Higher
CHEMISTRY

SECOND EDITION

Eric Allan and John Harris

HODDER
GIBSON
AN HACHETTE UK COMPANY

The Publishers would like to thank the following for permission to reproduce copyright material:

Photo credits Page 1 ASTRID & HANNS-FRIEDER MICHLER/SCIENCE PHOTO LIBRARY (left); BRITISH ANTARCTIC SURVEY/SCIENCE PHOTO LIBRARY (right); Page 67 NASA (left); © PHOTOTAKE Inc. / Alamy (right); Page 139 ROBERT BROOK/SCIENCE PHOTO LIBRARY (left); © Martin Shields / Alamy (centre); MOUNT STROMLO AND SIDING SPRING OBSERVATORIES/SCIENCE PHOTO LIBRARY (right).

Fig 1.16 ASTRID & HANNS-FRIEDER MICHLER/SCIENCE PHOTO LIBRARY; Fig 5.4 Axminster Power Tool Centre Ltd / www.axminster.co.uk; Fig 5.20 BRITISH ANTARCTIC SURVEY/SCIENCE PHOTO LIBRARY; Fig 7.1 D.BARNES/SCOTTISH VIEWPOINT; Fig 7.5 Charlie Neibergall/ AP/ PA Photos; Fig 7.7 © Car Culture/Corbis; Fig 7.9 Phil Degginger/ Photolibrary Group; Fig 7.10 NASA; Fig 8.5 NASA Goddard Space Flight Center (NASA-GSFC); Fig 10.6 Andy Sachs/ Stone/ Getty Images; Fig 11.3 Mark A. Leman/ Riser/ Getty Images; Fig 11.17 © Paul Rapson / Alamy; Fig 11.20 Polyval plc; Fig 12.1 John Pictor/ borderfields.co.uk; Fig 12.2 © J. Schwanke / Alamy; Fig 12.3 Mike Smith Photography/ borderfields.co.uk; Fig 12.17 © PHOTOTAKE Inc. / Alamy; Fig 13.4 MARTIN BOND/SCIENCE PHOTO LIBRARY; Fig 13.6 Rio Tinto Alcan; Fig 13.10 © Yann Arthus-Bertrand/CORBIS; Fig 13.12 © Leslie Garland Picture Library / Alamy; Fig 13.14 ROBERT BROOK/SCIENCE PHOTO LIBRARY; Fig 16.1 © Martin Shields / Alamy; Fig 17.6 Rio Tinto Alcan; Fig 18.6 STEVE ALLEN/SCIENCE PHOTO LIBRARY; Fig 18.7 Pat Groves/ Ecoscene; Fig 18.8 © Tony Marsh/Reuters/Corbis; Fig 18.10 EFDA-JET/ www.jet.efda.org; Fig 18.11 akg-images; Fig 18.13 MOUNT STROMLO AND SIDING SPRING OBSERVATORIES/SCIENCE PHOTO LIBRARY; Fig 18.14 CELESTIAL IMAGE CO./SCIENCE PHOTO LIBRARY. All other photos supplied by the authors.

Acknowledgements Extracts from past exam papers are reprinted with the permission of the Scottish Qualifications Authority.

Every effort has been made to trace all copyright holders, but if any have been inadvertently overlooked the Publishers will be pleased to make the necessary arrangements at the first opportunity.

Although every effort has been made to ensure that website addresses are correct at time of going to press, Hodder Gibson cannot be held responsible for the content of any website mentioned in this book. It is sometimes possible to find a relocated web page by typing in the address of the home page for a website in the URL window of your browser.

Hachette's policy is to use papers that are natural, renewable and recyclable products and made from wood grown in sustainable forests. The logging and manufacturing processes are expected to conform to the environmental regulations of the country of origin.

Orders: please contact Bookpoint Ltd, 130 Milton Park, Abingdon, Oxon OX14 4SB. Telephone: (44) 01235 827720. Fax: (44) 01235 400454. Lines are open 9.00–5.00, Monday to Saturday, with a 24-hour message answering service. Visit our website at www.hoddereducation.co.uk. Hodder Gibson can be contacted direct on: Tel: 0141 848 1609; Fax: 0141 889 6315; email: hoddergibson@hodder.co.uk

© Eric Allan and John Harris 1999, 2008
First published in 1999 by
Hodder Gibson, an imprint of Hodder Education,
an Hachette UK company,
2a Christie Street
Paisley PA1 1NB

This edition first published 2008

ISBN-13: 978 0340 959 114

Impression number	5	
Year	2012	2011

ISBN-13: 978 0340 959 121 (With Answers)

Impression number	7
Year	2014

Cover photo Albaimages/Alamy
Illustrations by Fakenham Prepress Solutions
Typeset in 11pt Minion by Fakenham Prepress Solutions, Fakenham, Norfolk N21 8NN
Printed and bound in Dubai

A catalogue record for this title is available from the British Library

Contents

Preface

This new edition is the first we have produced without the spur of a major revision of the course content. Our aim has been to make the book more attractive and accessible to students by the use of colour and more illustrations. We have, however, taken the opportunity to reflect changes in the style of examinations and to include recent questions from past SQA papers. Study question sections at the end of each chapter have received a facelift with the inclusion of sentence completion questions and/or multiple choice questions. Study questions which originated in past examination papers are marked with an asterisk. The End of Course section is entirely new, with multiple choice items in Part 1 and extended answer questions in Part 2. All of these questions come from Higher papers set on the current syllabus. The questions in Part 2 were chosen to provide further practice in calculations and problem-solving and also to give candidates experience in tackling questions that test more than one topic. We are grateful to SQA for permission to include past paper questions.

As we reviewed the text, it seemed appropriate to comment on matters that have become increasingly important to everyone, but which are not yet reflected by the Higher course content. When the previous edition was being prepared, the 'Greenhouse Effect' of some atmospheric gases was being discussed but global warming was not widely accepted. Now global warming is accepted as fact by most people and governments but it is also known that climate has fluctuated widely over historical time, even without the intervention of man and his burning of huge quantities of fossil fuels – as evidenced by the holding of 'Ice Fairs' on the frozen Thames at one period and the existence of vineyards in Yorkshire at another. What is most important is that consideration is now being given to the way we obtain our energy and the effects of this on the environment. The option of reducing the use of fossil fuels to produce less atmospheric carbon dioxide has the undoubted benefit of conserving a finite source of raw materials without which our petrochemical and plastics industries, to name only two, could not function.

The initial, understandable, rush to develop renewable energy from the wind is now being moderated by realisations that the wind is unreliable and can only produce a small proportion of our total energy requirements, at great expense, whilst having a huge impact, both visual and ecological, on the environment. The tourist industry, a major part of the economy in the same areas as favoured for wind farms, could be drastically affected. Similar considerations may apply to wave and tidal energy sources.

Until recently, it was almost unthinkable that nuclear power could once again play a major part in our energy production but new nuclear power stations are now being considered for the UK. Fusion power could now be an even greater prize than previously thought.

In terms of the detail of the Higher course, the degree of emphasis accorded certain fuels is worthy of comment. Diesel fuel was dropped from the course at the last revision but is now increasing in importance, whilst bio-diesel from oil-seed rape is seen as a way of combating global warming. Liquefied natural gas appears to be used less than before, but liquefied petroleum gas (LPG) is being used more. Hydrogen, as an energy transfer medium, has yet to make an impact.

Finally, but importantly, we wish to extend our thanks to Martin Armitage and former colleagues, Iain McGonigal, John Broadfoot, Johan Blaikie and Barbara Barrett for their advice and assistance. We are also indebted to our wives, Val and Chris, for their support and patience.

ERA, JHH, 2008

Energy Matters

The opening section of the Higher course shows the relevance of energy to reaction rates and introduces enthalpy change. Study of trends in the Periodic Table leads to a review of bonding, structure and properties of elements and compounds with an emphasis on intermolecular bonds. The final chapter reveals more facets of the mole, namely the Avogadro Constant and calculations involving volumes of gases.

1   Reaction Rate

From previous work you should know and understand the following:

★ A reaction can be speeded up by

- decreasing the *particle size* of any solid reactant

- increasing the *concentration* of a reactant in solution

- increasing the *temperature* at which the reaction occurs.

★ A **catalyst** is a substance which speeds up a reaction without being used up and can be recovered chemically unchanged at the end of the reaction.

★ A **mole** of a substance is its formula mass in grams (gram formula mass).

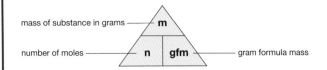

Hence, $n = \dfrac{m}{gfm}$ and $m = n \times gfm$

★ The volume of a liquid or a solution is measured in cubic centimetres (cm^3) or in litres (l).

$$1000\,cm^3 = 1\ litre$$

The concentration of a solution is measured in moles per litre, abbreviated to mol l^{-1}. You are also likely to come across the alternative term, **molarity**. A two-molar or 2 M solution has a concentration of 2 mol l^{-1}.

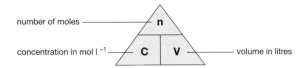

Hence, $n = C \times V$ and $C = \dfrac{n}{V}$ and $V = \dfrac{n}{C}$

Questions 1–3 are revision questions on the mole relevant to the work covered in this chapter.

Questions

1 Calculate the number of moles in

   a) 8 g of argon gas

   b) 48 g of oxygen gas

   c) 8 g of ammonium nitrate, NH_4NO_3

   d) 9.3 kg of calcium phosphate, $Ca_3(PO_4)_2$

2 Calculate the concentration in mol l^{-1} of each of the following solutions.

   a) 0.3 moles of NaCl in 600 cm^3 of solution

   b) 5 moles of KNO_3 in 2 litres of solution

   c) 3.31 g of $Pb(NO_3)_2$ in 200 cm^3 of solution

   d) 1 kg of NaOH in 5 litres of solution.

3 Calculate **i)** the number of moles and **ii)** the mass of solute in each of the following solutions:

   a) 100 cm^3 of 0.2 mol l^{-1} $AgNO_3$

   b) 1.5 litres of 4 mol l^{-1} Na_2CO_3

   c) 250 cm^3 of 2 M $Mg(NO_3)_2$

   d) 5 litres of 0.5 M $(NH_4)_2SO_4$.

Following the course of a reaction

In some reactions it is possible to take measurements as the reaction is taking place. A useful reaction to study is that between marble chips (calcium carbonate) and hydrochloric acid using the apparatus shown in Figure 1.1.

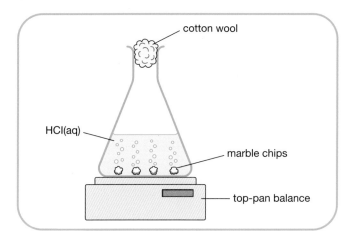

Figure 1.1

As the reaction proceeds, carbon dioxide gas is released and hence the mass of flask and contents decreases. A loose cotton wool 'plug' is used to prevent loss of acid spray during effervescence whilst allowing the gas to escape. The balanced equation is as follows:

$$CaCO_3(s) + 2HCl(aq) \rightarrow CaCl_2(aq) + CO_2(g) + H_2O(l)$$

Specimen results from an experiment in which 15 g of marble chips were added to 50 cm³ of 4 mol l⁻¹ hydrochloric acid are given in Table 1.1. Using these quantities ensures that the marble chips are present in

excess so the acid will eventually be neutralised. We shall reconsider the idea of excess reactant later in the chapter (page 10).

The decrease in mass is the mass of carbon dioxide released and this quantity can be plotted against time as shown in Figure 1.2. From the loss in mass it is also possible to carry out a mole calculation using the balanced equation to find the concentration of the acid at the various times. These calculated results are shown in the table and are plotted against time in Figure 1.3.

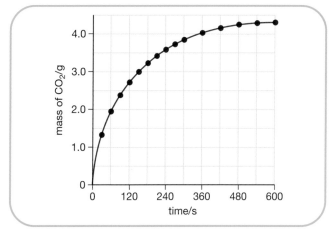

Figure 1.2 Mass of CO_2 against time

Time/s	Mass of flask and contents/g	Decrease in mass/g	Concentration of acid/mol l⁻¹
0	149.00	–	4.00
30	147.75	1.25	2.86
60	147.08	1.92	2.25
90	146.60	2.40	1.82
120	146.24	2.76	1.49
150	145.94	3.06	1.22
180	145.68	3.32	0.98
210	145.48	3.52	0.80
240	145.32	3.68	0.65
270	145.19	3.81	0.54
300	145.08	3.92	0.44
360	144.89	4.11	0.27
420	144.77	4.23	0.15
480	144.70	4.30	0.09
540	144.65	4.35	0.04
600	144.65	4.35	0.04

Table 1.1

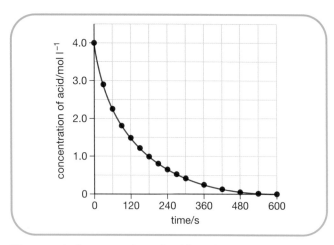

Figure 1.3 Concentration of acid against time

The rate of reaction is the change in concentration of reactants or products in unit time. As can be seen from Figures 1.2 and 1.3, the slope of the graph is steepest at the beginning of the reaction and levels off as time passes. This shows that the rate of reaction is greatest initially and decreases with time. This is true whether we consider the rate at which gas is released or the rate at which acid is consumed.

It is difficult to measure the actual rate at any one instant since the rate is always changing, but it is possible to calculate the average rate over a certain period of time. In this experiment the average rate would be calculated from the loss in mass or decrease in acid concentration which occurs in a certain time interval. The method of calculation is shown in the following worked example.

Worked Example 1.1

Use the data given in Table 1.1 to calculate the average rate of reaction during the second period of 30 seconds (i.e. between 30 s and 60 s) in terms of

   a) the mass of carbon dioxide produced,

   b) the decrease in the concentration of hydrochloric acid.

a) Mass of CO_2 released between 30 s and 60 s
$$= 1.92 - 1.25 = 0.67\,g$$

$$\text{Average rate} = \frac{\text{mass of } CO_2}{\text{time interval}}$$

$$\text{Average rate} = \frac{0.67}{30} = 0.022\,g\,s^{-1}$$

b) Decrease in concentration of HCl(aq) between 30 s and 60 s $= 2.86 - 2.25 = 0.61\,mol\,l^{-1}$

$$\text{Average rate} = \frac{\text{decrease in acid concentration}}{\text{time interval}}$$

$$\text{Average rate} = \frac{0.61}{30} = 0.020\,mol\,l^{-1}\,s^{-1}$$

Questions

4 Using data from Table 1.1 carry out similar calculations for the average rate of reaction during **a)** the first 30 s and **b)** the third period of 30 s (i.e. between 60 s and 90 s). Compare the results with those for the second period of 30 s calculated in Worked Example 1.1.

5 Taking the reaction shown in Figure 1.1 to be completed at 540 s, calculate the average rate per minute in terms of **a)** the mass of carbon dioxide produced and **b)** the decrease in concentration of hydrochloric acid. Use data from Table 1.1.

Factors affecting the rate of a reaction

Particle size

Reactions in which one of the reactants is a solid can be speeded up or slowed down by altering the particle size of the solid. The reaction discussed in the previous section provides a good example of this. If the marble chips are replaced by an equal mass of smaller pieces of marble, the reaction proceeds more quickly since the surface area exposed to the acid has increased. If the mass loss against time is compared with the results shown in Figure 1.3, then a graph with a steeper slope will be obtained (shown as a green line in Figure 1.4).

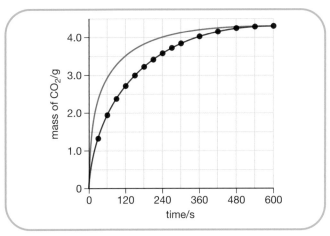

Figure 1.4 Effect of particle size on rate of reaction

When photographic film is exposed to light a chemical reaction occurs. The speed of the film is very important. The faster the film, i.e. the higher its ISO number, the smaller is the particle size of the light-sensitive material in the film. A fast film, e.g. ISO 400, means that good quality photographs can be obtained in poorer lighting conditions or that a very short exposure time can be used to enable photographs of fast-moving objects, such as racing cars, to be taken.

Concentration

You will already be aware that the concentration of a reactant affects the rate of a reaction. The reaction between marble and hydrochloric acid, for example, can be speeded up by increasing the concentration of the acid. When investigating the relationship between the rate of reaction and the concentration of a reactant, a '**clock reaction**' may be used. In a clock reaction, a time lapse occurs before a sudden end-point is reached.

Prescribed Practical Activity

We shall look at the reaction which occurs between hydrogen peroxide and acidified potassium iodide solution to see how the rate of this reaction depends on the concentration of iodide ions. The equation for the reaction is as follows:

$$H_2O_2(aq) + 2H^+(aq) + 2I^-(aq) \rightarrow 2H_2O(l) + I_2(aq)$$

Starch solution and sodium thiosulphate solution, $Na_2S_2O_3(aq)$, are also included in the reaction mixture. Iodine molecules produced in the reaction are immediately changed back into iodide ions by reacting

with thiosulphate ions according to the following equation:

$$I_2(aq) + 2S_2O_3^{2-}(aq) \rightarrow 2I^-(aq) + S_4O_6^{2-}(aq)$$

While this is happening the reaction mixture is colourless. When all of the thiosulphate ions have reacted, a blue–black colour suddenly appears as iodine – produced by the first reaction – is detected by starch.

As Figure 1.5 shows, potassium iodide solution, starch solution, sodium thiosulphate solution and dilute

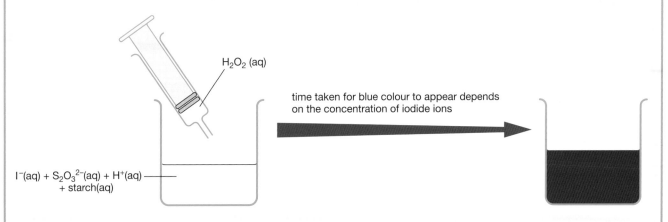

$I^-(aq) + S_2O_3^{2-}(aq) + H^+(aq) + starch(aq)$

$H_2O_2 (aq)$

time taken for blue colour to appear depends on the concentration of iodide ions

Figure 1.5

Prescribed Practical Activity (continued)

sulphuric acid are mixed. Hydrogen peroxide solution is added and the time taken for the mixture to turn blue–black is measured. The experiment is repeated using smaller volumes of the iodide solution but adding water so that the total volume of the reacting mixture is always the same. The concentrations and volumes of all other solutions are kept constant.

The number of moles of thiosulphate ions is the same in each experiment so that when the blue–black colour appears, the same extent of reaction has occurred. **Since rate is inversely proportional to time, the reciprocal of time (1/t) is taken to be a measure of the rate of the reaction.**

Specimen results are given in Table 1.2 and the graph of rate against volume of KI(aq) is shown in Figure 1.6.

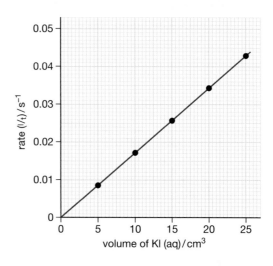

Figure 1.6

The graph of rate against volume of potassium iodide solution shows a straight line. Since the total volume is always the same, we can take the volume of KI(aq) to be a measure of iodide ion concentration.

The straight-line graph means that the rate of this reaction is directly proportional to the concentration of iodide ions. In other words, if the concentration of iodide ions is doubled then the rate of reaction doubles.

Volume of KI(aq)/ cm³	Volume of H_2O/cm³	Time (t)/s	Rate (1/t)/s⁻¹
25	0	23	0.043
20	5	29	0.034
15	10	39	0.026
10	15	60	0.017
5	20	111	0.009

Table 1.2

Temperature

From previous work you know that a reaction can be speeded up by raising the temperature at which it takes place.

Prescribed Practical Activity

The effect of changing the temperature on the rate of a reaction can be studied using the following reaction. Acidified potassium permanganate solution, which is purple due to the presence of permanganate ions (MnO_4^-), is decolourised by an aqueous solution of oxalic acid, $(COOH)_2$. This reaction is very slow at room temperature but is almost instantaneous above 80°C. The equation for this reaction is given below.

$$5(COOH)_2(aq) + 6H^+(aq) + 2MnO_4^-(aq) \rightarrow$$
$$2Mn^{2+}(aq) + 10CO_2(g) + 8H_2O(l)$$

This experiment is carried out at temperatures ranging from about 40°C to about 70°C. Volumes and concentrations of all the reactants are kept constant. As shown in Figure 1.7, the reaction starts when the oxalic acid is added to the permanganate solution previously acidified with dilute sulphuric acid. The time taken for the solution to become colourless is measured. The temperature is measured at the end-point of the reaction.

Since the number of moles of permanganate ions is the same in each experiment the same amount of reaction has occurred when the end-point has been reached.

Prescribed Practical Activity (continued)

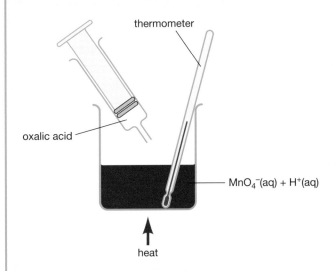

Figure 1.7

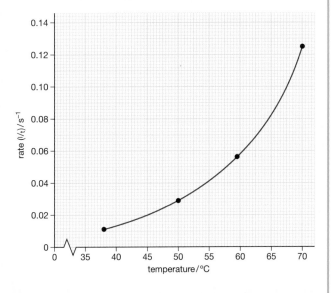

Figure 1.8

As in the previous experiment the reciprocal of the time taken to reach the end-point (1/t) is taken to represent the rate of reaction.

Specimen results for the experiment are given in Table 1.3 and the graph of rate against temperature is shown in Figure 1.8.

Temperature/ °C	Time (t)/s	Rate (1/t)/s^{-1}
38	87	0.011
50	35	0.029
59	18	0.056
70	8	0.125

Table 1.3

As expected, the rate of reaction increases with rising temperature. However, since the graph of rate against temperature is a curve, as can be seen from Figure 1.8, the rate is not directly proportional to the temperature. In fact it can be seen from the graph that the rate of reaction doubles if there is a temperature rise of about 10°C.

There are many applications of the fact that a small change in temperature has a marked effect on the rate of a reaction. For example, roasting a chicken weighing 2 kg takes about 1 hour 30 mins in a hot oven (200°C) and twice as long in a moderate oven (150°C). If a slow cooking pot is used, the cooking time increases to several hours, since the temperature is not much over 100°C. Food can be stored in a freezer for much longer than in a domestic fridge due to lower temperatures which slow down decomposition reactions.

Temperature control is often essential in industrial processes. In the manufacture of nitric acid by the

Ostwald process, ammonia is catalytically oxidised to produce nitrogen monoxide according to the following equation:

$$4NH_3(g) + 5O_2(g) \rightarrow 4NO(g) + 6H_2O(g)$$

The reaction is operated at about 900°C and approximately 96% conversion of ammonia occurs. The reaction is highly exothermic and, if the temperature rises too much, damage to the platinum–rhodium catalyst may occur. A higher temperature also increases the chance of ammonia being converted to nitrogen. This would decrease the yield of nitrogen monoxide and consequently affect the production of nitric acid.

Collision theory

From an early stage in studying science, you will have been aware that all substances are made up of very small particles which are called atoms, ions or molecules. Furthermore, these particles are continually moving, the speed and extent of the motion depending on whether the substance is a gas, a liquid, a solid or in solution. This description is often referred to as the 'kinetic model of matter'.

For a chemical reaction to occur, the reactants must be brought together in some way so that their particles will collide. This is the basis of the **collision theory**. Any factor which increases the number of collisions per second between the particles of the reactants is likely to increase the rate of reaction. More collisions occur if the *particle size* of a solid reactant is decreased, since its overall surface area is increased. Similarly, if the *concentration* of a reactant is increased, more collisions between particles will taken place. These points are illustrated in Figure 1.9.

Questions

6 The graph shows how the volume of nitrogen dioxide increases with time when 2 g of copper turnings react with excess concentrated nitric acid. Copy the graph and add similar curves for the reaction between concentrated nitric acid and **a)** 1 g of copper powder, **b)** a 2 g piece of copper foil. Label the curves **a)** and **b)** as appropriate.

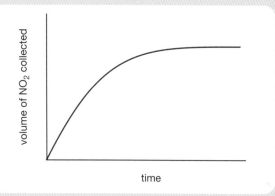

7* Excess zinc was added to 2 mol l^{-1} sulphuric acid at room temperature and the volume of hydrogen produced was plotted against time as shown.

a) Why does the gradient of the curve decrease as the reaction proceeds?

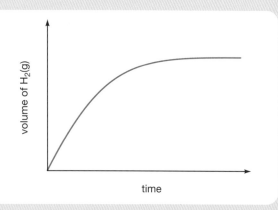

b) Copy the graph and add corresponding curves obtained when the reaction is repeated **i)** at a higher temperature, **ii)** using an equal volume of 1 mol l^{-1} sulphuric acid. Label the curves **i)** and **ii)** as appropriate.

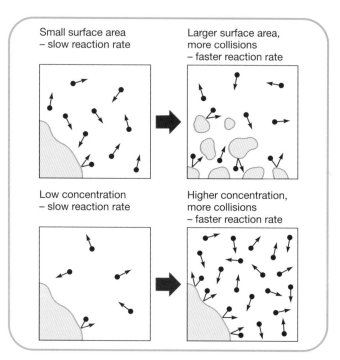

Figure 1.9 Effect of particle size and concentration on rate of reaction

Raising the *temperature* at which the reaction occurs does more than merely increase the number of collisions between particles. Temperature can be regarded as a measure of the average kinetic energy of the particles in a substance. Hence, at a higher temperature the particles have greater kinetic energy and will collide with greater force.

Reactions occur, then, when reactant particles collide. However, it would appear that not all collisions result in a successful reaction. If they did, all reactions would be virtually instantaneous. Reactions in which covalent substances take part are often slow, even when the substances are gases as in the case of hydrogen and oxygen.

Unless ignited by a flame, a mixture of hydrogen and oxygen will not react to any appreciable extent at room temperature. This is despite the fact that, as gases, their molecules are separate and will mix rapidly by diffusion and many collisions between molecules will occur per second.

Not all reactions involving covalent substances are slow. The colourless gas nitrogen monoxide combines rapidly with oxygen, even at room temperature, to form brown fumes of nitrogen dioxide.

$$2NO(g) + O_2(g) \rightarrow 2NO_2(g)$$

Reactions which involve separate ions in solution are often very fast if not instantaneous. When an acid and alkali are mixed, large numbers of the reacting particles, i.e. H^+ and OH^- ions, collide at the moment of mixing and rapidly combine to form water molecules. Similarly, mixing solutions of barium chloride and sodium sulphate brings large numbers of barium ions and sulphate ions together and insoluble $BaSO_4$ is rapidly precipitated.

Ionic reactions involving a solid may be slow. For example, large marble chips react slowly with dilute acid at room temperature. As you are already aware, this reaction can be accelerated by decreasing the size of the marble chips, increasing the concentration of the acid, raising the temperature or, indeed, using any combination of these factors.

Activation energy and energy distribution

In 1889 a Swedish chemist by the name of Arrhenius put forward the idea that, for a reaction to occur, the colliding particles must have a minimum amount of kinetic energy, called the **activation energy**. The activation energy required varies from one reaction to another. If in a certain reaction the activation energy is high, only a few particles will have enough energy for collisions between them to be successful and hence the reaction will normally be slow. Conversely, a reaction with a low activation energy will be fast under normal conditions.

Questions

8 Refer to the information given on this page about the reactions involving **a)** hydrogen and oxygen, and **b)** nitrogen monoxide and oxygen. Which of these reactions is likely to have the higher activation energy?

9 A piece of phosphorus ignites when touched with a hot wire, while magnesium ribbon needs strong heating before it will catch fire.

What does this suggest about the activation energies of these two reactions?

At a given temperature individual molecules of a gas have widely different kinetic energies. Most molecules will have energy near to the average value, but some will be well below average, while others will be well above. The distribution of kinetic energy values is illustrated in Figure 1.10. The kinetic energy of individual molecules will continually change due to collisions with other molecules. However, at constant temperature the overall distribution of energies remains the same.

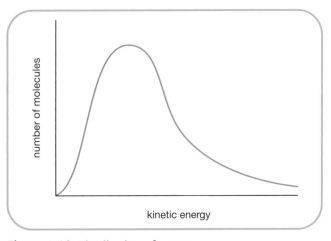

Figure 1.10 Distribution of energy

Figure 1.11 shows the same distribution of kinetic energy but it also incorporates the activation energy, E_A. The shaded area represents all of the molecules which have energy greater than the activation energy, i.e. the proportion of molecules which have sufficient energy to react. If the activation energy is greater then the shaded area would be smaller thus representing a smaller proportion of the total number of molecules.

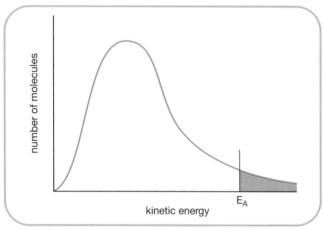

Figure 1.11 Distribution of energy including activation energy

The distribution of energy changes when the temperature changes. The effect of a small rise in temperature, from T_1 to T_2, is shown in Figure 1.12.

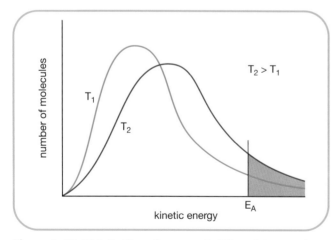

Figure 1.12 Distribution of energy at different temperatures

The average energy is increased but the most significant feature is the considerable increase in the area that is shaded. In other words at a higher temperature there are many more molecules with energy equal to or greater than the activation energy.

This is the real reason why a small change in temperature can have such a marked effect on the rate of a reaction. A small rise in temperature causes a significant increase in the number of molecules which have energy greater than the activation energy.

Photochemical reactions

In some chemical reactions light energy is used to increase the number of molecules which have energy greater than the activation energy. The most important photochemical reaction is arguably photosynthesis in which light energy is absorbed by chlorophyll to convert carbon dioxide and water into glucose and oxygen.

The use of photographic film provides another important example of a photochemical reaction. When a black-and-white film is exposed to light, silver ions are reduced to silver atoms. The darkest parts on the negative are those which have been exposed to the strongest light.

$$Ag^+ + e^- \rightarrow Ag$$

In the examples given above, light is needed to sustain the reaction. Some reactions are 'set off' or initiated by light. A mixture of hydrogen and chlorine gases explodes when exposed to a light source of high enough energy. In this case, the light energy breaks the bonds in the chlorine molecules to form highly reactive chlorine atoms, which then set off a very rapid chain reaction.

$$H_2 + Cl_2 \rightarrow 2HCl$$

Excess reactant

In previous work in chemistry you will have used balanced equations to help you calculate the mass of a product from the mass of a reactant, or vice versa. In reactions involving two reactants this calculation can only be done if there is more than enough of the other reactant, i.e. it is present in excess. Questions 10 and 11 on page 11 provide revision of this type of calculation. Note the use of the word 'excess' in question 10 and that the phrase 'complete reduction' in question 11 implies that there is excess reducing agent present.

In the experiment described on page 3 calcium carbonate, in the form of marble chips, was reacted with hydrochloric acid. The quantities were chosen to ensure that excess marble was used so that the acid would be completely neutralised. Worked Example 1.2 shows by calculation a) that the marble is present in excess, and b) that the mass of carbon dioxide

produced depends on the number of moles of the reactant which is *not* in excess, i.e. the acid.

Questions 12–14 give you some practice in this type of calculation. In each question the balanced equation for the reaction is provided.

Worked Example 1.2

15 g of calcium carbonate were reacted with 50 cm^3 of 4 mol l^{-1} hydrochloric acid.

a) Show by calculation which reactant was present in excess.

b) Calculate the mass of carbon dioxide produced.

$CaCO_3 + 2HCl \rightarrow CaCl_2 + CO_2 + H_2O$
1 mol   2 mol   1 mol
100 g   44 g

a) Number of moles of $CaCO_3$,

$$n = \frac{m}{gfm} = \frac{15}{100} = 0.15 \text{ mol}$$

Number of moles of HCl,

$$n = C \times V = 4 \times \frac{50}{1000} = 0.2 \text{ mol}$$

According to the equation, 1 mol of $CaCO_3$ neutralises 2 mol of HCl.

Hence, 0.1 mol of $CaCO_3$ neutralises 0.2 mol of HCl.

Since there is more than 0.1 mol of $CaCO_3$ present, this reactant is in excess.

b) To calculate the mass of carbon dioxide produced we use the quantity of the reactant which is completely reacted, i.e. the acid, and not the one which is present in excess.

According to the equation, 2 mol of HCl produce 1 mol of CO_2.

Hence, 0.2 mol of HCl produce 0.1 mol of CO_2.

0.1 mol = 0.1 × 44 = 4.4 g

4.4 g of carbon dioxide were produced.

Questions

10 4.46 g of lead(II) oxide were reacted with excess dilute nitric acid. Calculate the mass of lead(II) nitrate produced.

$$PbO + 2HNO_3 \rightarrow Pb(NO_3)_2 + H_2O$$

11 Calculate the mass of iron(III) oxide which, on complete reduction by carbon monoxide, will produce 558 tonnes of iron.

$$Fe_2O_3 + 3CO \rightarrow 2Fe + 3CO_2$$

12 A piece of magnesium ribbon weighing 0.6 g was added to 40 cm^3 of 2 mol l^{-1} hydrochloric acid.

$$Mg + 2HCl \rightarrow MgCl_2 + H_2$$

a) Show by calculation that excess acid has been used.

b) Calculate the number of moles of excess acid.

c) Describe briefly how the result in b) could be checked by experiment.

13 2.6 g of zinc dust were added to 30 cm^3 of 1.0 mol l^{-1} copper(II) sulphate solution.

$$Zn + CuSO_4 \rightarrow ZnSO_4 + Cu$$

a) Show by calculation that the zinc was in excess.

b) Calculate the mass of copper produced.

c) Describe briefly how the excess zinc could be removed and the copper obtained to check this result.

14 20 cm^3 of 0.2 mol l^{-1} lead(II) nitrate solution and 10 cm^3 of 0.6 mol l^{-1} potassium iodide solution were mixed together.

$$Pb(NO_3)_2(aq) + 2KI(aq) \rightarrow PbI_2(s) + 2KNO_3(aq)$$

a) Show by calculation which reactant was in excess.

b) Calculate the mass of precipitate formed.

Catalyst	Process	Reaction	Importance
Vanadium(V) oxide	Contact	$2SO_2 + O_2 \rightarrow 2SO_3$	Manufacture of sulphuric acid
Iron	Haber	$N_2 + 3H_2 \rightarrow 2NH_3$	Manufacture of ammonia
Platinum	Catalytic oxidation of ammonia	$4NH_3 + 5O_2 \rightarrow$ $4NO + 6H_2O$	Manufacture of nitric acid
Nickel	Hydrogenation	Unsaturated oils + $H_2 \rightarrow$ Saturated fats	Manufacture of margarine
Aluminium silicate	Catalytic cracking	Breaking down long-chain hydrocarbon molecules	Manufacture of fuels and monomers for the plastics industry

Table 1.4 Catalysts used in industrial processes

Catalysts

A **catalyst** is a substance which alters the rate of a reaction without being used up in the reaction. A simple example of catalysis can be shown using hydrogen peroxide. A solution of hydrogen peroxide evolves oxygen very slowly, even on heating. Oxygen is much more rapidly released when manganese(IV) oxide is added.

$$2H_2O_2(aq) \rightarrow 2H_2O(l) + O_2(g)$$

Catalysts play an important part in many industrial processes, some of which you will have encountered in previous years. Table 1.4 summarises some of these processes.

The catalysts listed in Table 1.4, along with manganese(IV) oxide in the decomposition of hydrogen peroxide, are said to be **heterogeneous catalysts**, since they are in a different physical state from the reactants. In the Contact Process, vanadium(V) oxide, in the form of solid pellets, is a heterogeneous catalyst in a reaction involving gaseous reactants, i.e. sulphur dioxide and oxygen.

A catalyst which is in the same physical state as the reactants is said to be **homogeneous**. An example of this is illustrated in Figure 1.13. The reaction between aqueous solutions of Rochelle salt (potassium sodium tartrate) and hydrogen peroxide is slow even when the mixture is heated. It is catalysed when an aqueous

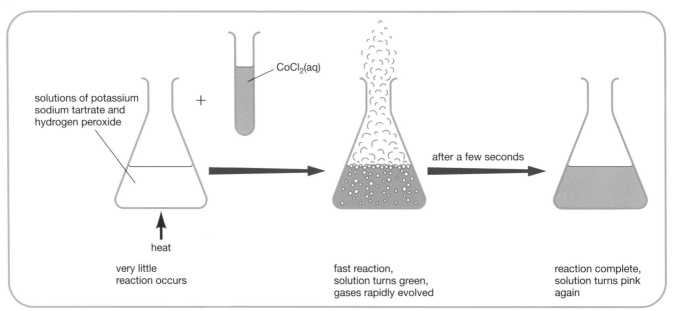

Figure 1.13 An example of homogeneous catalysis

solution containing cobalt(II) ions is added. The immediate colour change to green is thought to be due to oxidation to form cobalt(III) ions which then catalyse the reaction. The return of the pink colour at the end of the reaction shows that cobalt(II) ions have been reformed. This sequence of colour changes shows that a catalyst may undergo a temporary chemical change during its catalytic activity.

Another example of homogeneous catalysis can be shown in the reaction between potassium permanganate solution, acidified with dilute sulphuric acid, and oxalic acid solution. This reaction is very slow at room temperature but it can be catalysed by adding an aqueous solution of manganese(II) ions. This speeds up the decolouration of permanganate ions by oxalic acid.

Most of the catalysts referred to so far are either transition metals or compounds of these metals. Catalytic behaviour is just one of the characteristic properties of transition metals. From previous work you will also be aware that transition metals have variable valency and that they tend to form coloured ions.

How heterogeneous catalysts work

In a reaction in which a heterogeneous catalyst is used it is an advantage if the catalyst has a large surface area. It is believed that catalysis occurs on the surface of the catalyst at certain points called **active sites**. At these sites molecules of at least one of the reactants are **ad**sorbed. How a heterogeneous catalyst works can be represented in three stages as shown in Figure 1.14.

This can also help us to understand how a catalyst can become poisoned. **Poisoning** will occur if certain molecules are preferentially adsorbed or even permanently attached to the surface of the catalyst. This will reduce the number of active sites available for the adsorption of reactant molecules and will render the catalyst ineffective.

Traces of hydrogen sulphide, along with arsenic compounds, should be removed from the reacting gases in the Contact Process before passing over the catalyst. Carbon monoxide is a catalyst poison in the Haber Process. If the reactants in a catalysed industrial process contain impurities, additional costs will arise from regenerating or renewing the catalyst.

During the catalytic cracking of long-chain hydrocarbons, carbon is a by-product and is deposited on the surface of the catalyst thus reducing its efficiency. In this case the catalyst can be regenerated by burning off the carbon in a plentiful supply of air. The spent catalyst is removed from the reactor and returned after regeneration; see Chapter 13, page 146.

Catalytic converters

In a car engine which burns petrol, the main products of combustion are carbon dioxide and water. However, sparking of the air–fuel mixture produces some carbon

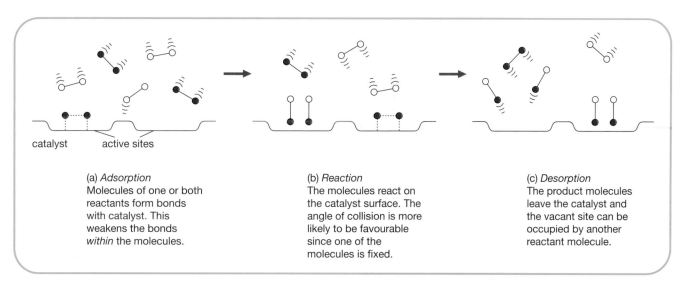

catalyst   active sites

(a) *Adsorption*
Molecules of one or both reactants form bonds with catalyst. This weakens the bonds *within* the molecules.

(b) *Reaction*
The molecules react on the catalyst surface. The angle of collision is more likely to be favourable since one of the molecules is fixed.

(c) *Desorption*
The product molecules leave the catalyst and the vacant site can be occupied by another reactant molecule.

Figure 1.14 Heterogeneous catalysis

monoxide as well as oxides of nitrogen, NO_x. These compounds are harmful to the environment since carbon monoxide is highly poisonous and nitrogen oxides contribute to the problem of acid rain.

Cars with petrol engines can have **catalytic converters** fitted as part of their exhaust systems. The converter contains a ceramic support material covered with expensive metals such as platinum and rhodium which catalyse the conversion of carbon monoxide to carbon dioxide and nitrogen oxides to nitrogen as shown in Figure 1.15. Catalytic converters could not be fitted to exhaust systems of cars which ran on leaded petrol, otherwise the lead compounds produced during combustion would poison the catalyst.

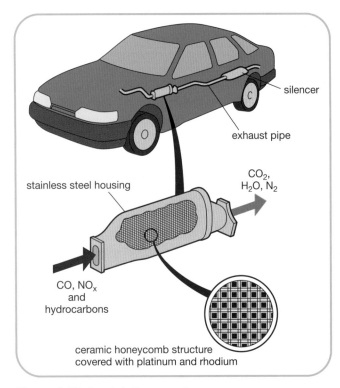

silencer

exhaust pipe

stainless steel housing

CO_2, H_2O, N_2

CO, NO_x and hydrocarbons

ceramic honeycomb structure covered with platinum and rhodium

Figure 1.15 A catalytic converter

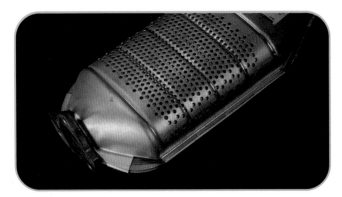

Figure 1.16 A catalytic converter

Enzymes

Many biochemical reactions in the living cells of plants and animals are catalysed by **enzymes**. Examples of enzymes include i) amylase, which catalyses the hydrolysis of starch, and ii) catalase, which catalyses the decomposition of hydrogen peroxide. Catalase is present in blood and helps to prevent build-up of hydrogen peroxide, a powerful oxidising agent, in the body.

An enzyme has a complex molecular structure and its molecular shape usually plays a vital role in its function as a catalyst. It operates most effectively at a certain optimum temperature and within a narrow pH range.

Enzymes are usually highly *specific*. Maltose and sucrose are disaccharides but are hydrolysed by different enzymes, the former by an enzyme called maltase, the latter by invertase. The hydrolysis of sucrose to form glucose and fructose is used commercially in the production of soft-centred chocolates and other confectionery.

The effect of invertase on sucrose solution can be shown experimentally as illustrated in Figure 1.17. The filtrate reacts with Benedict's solution producing an orange-red precipitate. This shows that sucrose has been hydrolysed to form a reducing sugar.

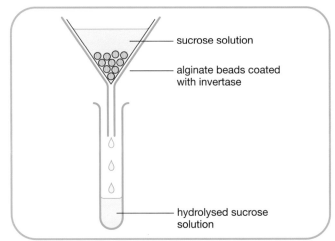

sucrose solution

alginate beads coated with invertase

hydrolysed sucrose solution

Figure 1.17 Enzyme-catalysed hydrolysis

There are many applications of enzymes in industry. In earlier work you have come across the use of yeast to provide enzymes for the fermentation of glucose to produce ethanol. Other examples include

♦ lipase to enhance flavour in ice-cream, cheese and chocolate

♦ rennin in cheese production

♦ proteases to tenderise meat

♦ amylase in removing starch from fabrics, a process called desizing.

We shall come across enzymes again in Chapter 12.

Question

15 In each of the following examples decide whether the catalyst is homogeneous or heterogeneous. Explain your choice

Reactants	Catalyst
a) Methane and steam	Nickel
b) $C_2H_5OH(l)$ and $CH_3COOH(l)$	Conc. $H_2SO_4(l)$
c) Ethene and steam	Conc. $H_3PO_4(l)$

Study Questions

In questions 1–4 choose the correct word from the following list to complete the sentence.

concentration	divided	heterogeneous
homogeneous	multiplied	sulphur
temperature	zinc	

1 The kinetic energy of particles taking part in a reaction can be altered by changing the _____ .

2 The reaction involving car exhaust gases when they pass through a catalytic converter is an example of _____ catalysis.

3 In a reaction where a gas is released, the average rate of reaction over a period of time can be calculated from the loss of mass of gas _____ by the time interval.

4 Zinc and sulphur react according to the equation: $Zn + S \rightarrow ZnS$. If equal masses of these elements react together, the excess reactant is _____ .

5* The graph shows the change in the concentration of a reactant with time for a given chemical reaction.

What is the average rate of this reaction, in $mol\ l^{-1}\ s^{-1}$, between 10 s and 20 s?

A 1.0×10^{-2} B 1.0×10^{-3}
C 1.5×10^{-2} D 1.5×10^{-3}

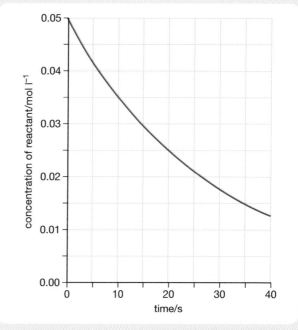

6 In which process is a homogeneous catalyst used?

A Synthesis of ammonia from nitrogen and hydrogen using iron

B Decomposing potassium chlorate crystals using MnO_2 powder

C Catalytic cracking of hydrocarbons using Al_2O_3 powder

D Manufacture of fats from vegetable oils using nickel

Study Questions (continued)

7 $Mg + H_2SO_4 \rightarrow MgSO_4 + H_2$

A student added 8 g of magnesium to 200 cm³ of 1.0 mol l⁻¹ sulphuric acid.

Which statement is true for this experiment?

A Excess acid has been used.

B 0.4 mol of magnesium sulphate is produced.

C All of the magnesium reacts.

D 0.4 g of hydrogen gas is released.

8* A student was asked to write a plan of the procedure for an investigation. The entry made in a laboratory notebook is shown.

Aim

To find the effect of concentration on the rate of the reaction between hydrogen peroxide and an acidified solution of iodide ions.

$H_2O_2(aq) + 2H^+(aq) + 2I^-(aq) \rightarrow 2H_2O(l) + I_2(aq)$

Procedure

1 Using a 100 cm³ measuring cylinder, measure out 10 cm³ of sulphuric acid, 10 cm³ of sodium thiosulphate solution, 1 cm³ of starch solution and 25 cm³ of potassium iodide solution into a dry 100 cm³ glass beaker and place the beaker on the bench.

2 Measure out 5 cm³ of hydrogen peroxide solution and start the timer.

3 Add the hydrogen peroxide solution to the beaker. When the blue/black colour just appears, stop the timer and record the time (in seconds).

4 Repeat this procedure four times but using different concentrations of potassium iodide solution. This is achieved by adding 5 cm³, 10 cm³, 15 cm³, 20 cm³ of water to the 25 cm³ of potassium iodide solution before adding it to the glass beaker.

a) Why is instruction 4 **not** the best way of altering the concentration of the potassium iodide solution?

b) State **two** other ways of improving the student's plan of this investigation procedure.

9* Hydrogen peroxide can be used to clean contact lenses. In this process, the enzyme catalase is added to break down hydrogen peroxide. The equation for the reaction is:

$$2H_2O_2 \rightarrow 2H_2O + O_2$$

The rate of oxygen production was measured in three laboratory experiments using the same volume of hydrogen peroxide at the same temperature.

Experiment	Concentration of $H_2O_2/$ mol l⁻¹	Catalyst used
A	0.2	Yes
B	0.4	Yes
C	0.2	No

The curve obtained for experiment A is shown.

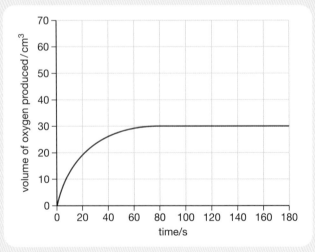

a) Calculate the average rate of reaction over the first 40 s.

b) Copy the graph and add curves to show the results of experiments B and C. Label each curve clearly.

Study Questions (continued)

10 For each of the following examples of neutralisation:

i) show by calculation which reactant is present in excess, and

ii) calculate the mass of salt produced.

a) 3.0 g lead(II) carbonate were added to 40 cm^3 of 0.5 mol l^{-1} nitric acid.

$$PbCO_3 + 2HNO_3 \rightarrow Pb(NO_3)_2 + CO_2 + H_2O$$

b) 8.1 g of aluminium were added to 250 cm^3 of 4 mol l^{-1} hydrochloric acid.

$$2Al + 6HCl \rightarrow 2AlCl_3 + 3H_2$$

2 Enthalpy

From previous work you should know and understand the following:

★ An **exothermic reaction** gives out energy when it takes place.

★ **Combustion** is the burning of a substance during which it combines with oxygen.

★ **Neutralisation** occurs when an acid reacts with an alkali or other base.

Potential energy

Exothermic and endothermic reactions

During your study of chemistry you will often have observed that when chemical reactions occur they are accompanied by a significant change in energy. Most of the reactions that you have come across will have involved a release of energy to the surroundings, usually in the form of heat, and are thus said to be exothermic.

Examples of such reactions include:

♦ combustion of elements, carbon compounds and other fuels

♦ neutralisation of acids by alkalis and reactive metals

♦ displacement of less reactive metals.

Energy may also be released in a chemical reaction in other forms, such as light, e.g. when magnesium burns, or sound, e.g. hydrogen–oxygen and hydrogen–chlorine explosions.

Reactions in which heat is absorbed from the surroundings are said to be endothermic. Although less frequent, such reactions do occur, and examples include:

♦ dissolving certain salts in water (e.g. ammonium nitrate, potassium nitrate)

♦ neutralising ethanoic acid with ammonium carbonate or sodium hydrogencarbonate

♦ making a fuel called 'water gas' by reacting steam with hot coke.

$$C(s) + H_2O(g) \rightarrow \underbrace{CO(g) + H_2(g)}_{\text{'water gas'}}$$

Your teacher may demonstrate the endothermic reaction illustrated in Figure 2.1. Powdered samples of hydrated barium hydroxide and ammonium thiocyanate are mixed together in a beaker on a wet wooden block. The temperature of the reacting mixture falls well below 0°C, sufficient to freeze the water and cause the beaker to stick to the block.

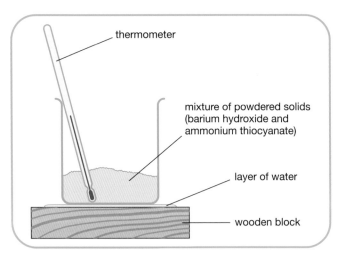

Figure 2.1 An endothermic reaction

During an exothermic reaction, energy possessed by the reactants – potential energy – is released to the surroundings. Hence the products of an exothermic reaction have less potential energy than the reactants. This can be illustrated in a potential energy diagram (Figure 2.2) which shows the energy pathway as the reaction proceeds from reactants to products.

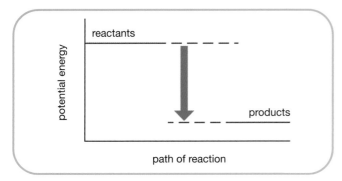

Figure 2.2 Exothermic reaction

Conversely, in an endothermic reaction, the reactants absorb energy from the surroundings so that the products possess more energy than the reactants.

The potential energy diagram in Figure 2.3 shows this.

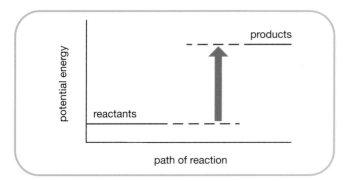

Figure 2.3 Endothermic reaction

Enthalpy change

The potential energy diagrams in the previous section – Figures 2.2 and 2.3 – show the change in energy during exothermic and endothermic reactions. The difference in potential energy between reactants and products is called the enthalpy change, denoted by the symbol ΔH. Enthalpy changes are usually quoted in kilojoules per mole of reactant or product, abbreviated to kJ mol^{-1}.

Since the reactants lose energy in an exothermic reaction, ΔH is said to be negative, as is shown in Figure 2.4. In an endothermic reaction, the reactants take in heat from the surroundings, so that the products possess more energy than the reactants. As a result, an endothermic change has a positive ΔH value, as shown in Figure 2.5.

It is important to note that, although it is essential to include the minus sign in front of the numerical value for the enthalpy change if the reaction is exothermic, it

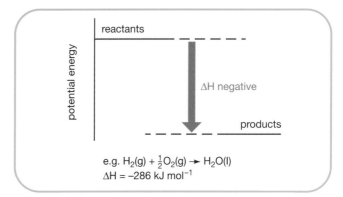

Figure 2.4 ΔH for an exothermic reaction

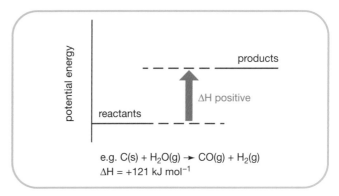

Figure 2.5 ΔH for an endothermic reaction

is not necessary to include the plus sign in the case of an endothermic reaction. The absence of a sign from the ΔH value will be taken to indicate that the reaction is endothermic.

Activation energy and activated complex

In the previous chapter we introduced the idea of activation energy, defining it as the minimum kinetic energy required by colliding molecules for a reaction to occur. We can also consider activation energy from the point of view of potential energy. In the potential energy diagrams shown in Figures 2.6 and 2.7, the activation energy appears as an 'energy barrier' which has to be overcome as the reaction proceeds from reactants to products. Whether a reaction is fast or slow will depend on the height of this barrier. The higher the barrier, the slower will be the reaction. It is worth emphasising at this point that **the rate of reaction does *not* depend on the *enthalpy* change**.

As the reaction proceeds from reactants to products, an intermediate stage is reached at the top of the activation energy barrier at which a highly energetic

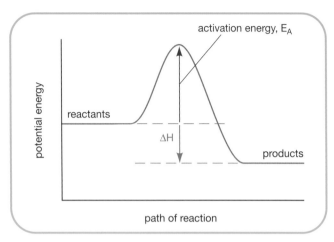

Figure 2.6 The activation energy for an exothermic reaction

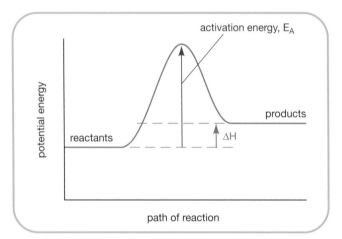

Figure 2.7 The activation energy for an endothermic reaction

species called an **activated complex** is formed. This is illustrated in Figure 2.8, which also shows that the activation energy can be redefined as the energy needed by colliding particles to form the activated complex.

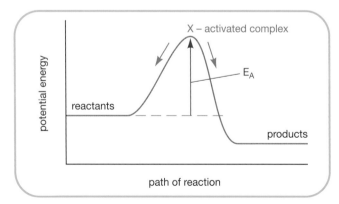

Figure 2.8 Activated complex

Activated complexes are very unstable and only exist for a very short time. From the peak of the energy

barrier the complex can lose energy in one of two ways to form stable substances, i.e. either to yield the products or to form the reactants again.

The addition reaction between ethene and bromine is believed to go via the activated complex shown in Figure 2.9 when the reaction is carried out under certain conditions. The dotted lines in the structural formula of the complex indicate partial bonding between atoms. The first step shows the formation of the activated complex and a bromide ion which combine in the second step to form dibromoethane. The first step can, however, be reversed, i.e. the complex can break down to reproduce the reactants, namely ethene and bromine.

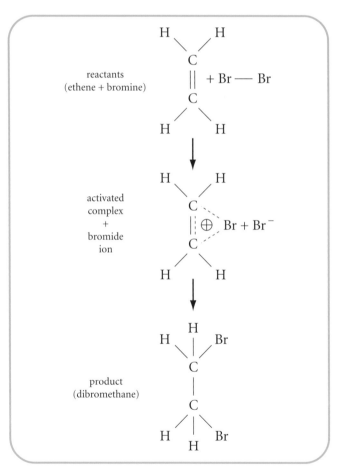

Figure 2.9 Reaction showing the formation of an activated complex

Potential energy diagrams give useful information about the energy profile of a reaction. When drawn to scale they can also be used to calculate the enthalpy change and/or the activation energy of a reaction as shown in Question 1.

Question

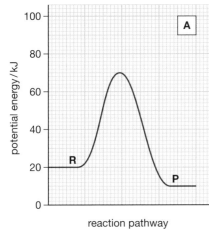
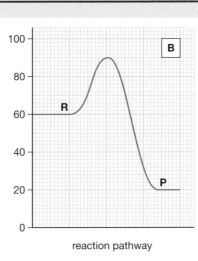
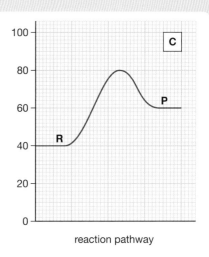

Key: **R** = reactants   **P** = products

1   Potential energy diagrams of three different reactions are shown above.

   a) Identify the endothermic reaction(s).

   b) For each reaction calculate **i)** its activation energy, E_A and **ii)** its enthalpy change, ΔH.

   c) Which reaction is likely **i)** to be the slowest, **ii)** to have the least stable activated complex?

Catalysts

In the previous chapter we considered the function of a catalyst. We noted how a heterogeneous catalyst works by providing an alternative route for the reaction which needs less energy of collision between particles to produce a successful outcome. Although homogeneous catalysts operate in a different way, it can still be said that in general catalysts provide alternative reaction pathways involving less energy. In other words a catalyst *lowers the activation energy of a reaction*. This is illustrated in the potential energy diagram shown in Figure 2.10.

It is worth drawing attention to the contrasting ways in which the use of a catalyst and the use of heat affect the rate of a reaction. **Heating speeds up a reaction by increasing the number of molecules which have energy greater than the activation energy. A catalyst speeds up a reaction by lowering the activation energy.** The former provides energy to overcome the energy barrier; the latter lowers the barrier. This confirms the important role played by catalysts in saving energy in many industrial processes.

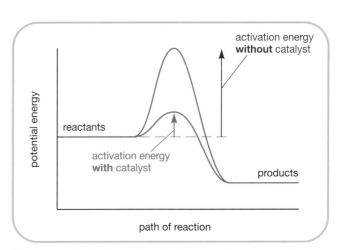

Figure 2.10 The lowering of the activation energy by a catalyst

Experimental measurement of enthalpy changes
Enthalpy of combustion

The enthalpy of combustion of a substance is the enthalpy change when one mole of the substance is burned completely in oxygen. Equations for the complete combustion of propane and methanol, for example, are given below along with their enthalpies of combustion.

$$C_3H_8(g) + 5O_2(g) \rightarrow 3CO_2(g) + 4H_2O(l)$$

$$\Delta H = -2220\,kJ\,mol^{-1}$$

$$CH_3OH(l) + \tfrac{3}{2}O_2(g) \rightarrow CO_2(g) + 2H_2O(l)$$

$$\Delta H = -727\,kJ\,mol^{-1}$$

Note that

♦ there is a negative sign in the ΔH values since combustion is an exothermic reaction

♦ it is usual to write the equation showing one mole of substance that is burning.

The equation for the combustion of methanol shows that 1.5 moles of oxygen are needed per mole of methanol. If the equation is doubled to remove half moles of oxygen, then the ΔH value is doubled. The units are then kJ, not $kJ\,mol^{-1}$, since the energy released relates to the complete combustion of 2 moles of methanol.

Prescribed Practical Activity

The enthalpy of combustion of a simple alkanol can be determined by experiment using apparatus like that shown in Figure 2.11. The burner containing the alkanol is weighed before and after burning. The alkanol is allowed to burn until the temperature of the water in the copper can has been raised by, say, 10°C before extinguishing the flame.

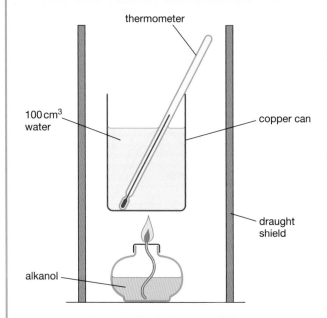

thermometer

100 cm³ water

copper can

draught shield

alkanol

Figure 2.11

The heat energy (E_h) gained by the water in the copper can can be calculated from the formula:

$$E_h = cm\,\Delta T$$

Where

c is the specific heat capacity of water, $4.18\,kJ\,kg^{-1}\,°C^{-1}$

m is the mass of water heated, in kg

ΔT is the change in temperature of the water.

This is a measure of the heat released by the burning alkanol and from this the enthalpy of combustion of the alkanol can be calculated. The method of calculation is shown in Worked Example 2.1 using specimen data for the burning of methanol, which gives a result close to the accepted figure given in your Data Book. The result obtained using the above apparatus will be considerably less. It is useful to consider the possible sources of error and how they could be minimised.

Worked Example 2.1

Enthalpy of combustion of methanol, CH_3OH

Data:

Mass of burner + methanol
before burning $= 53.65\,g$

Mass of burner + methanol
after burning $= 53.46\,g$

Mass of water heated, m $= 100\,g = 0.1\,kg$

Temperature rise of water, ΔT $= 10°C$

Calculation:

Heat energy released, $E_h = cm\Delta T$
$= 4.18 \times 0.1 \times 10$
$= 4.18\,kJ$

Gram formula mass of methanol,
CH_3OH $= 32\,g$

Mass of methanol burned $= 0.19\,g$

Number of moles of methanol
burned, n $= \dfrac{0.19}{32}$

Heat energy released per mole, $\dfrac{E_h}{n} = \dfrac{4.18}{n}$

$= \dfrac{4.18 \times 32}{0.19}$

$= 704\,kJ$

Enthalpy of combustion of
methanol, ΔH $= -704\,kJ\,mol^{-1}$

Note that, since the reaction is exothermic, it is necessary to insert a negative sign in the final result.

Questions

2   0.01 moles of methane when burned raised the temperature of 200 cm^3 of water by 10.5°C. Calculate the enthalpy of combustion of methane.

3   Mass of burner + propanol before burning $= 84.25\,g$
Mass of burner + propanol after burning $= 83.95\,g$
Mass of water heated $= 120\,g$
Temperature of water before heating $= 19.0°C$
Temperature of water after heating $= 31.5°C$

The data given above were obtained by some pupils using apparatus similar to that shown in Figure 2.11. The alkanol used was propanol, C_3H_7OH.

a) Use the data to calculate the enthalpy of combustion of propanol obtained in this experiment.

b) Compare this result with the Data Book value and describe what you consider to be the main sources of experimental error.

Enthalpy of solution

The enthalpy change of solution of a substance is the enthalpy change when one mole of the substance dissolves in water.

The enthalpy of solution of a soluble substance can be determined experimentally as illustrated in Figure 2.12. The temperature of the water before adding a weighed amount of solute is measured along with the temperature of the final solution. A thermometer reading to the nearest 0.1 or 0.2°C will enable more accurate results to be obtained. The method of calculating the enthalpy change is shown in Worked Example 2.2.

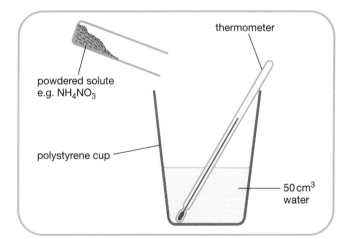

thermometer

powdered solute
e.g. NH_4NO_3

polystyrene cup

$50\,cm^3$
water

Figure 2.12 Enthalpy of solution

Worked Example 2.2

Enthalpy of solution of NH_4NO_3

Data:

Mass of solute (ammonium
nitrate) $= 1.00\,g$

Mass of water used, m $= 50\,g = 0.05\,kg$

Temperature of water initially $= 20.4°C$

Temperature of solution $= 18.8°C$

Calculation:

Temperature fall, ΔT $\qquad = 1.6°C$

Heat energy absorbed, $E_h = cm\Delta T$
$\qquad\qquad\qquad\qquad\quad = 4.18 \times 0.05 \times 1.6\ kJ$

Gram formula mass of ammonium
nitrate $\qquad\qquad\qquad\quad = 80\ g$

Number of moles of solute
used, n $\qquad\qquad\quad = \dfrac{1.00}{80} = 0.0125$

Heat energy absorbed
per mole, $\dfrac{E_h}{n}$ $\qquad = \dfrac{4.18 \times 0.05 \times 1.6}{0.0125}$

$\qquad\qquad\qquad\qquad\quad = 26.8\ kJ$

Enthalpy of solution of
ammonium nitrate, $\Delta H \quad = 26.8\ kJ\ mol^{-1}$

Note that this reaction is endothermic. As mentioned earlier it is not essential to include a plus sign before the numerical value. Note also that in this calculation (as well as in Worked Example 2.3) two approximations are being made, namely that

★ the density of a dilute aqueous solution is the same as that of water, i.e. $1\ g\ cm^{-3}$ at room temperature

★ the specific heat capacity of a dilute aqueous solution is the same as that of water, i.e. $4.18\ kJ$ $kg^{-1}\ °C^{-1}$.

Questions

4 $3.03\ g$ of potassium nitrate, KNO_3, were dissolved in $100\ cm^3$ of water. The temperature of the water fell by $2.5°C$. Calculate the enthalpy of solution of this salt.

5 Calculate the enthalpy of solution of sodium hydroxide from the following data.

Mass of solute, NaOH $\qquad = 0.80\ g$
Volume of water used $\qquad = 40\ cm^3$
Temperature of water initially $\quad = 20.2°C$
Temperature of solution $\qquad = 25.0°C$

Enthalpy of neutralisation

The enthalpy of neutralisation of an acid is the enthalpy change when the acid is neutralised to form one mole of water.

As you already know, when any acid such as hydrochloric acid is neutralised by any alkali such as sodium hydroxide, a salt – in this case sodium chloride – and water are formed.

$$HCl(aq) + NaOH(aq) \rightarrow NaCl(aq) + H_2O(l)$$

When an acid is neutralised by an alkali the reaction can be expressed by the following equation (with spectator ions omitted).

$$H^+(aq) + OH^-(aq) \rightarrow H_2O(l)$$

The enthalpy of neutralisation of an acid by an alkali can be found by experiment as shown in Figure 2.13. The temperature of each reactant is measured before mixing so that the average initial temperature can be calculated. The solutions are then mixed and the highest temperature of the neutral solution is noted. The method of calculating the enthalpy change is shown in Worked Example 2.3.

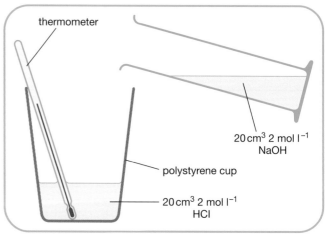

thermometer

$20\ cm^3\ 2\ mol\ l^{-1}$ NaOH

polystyrene cup

$20\ cm^3\ 2\ mol\ l^{-1}$ HCl

Figure 2.13 Enthalpy of neutralisation

Worked Example 2.3

Enthalpy of neutralisation of HCl(aq) by NaOH(aq)

Data:

Solutions used: 20 cm^3 2 mol l^{-1} HCl and 20 cm^3 2 mol l^{-1} NaOH

Temperature of acid before mixing	= 19.5°C
Temperature of alkali before mixing	= 18.5°C
Temperature of solution after mixing	= 32.5°C

Calculation:

Average initial temperature $= \dfrac{19.5 + 18.5}{2} = 19.0°C$

Temperature rise, ΔT $= 32.5 - 19.0 = 13.5°C$

Total volume of solution $= 40$ cm^3

Mass of solution heated $= 40$ g $= 0.04$ kg

Heat energy released E_h $= cm\Delta T = 4.18 \times 0.04 \times 13.5$ kJ

Number of moles of acid, n $= C \times V$

   $= 2 \times 0.02 = 0.04$

Heat energy released per mole, $\dfrac{E_h}{n}$ $= \dfrac{4.18 \times 0.04 \times 13.5}{0.04} = 56.4$ kJ mol^{-1}

Enthalpy of neutralisation, ΔH $= -56.4$ kJ mol^{-1}

In the calculation above 0.04 moles of HCl are neutralised by 0.04 moles of NaOH producing 0.04 moles of water. Hence, in the calculation, cmΔT is divided by n = 0.04 so that the ΔH value obtained refers to one mole of acid being neutralised to form one mole of water.

Questions

6 Calculate the enthalpy of neutralisation of nitric acid by potassium hydroxide given that a temperature rise of 6.5°C was observed on mixing 10 cm^3 of 1 mol l^{-1} HNO$_3$ and 10 cm^3 of 1 mol l^{-1} KOH.

7 Calculate the enthalpy of neutralisation of sulphuric acid by sodium hydroxide from the following data. Solutions used: 20 cm^3 of 1 mol l^{-1} H$_2$SO$_4$ and 20 cm^3 of 2 mol l^{-1} NaOH

Temperature of acid before mixing	= 20.0°C
Temperature of alkali before mixing	= 20.8°C
Temperature of solution after mixing	= 34.0°C

Remember that this enthalpy change relates to the formation of one mole of water.

In question 7 sulphuric acid, a dibasic acid, is used but the enthalpy of neutralisation is about the same as in question 6 since one mole of sulphuric acid produces *two* moles of water when completely neutralised by sodium hydroxide. This reaction is shown by the following equation:

$$H_2SO_4(aq) + 2NaOH(aq) \rightarrow Na_2SO_4(aq) + 2H_2O(l)$$

Monobasic or monoprotic acids are those which can yield one mole of hydrogen ions per mole of acid. Examples include hydrochloric acid (HCl), nitric acid (HNO$_3$) and ethanoic acid (CH$_3$COOH). Sulphuric acid (H$_2$SO$_4$) is an example of a dibasic or diprotic acid, since it can yield two moles of hydrogen ions per mole of acid, while phosphoric acid (H$_3$PO$_4$) is a tribasic or triprotic acid.

Study Questions

In questions 1–4 choose the correct word from the following list to complete the sentence.

endothermic	higher	exothermic	lower
salt	stable	water	unstable

1   A catalyst provides a reaction with an alternative route which has a _____ activation energy than when no catalyst is used.

2   A reaction is _____ if its enthalpy change has a positive value.

3   The enthalpy of neutralisation of an acid is the enthalpy change when the acid is neutralised to form one mole of _____ .

4   The _____ arrangement of atoms formed at the top of the potential energy barrier during a reaction is called the activated complex.

5   The enthalpy of combustion of ethene relates to which of the following equations?

A $C_2H_4(g) + 2O_2(g) \rightarrow 2CO_2(g) + 2H_2(g)$

B $C_2H_4(g) + 3O_2(g) \rightarrow 2CO_2(g) + 2H_2O(l)$

C $C_2H_4(g) + 2O_2(g) \rightarrow 2CO(g) + 2H_2O(l)$

D $C_2H_4(g) + O_2(g) \rightarrow 2CO(g) + 2H_2(g)$

6   When a catalyst is used in a reaction

A   the enthalpy change decreases

B   the activation energy decreases

C   the enthalpy change increases

D   the activation energy increases.

7   The graph in the next column indicates an energy diagram for the decomposition of ethanal (CH_3CHO) vapour according to the equation:

$$CH_3CHO(g) \rightarrow CH_4(g) + CO(g)$$

a) What is the value for the activation energy of the reaction?

b) What is the enthalpy change for the reaction? Is the reaction exothermic or endothermic?

c) Iodine vapour catalyses the above reaction. Copy the graph and on it indicate by means of a dotted line the reaction pathway for a catalysed reaction.

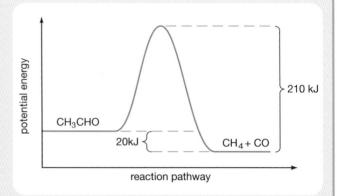

8* The reaction between hydrazine and hydrogen peroxide can be used for propelling rockets.

$$N_2H_4(l) + 2H_2O_2(l) \rightarrow N_2(g) + 4H_2O(g)$$
$$\Delta H = -685 \text{ kJ mol}^{-1}$$

From the information given suggest *two* reasons for its suitability.

9

Reactant	Hydrochloric acid	Ammonia solution
Concentration/ mol l^{-1}	1.0	2.0
Volume/cm^3	40.0	20.0
Temperature/°C	20.5	20.5

The highest recorded temperature after mixing these solutions was 28.5°C.

Use the data given above to calculate the enthalpy of neutralisation of hydrochloric acid by ammonia solution. Ammonia solution, $NH_3(aq)$, reacts as $NH_4OH(aq)$ when it neutralises an acid. The equation for the reaction with hydrochloric acid is

$$NH_4OH(aq) + HCl(aq) \rightarrow NH_4Cl(aq) + H_2O(l)$$

10   A polystyrene cup contained 50 cm^3 of water at 21.2°C. 2.53 g of powdered ammonium chloride was added and dissolved with stirring. Calculate the lowest temperature which the solution should reach given that the enthalpy of solution of NH_4Cl is + 15.0 kJ mol^{-1}.

Study Questions (continued)

11 Two litres of water, initially at 25°C, were heated using a butane burner as shown below.

aluminium pot
containing water

BUTANE

a) Using appropriate data from your Data Book, calculate the number of moles of butane needed to boil this quantity of water.

b) In practice, some heat is lost to the surroundings. Calculate the mass of butane burned if 80% of the heat produced by the burner is transferred to the water.

12* Methanoic acid, HCOOH, can break down to carbon monoxide and water by two different reactions, **A** and **B**.

Reaction A (catalysed)

$$HCOOH\ (aq) + H^+(aq) \rightarrow CO(g) + H_2O(l) + H^+(aq)$$

Reaction B (uncatalysed)

$$\overset{\text{heat}}{HCOOH\ (aq) \rightarrow CO(g) + H_2O(l)}$$

a) i) What is the evidence in the equation for Reaction **A** that the $H^+(aq)$ ion acts as a catalyst?

ii) Explain whether Reaction **A** is an example of heterogeneous or homogeneous catalysis.

b) The energy diagram for the **catalysed** reaction is:

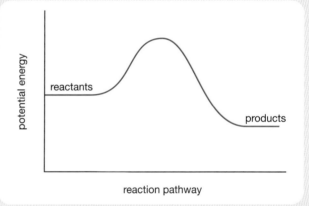

Copy this diagram and draw a line on it to show the reaction pathway for the **uncatalysed** reaction.

3   Patterns in the Periodic Table

From previous work you should know and understand the following:

★ Atomic structure, atomic number and atomic mass
   (Summary in Chapter 18, page 193)

★ Electron arrangement and energy levels

★ Valency.

The development of the Periodic Table

The Periodic Table will already be familiar to you but more detailed study is now necessary. It is worthwhile seeing how the shape of the Periodic Table has developed. Nineteenth century chemists had a bewildering mass of chemical information available to them. To try to simplify this information, and find some basic underlying principles, they devised various classifications of the elements. At that time the only apparently fundamental information available for atoms of different elements was the atomic mass, and this was made the basis of classification. Various groupings were tried, but most were unsuccessful or were derided. One example was an idea of Newlands that, when arranged in order of atomic mass, every eighth element showed similarities like musical notes in a scale. He called his idea the 'Law of Octaves'.

In the mid-nineteenth century, Lothar Meyer began to accumulate data on the elements and to investigate the variation with atomic mass. He discovered that many important quantities which could be measured accurately varied periodically, i.e. in a regular way, when plotted against atomic mass. Quantities in this category include melting point, boiling point and density. A graph of this type (but using atomic number instead of atomic mass as the independent variable) is shown in Figure 3.1. It can be seen that alkali metals, halogens and noble gases lie at characteristic positions on the series of undulations produced.

Note: the properties plotted on graphs of this type are not continuous functions. Bar graphs would be a more correct way of displaying the data, but in the form used it is easier to label specific features.

The greatest step in the progress towards a Periodic Law was taken by Mendeleev in 1869. His main points are summarised below.

♦ The elements fall into a repeating pattern of similar properties if arranged in order of increasing atomic mass.

♦ The list was arranged into vertical and horizontal sequences, called **groups** and **periods**. The groups contained elements which were chemically similar.

♦ Blanks were left for as yet unknown elements if it appeared that without these blanks dissimilar elements would be thrown together. Mendeleev predicted the properties for the missing elements and when they were discovered later, including gallium and germanium, the predictions were accurate. When the noble gases were later discovered, they fitted into the Periodic Table as a separate column.

♦ Errors in atomic mass determinations, particularly of beryllium, were probable. Despite the great advance made, and the ready acceptance of the Periodic Law, there were still anomalies.

♦ Certain elements in the table were in reverse order of atomic mass.

♦ There was no easy way of placing the 'rare earths' – the elements lanthanum to lutetium.

The true significance of the Periodic Law was seen after the discovery of 'atomic number' by Moseley in the early twentieth century. The atoms of the elements have an increasing number of protons and electrons which largely determine the chemical properties of the

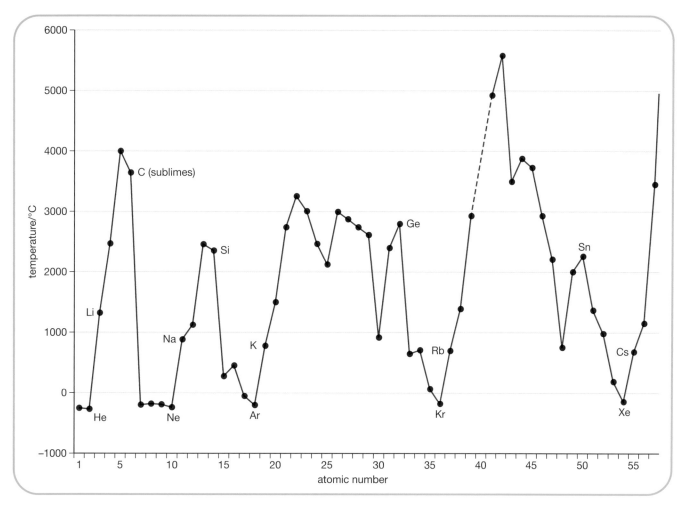

Figure 3.1 Variation of boiling point with atomic number

elements. Most of the anomalies of the Periodic Table were removed when classification was made by atomic number.

In the modern Periodic Table, the elements are arranged by increasing atomic number, each new horizontal row – period – commencing when a new layer of electrons starts to fill in the atom. Each main vertical column contains elements with the same number of electrons in the outermost layer. Since valency depends on the number of unpaired outermost electrons, the main vertical groups contain elements with the same main valency, e.g. group I contains the alkali metals of valency 1 and group VII contains the halogens of valency 1.

We shall be making use of the ideas of Lothar Meyer and Mendeleev to try to achieve a better understanding of the relationships between the first 20 elements.

Trends in physical properties of the elements

Melting and boiling points

In the Lothar Meyer curves (Figures 3.1 and 3.2) the repeating pattern of high and low values of melting and boiling points can be seen. Generally speaking, the high values occur for elements just starting a new layer of electrons, with a decrease as the layer fills. High values indicate that a large energy input is necessary to separate the particles of the elements sufficiently so that they first become liquid and then vapour. In other words, the bonding between the particles of the elements on the left of the Periodic Table is stronger, or more extensive, than that between the particles of the right-hand elements.

Melting and boiling points *decrease* in group I of the Periodic Table as the atomic number *increases*. The reason is the decrease of the force of attraction between

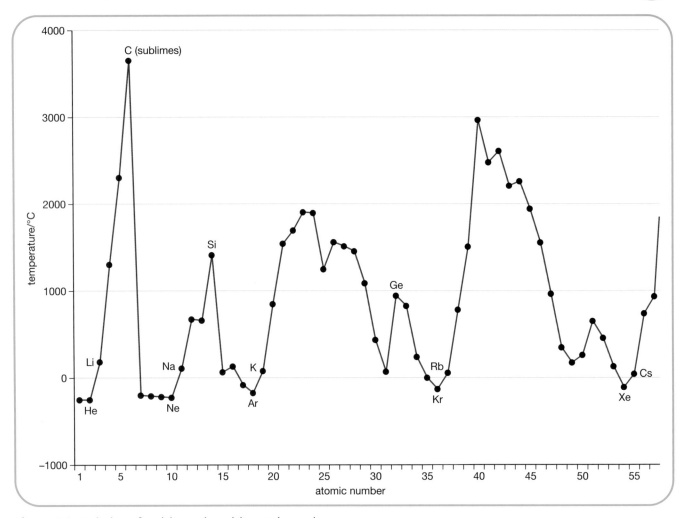

Figure 3.2 Variation of melting point with atomic number

the particles of the element as the atomic number increases. In group VII, melting and boiling points increase *with* atomic number, since the force of attraction between molecules increases with atomic number.

Density

A plot of density in gram cm^{-3} (Figure 3.3) shows a typical Lothar Meyer graph. In any period of the Periodic Table, density first increases from group I to a maximum in the centre of the period, and then decreases again towards group 0. In any group the density tends to increase as atomic number increases.

Atomic size

A measure of atomic size is the covalent atomic radius which is defined as half the distance between the nuclei of two bonded atoms of the element. Internuclear distances can be measured accurately by X-ray

diffraction, whereas the boundaries of atoms which are defined by electron clouds are not clearly observable. The values for some covalent radii are quoted in Table 3.1 (units are picometres, i.e. 10^{-12} m). The graph (Figure 3.4) shows their periodic variation.

Two quite clear trends are discernible in the table shown:

1   In a horizontal row (period) of the Periodic Table, covalent radii decrease because the atoms being considered all have the same number of occupied energy levels, whilst there is an increase of one proton in the nucleus from one element to the next. This increase in the nuclear positive charge exerts an increasing attraction on the electrons, resulting in the outer layer decreasing in size. This means the covalent radius becomes smaller.

2   In any vertical column (group) all the elements have the same number of outer electrons, but one more

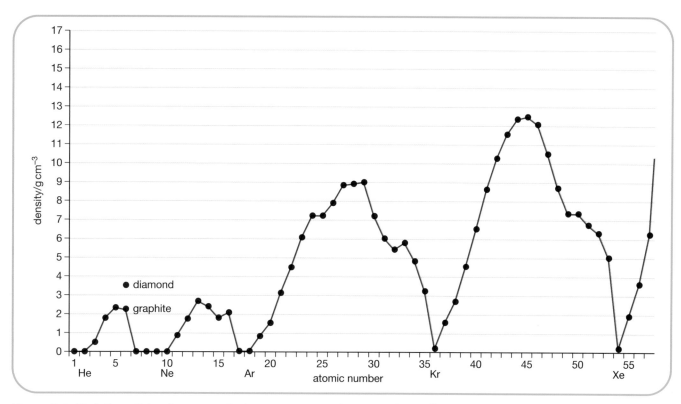

Figure 3.3 Variation of density at standard temperature and pressure ($g\,cm^{-3}$) with atomic number

Li	Be	B	C	N	O	F	Ne	
134	129	90	77	75	73	71	—	
Na	Mg	Al	Si	P	S	Cl	Ar	increase
154	145	130	117	110	102	99	—	in value
K	Ca	Ga	Ge	As	Se	Br	Kr	in each
196	174	120	122	121	117	114	—	column
Rb	Sr	In	Sn	Sb	Te	I	Xe	
216	191	150	140	143	135	133	—	

——————————————— decrease in value in each row ———————————————→

Table 3.1 Covalent atomic radii (pm)

energy level is occupied by electrons in each succeeding element. Although nuclear charge also increases, its effect is outweighed by the much greater radius of successive electron layers and hence covalent radius increases down a group.

Bond lengths between two atoms in a covalent or polar covalent compound are approximately the sum of the two appropriate covalent radii.

Question

1 Explain why
   a) the potassium atom is larger than the sodium atom.
   b) the chlorine atom is smaller than the sodium atom.

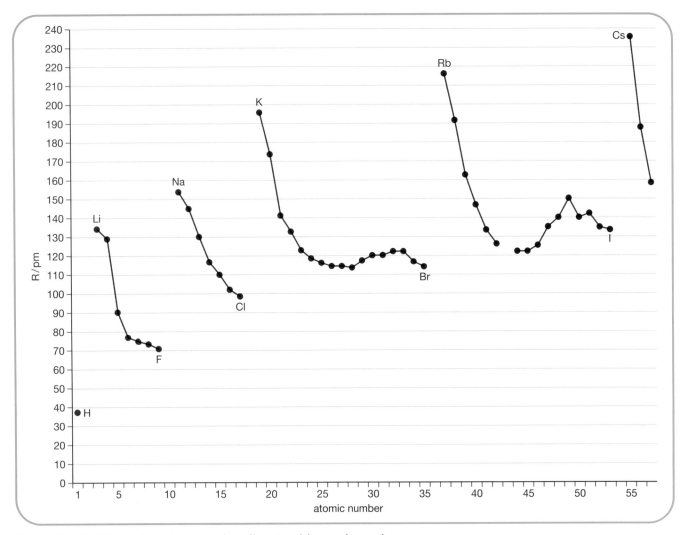

Figure 3.4 Variation of covalent atomic radius, R, with atomic number

First ionisation energy (or enthalpy)

In the formation of an ionic bond, one important factor is the energy change involved in creating positive ions from neutral isolated atoms, i.e. atoms considered to be in the gaseous state. The change, as with all energy values in chemistry, is measured per mole and is called the **ionisation energy**. It is an enthalpy change and is represented by a ΔH value.

$$Na(g) \rightarrow Na^+(g) + e^- \quad \Delta H = (+)502 \, kJ \, mol^{-1}$$

For sodium, 502 kJ of energy are required to remove the first electron from each of one mole of sodium atoms in the gaseous phase. This is properly called the first ionisation energy of sodium since it is possible to measure the energy required to remove successive electrons from sodium atoms.

For example, the second ionisation energy is the enthalpy change associated with

$$Na^+(g) \rightarrow Na^{2+}(g) + e^- \quad \Delta H = (+)4560 \, kJ \, mol^{-1}$$

The values for some first ionisation energies are quoted in Table 3.2 (units are $kJ \, mol^{-1}$).

These values are plotted, together with second ionisation energies, for the first 20 elements in Figure 3.5.

It is not necessary to explain all the minor features of the graph, but some major points can be made. Firstly, in each group of elements, there is a decrease of first ionisation energy as the group descends. The electron is being removed from the outermost layer of electrons. This layer is increasingly distant from the nuclear attraction and hence, although the nuclear charge is

Li	Be	B	C	N	O	F	Ne	
526	905	807	1090	1410	1320	1690	2090	
Na	Mg	Al	Si	P	S	Cl	Ar	decrease
502	744	584	792	1020	1010	1260	1530	down
K	Ca	Ga	Ge	As	Se	Br	Kr	group
425	596	577	762	947	941	1150	1350	
Rb	Sr	In	Sn	Sb	Te	I	Xe	
409	556	556	709	834	870	1020	1170	

———— overall increase along period ————→

Table 3.2 First ionisation energies (kJ mol^{-1})

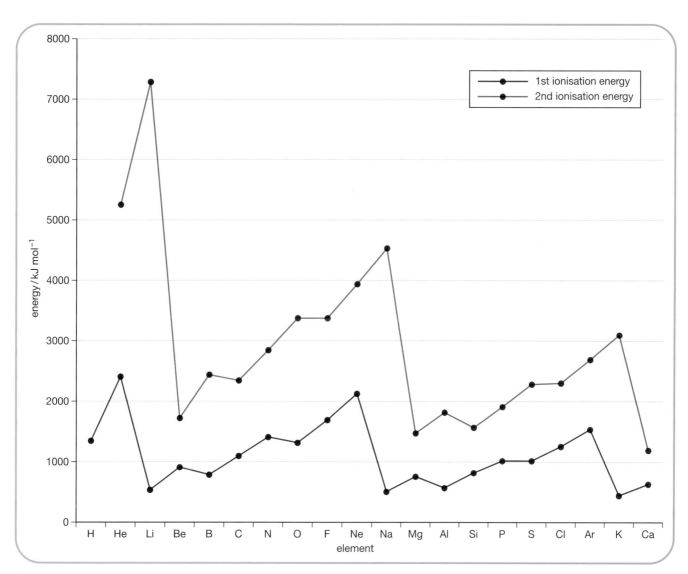

Figure 3.5 First and second ionisation energies for the first 20 elements

also increasing, less energy is required to remove an electron.

An additional factor is the screening effect of electrons in inner orbitals. These inner electrons reduce the attraction of the nucleus for outermost electrons, hence reducing the ionisation energy.

In each period, the pattern is less straightforward, but there is an overall increase. The electron being removed is in the same layer for any element in the same period e.g. Li–Ne or Na–Ar. As already pointed out, the nuclear charge is increasing along each period, resulting in a contraction in the sizes of the electron orbitals. The outermost electrons are therefore more strongly held and so the energy required to remove them, the ionisation energy, increases along each period.

Finally it is worth noting that within each period, the noble gas has the highest value for first ionisation energy. This goes some way to explaining the great stability of filled orbitals and the resistance of the noble gases to formation of compounds. It should be noted, however, that electrons *can* be removed from noble gas atoms. If some other change can compensate for the energy required then ionic compounds of the noble gases *can* be made.

Questions

2 Suggest a reason for the large difference between 1st and 2nd ionisation energies for each of the group I elements.

3 Use your Data Book to calculate the energy required for the following changes:

   a) $Ca(g) \rightarrow Ca^{2+}(g) + 2e^-$

   b) $Al(g) \rightarrow Al^{3+}(g) + 3e^-$

4 Xe is the noble gas whose compounds were prepared first. Suggest why this was the case.

Electronegativity

In a covalent bond, formed by the sharing of an electron pair between two atoms, the attraction of the atoms for these electrons depends on the elements concerned. The relative powers of the atoms in a molecule to attract bonding electrons to themselves are defined as their electronegativities. This has been measured in a quantitative fashion, but it is generally agreed that too great an importance should not be attached to the values. Electronegativity values for some elements are, however, given in Table 3.3.

H							
2.2							
Li	Be	B	C	N	O	F	
1.0	1.5	2.0	2.5	3.0	3.5	4.0	
Na	Mg	Al	Si	P	S	Cl	decrease
0.9	1.2	1.5	1.9	2.2	2.5	3.0	down
K	Ca	Ga	Ge	As	Se	Br	group
0.8	1.0	1.6	1.8	2.2	2.4	2.8	
Rb	Sr	In	Sn	Sb	Te	I	
0.8	1.0	1.7	1.8	2.1	2.1	2.6	
Cs	Ba						
0.8	0.9						

increase
across period

Table 3.3 Electronegativity values

In general, the electronegativity increases from left to right along a period since nuclear charge increases in the same direction, and the electronegativity decreases down a group of the Periodic Table since the atomic size increases down the group.

The difference in electronegativity values for the atoms joined gives an indication of the relative degrees of polarity in covalent bonds.

Study Questions

1 Which of the following increases when a chlorine atom changes to a chloride ion?

   A The atomic number

   B The charge of the nucleus

   C The mass number

   D The number of full energy levels

2 When compared with potassium, lithium has

   A a lower first ionisation energy

   B a greater electronegativity

   C a lower melting point

   D a greater covalent radius

3 Which of the following statements is true?

   A Magnesium is more electronegative than sulphur.

   B Sulphur has a greater first ionisation energy than magnesium.

   C Magnesium has a smaller covalent radius than sulphur.

   D Sulphur has a lower relative atomic mass than magnesium.

4 $Fe(g) \rightarrow Fe^{2+}(g) + 2e^-$

   The enthalpy change for the reaction shown above in $kJ\ mol^{-1}$ is

   A 1532 B 1570 C 2336 D 3140.

5* Graph 1 shows the boiling points of the group VII elements.

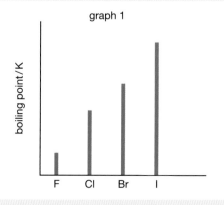

graph 1

a) Why do the boiling points increase down group VII?

b) Graph 2 shows the melting points of elements from lithium to neon across the second period.

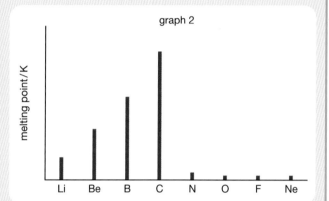

graph 2

Give a reason for the high melting points of boron and carbon.

c) Graph 3 shows the first ionisation energies of the group I elements.

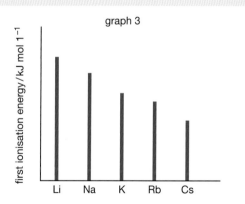

graph 3

Explain why there is a decrease in first ionisation energy down this group.

d) Over.

Study Questions (continued)

d) Graph 4 shows the first ionisation energies of successive elements with increasing atomic number.

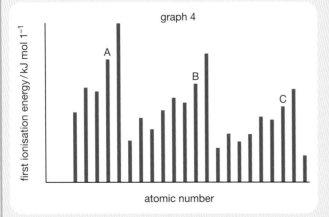

graph 4

Elements A, B and C belong to the same group of the Periodic Table. Identify the group.

6* Ionisation energies can be found by applying an increasing voltage across test samples of gases until the gases ionise. The results below were obtained from experiments using hydrogen atoms and then helium atoms.

Element	Voltage at which an atom of gas ionises/V	
Hydrogen	13.6	no further change
Helium	24.6	54.5

a) Why are there two results for helium but only one for hydrogen?

b) i) Write an equation which would represent the first ionisation energy of helium gas.

   ii) Why is the first ionisation energy of helium higher than that of hydrogen?

7* a) Atoms of different elements have different attractions for bonded electrons. What term is used as a measure of the attraction an atom involved in a bond has for the electrons of the bond?

b) Atoms of different elements are different sizes. What is the trend in atomic size across the period from sodium to argon?

c) Atoms of different elements have different ionisation energies. **Explain clearly** why the first ionisation energy of potassium is less than the first ionisation energy of sodium.

8* The graph below relates the ionic radii of some elements to their atomic numbers.

a) Copy the graph and, plot the ionic radii you would predict for the ions of the elements with atomic numbers 13 and 15. You may wish to refer to a Periodic Table in the Data Book to help you.

b) The value quoted for hydrogen is for the hydride ion (H^-).

   i) Why is no value quoted for the H^+ ion?

   ii) Why is the H^- ion larger than the Li^+ ion?

c) Why is there a large increase in ion size from boron to nitrogen?

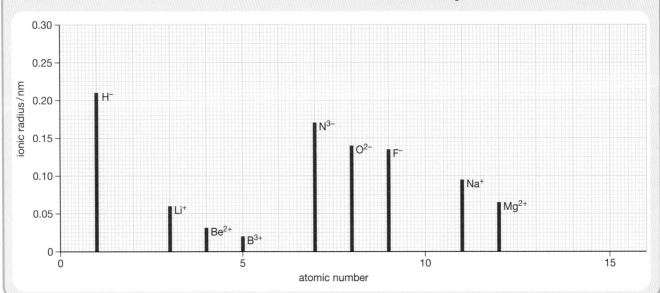

Bonding, Structure and Properties of Elements

From previous work you should know and understand the following:

★ Atom structure

★ Periodicity

★ Covalent bonding

★ Physical properties of metals.

In this chapter, you will see that the properties and structures of elements are related to the types of bonding present. Bonding is a term describing the mechanism by which atoms join together. **Structure** describes the way in which the atoms, or particles derived from them, are arranged. The resulting characteristics, whether physical or chemical, of the substances are their **properties**. The complexities of compounds will be considered in the next chapter.

Types of bonding in elements
Metallic bonding

Metallic bonding is the electrostatic force of attraction between positively-charged ions, formed by the loss of the outer electrons of metal atoms, and these electrons. The electrons are delocalised, i.e. they are held in common by all the ions. Although the metal ions pack together in various specific ways, resulting in the metals having crystalline structures, metallic bonding is not directional.

Covalent bonding

Covalent bonding is also electrostatic, but this time the atoms are held together by the attraction between their positive nuclei and negatively-charged shared pairs of electrons. Each shared pair of electrons constitutes a covalent bond. Since all the atoms in an element are alike in terms of protons and electrons, the bonding electrons are shared equally. In the atoms from which the two electrons come to make the shared pair, the electrons are held in orbitals which have a fixed orientation to each other. When more than one covalent bond is formed by an atom, the bonds are therefore orientated in specific directions to each other.

> **Question**
>
> 1   What term describes the property of atoms which results in equal sharing of electrons in a covalently bonded molecule?

Van der Waals' bonding

Van der Waals' bonding is very weak bonding between molecules. Its mechanism will be explained later in this chapter.

With these types of bonding in mind, the bonding, structure and properties of the first 20 elements will be examined.

Bonding in the first 20 elements
The 'noble gases'

The bonding in the elements is least complex at the right-hand side of the Periodic Table, i.e. for the 'noble gases'. These elements, with the exception of helium, always have an outer layer of eight electrons which is an especially stable arrangement. Because of the stability of the outer electrons, the 'noble gases' do not form either covalent or ionic bonds between their atoms. They are 'monatomic', i.e. their molecules consist of only one atom.

Although there are no ionic or covalent bonds between the atoms, there are other very weak forces. These forces are caused by the uneven distribution of the constantly moving electrons around the nuclei of the atoms. This causes the formation of **temporary dipoles** on the atoms; see Figure 4.1. The atoms then attract each other. The dipoles are constantly changing, but there are always some in existence.

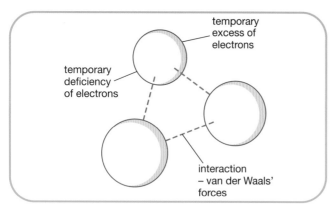

Figure 4.1 Van der Vaals' forces

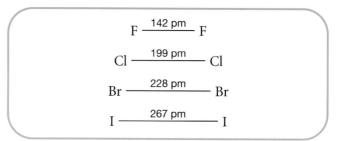

Figure 4.2 Bonding in the halogen molecules

These forces are called van der Waals' forces and are very weak compared with ionic and covalent bonds. Although weak, the van der Waals' forces are strong enough to allow the 'noble gases' to liquefy and solidify if they are cooled enough to remove the thermal kinetic energy of the atoms. Not surprisingly, helium, with only two electrons per atom, has the weakest van der Waals' forces between its atoms and is the most difficult element to condense and freeze. It only freezes in temperatures near to absolute zero (i.e. $-273°C$).

Because the other noble gases have increasing numbers of electrons, the van der Waals' forces, and hence melting and boiling points, increase down the group.

Groups VII, VI and V

In groups VII, VI and V, the structures of the elements are based on covalent bond formation to achieve eight outer electrons.

Group VII elements

The halogens with one unpaired outer electron can form one covalent bond and, as a result, diatomic molecules F_2, Cl_2, Br_2 and I_2 are formed. These molecules interact only weakly by the van der Waals' mechanism so that all the elements are volatile, and fluorine and chlorine are gaseous.

Group VI elements

Oxygen
Each oxygen atom uses its two unpaired electrons to form two covalent bonds with one other oxygen atom (except when the rarer form ozone, O_3, is formed). The molecules interact by van der Waals' bonding but since the interaction is weak, O_2 is gaseous.

Sulphur
When they have the ability to form two or more bonds, atoms can bond to more than one other atom. In the case of sulphur, closed eight-membered puckered rings are found in the crystalline forms, and zig-zag chains are found in plastic sulphur. Van der Waals' forces between the molecules are strong enough to make sulphur solid at room temperature.

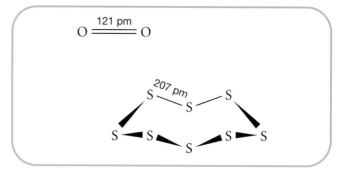

Figure 4.3 Bonding in oxygen and sulphur

Group V elements

Nitrogen
Nitrogen atoms form diatomic molecules with a triple bond, and only weak van der Waals' interactions.

Phosphorus
Phosphorus makes use of single bonds to three other atoms to form tetrahedral P_4 molecules. Fewer electrons in P_4 than in S_8 make the van der Waals' forces weaker and P_4 has a lower melting point than S_8.

Question

2 Why do the melting points of the halogens increase from fluorine to iodine?

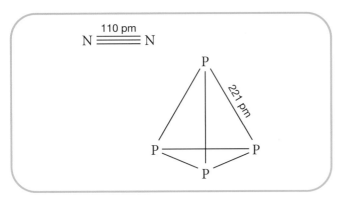

Figure 4.4 Bonding in nitrogen and phosphorus

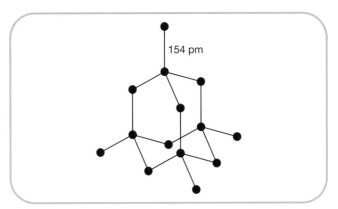

Figure 4.5 Diamond structure

In the elements of groups VII, VI and V, the intramolecular forces, i.e. the bonds *within* the molecule, are covalent. The intermolecular forces, those *between* the molecules, are the very weak van der Waals' forces. Most of these elements are therefore quite volatile, even if solid at room temperature, since only the intermolecular forces have to be broken to melt and boil them.

Group IV elements

The standard structure of the group is an infinite three dimensional network or lattice as in diamond and silicon. Each atom bonds covalently to four other atoms. The resultant structure is exceptionally hard and rigid. There are no discrete molecules, each atom being linked ultimately to each other atom in the piece of the element. There are no free electrons to allow conduction but in diamonds, for example, 'tunnels' between the atoms allow light to pass through, thus making them transparent.

Figure 4.6 Diamond model

Graphite, the other well-known variety of carbon, shown in Figure 4.7, has a structure based on three covalent bonds from each atom in one plane, forming layers of hexagonal rings. Each carbon atom contributes its fourth unpaired electron to delocalised orbitals extending over the layers. The result is strong bonding within the layers but only weak interaction between the layers. Since the delocalised electrons are held quite weakly, they can flow across the layers. Graphite therefore conducts, similarly to a metal. The layers separate easily so graphite is flaky, but because the layers are offset with respect to each other, light cannot pass through, so graphite is opaque.

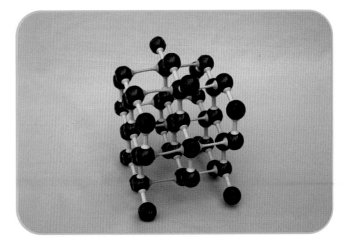

Figure 4.7 Graphite model

The fullerenes, discovered in 1985, are discrete, covalently bonded molecules. The smallest is spherical and is named buckminsterfullerene after an architect who designed large geodesic dome structures consisting

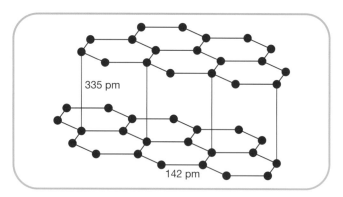

Figure 4.8 Graphite structure

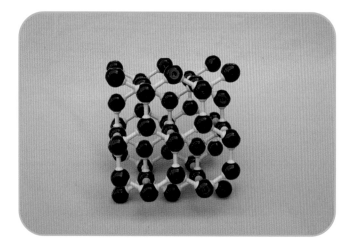

Figure 4.9 The alternate layers in graphite are offset

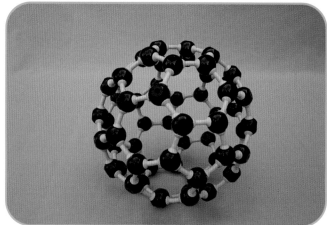

Figure 4.10 Fullerene, C_{60}, model

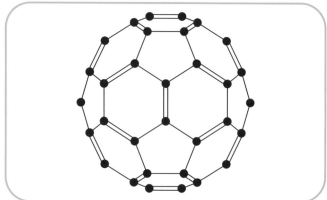

Figure 4.11 Fullerene structure

of 5- and 6-sided panels. The spherical molecule, C_{60}, has 5- and 6-membered rings of carbon atoms producing the overall shape. Other molecules with elongated shapes can exist, e.g. C_{70}, and there are also much longer 'nanotubes', but all contain 5- and 6-membered carbon atom rings. The properties of these molecules are currently under intensive investigation with a view to discovering applications of them. Amongst other reactions, fullerenes will form addition compounds with halogens, polymers (some containing palladium will behave as catalysts in alkene hydrogenation) and compounds with metals.

Groups I, II and III

Elements in these groups have insufficient electrons to allow the achievement of an octet of electrons in their outer layer by covalent bonding. Generally the elements contribute their outer electrons to a common 'pool' of delocalised electrons which act as a binding medium for the resultant positive ions. The bonding is less directional than covalent bonding, and the metals are therefore malleable and ductile. The electrons are capable of easy movements and hence the elements are electrical conductors. They are typical metals, and the above describes metallic bonding, as shown in Figure 4.12.

The one exception of any note in Groups I, II and III is boron. This forms a structure made up of B_{12} groups, which are interbonded with other groups. The result is an element almost as hard as diamond. The explanation of the bonding in boron is complex, as is that of the bonding of boron compounds.

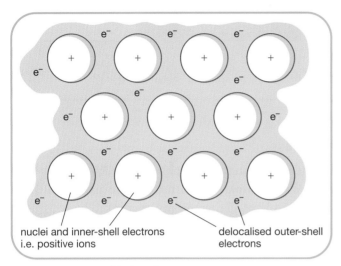

nuclei and inner-shell electrons
i.e. positive ions

delocalised outer-shell
electrons

Figure 4.12 Metallic bonding

Question

3   Suggest why aluminium conducts better than sodium.

Some specific properties of elements related to bonding

If we now look at some of the physical properties of these first 20 elements, we can see how these properties relate to the bond types present.

Melting and boiling points

Figure 4.14 shows that where the elements consist of discrete molecules (the monatomic and diatomic gases and P_4 and S_8), the melting and boiling points are low. This is because only the weak intermolecular van der Waals' forces have to be overcome in melting and boiling the element. The strong, covalent, intramolecular forces are unaffected.

In the covalent network solids carbon and silicon, covalent bonds must be broken when melting or boiling takes place – a much more difficult undertaking. Melting and boiling points are therefore much higher.

Similarly for group I, II and III elements, their very strong metallic bonds have to be overcome and these elements also have high melting and boiling points compared with covalent molecular elements.

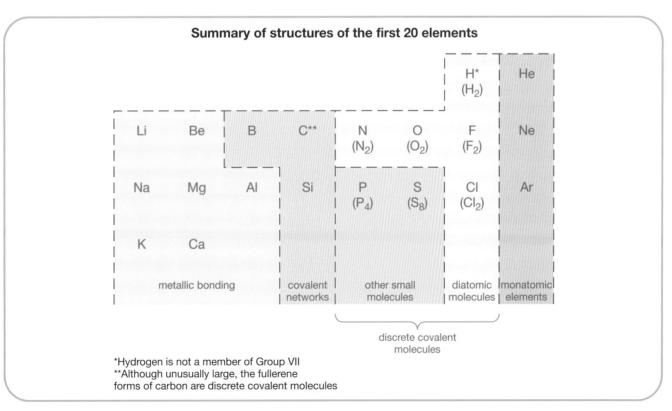

Summary of structures of the first 20 elements

						H* (H_2)	He
Li	Be	B	C**	N (N_2)	O (O_2)	F (F_2)	Ne
Na	Mg	Al	Si	P (P_4)	S (S_8)	Cl (Cl_2)	Ar
K	Ca						

metallic bonding | covalent networks | other small molecules | diatomic molecules | monatomic elements

discrete covalent molecules

*Hydrogen is not a member of Group VII
**Although unusually large, the fullerene forms of carbon are discrete covalent molecules

Figure 4.13

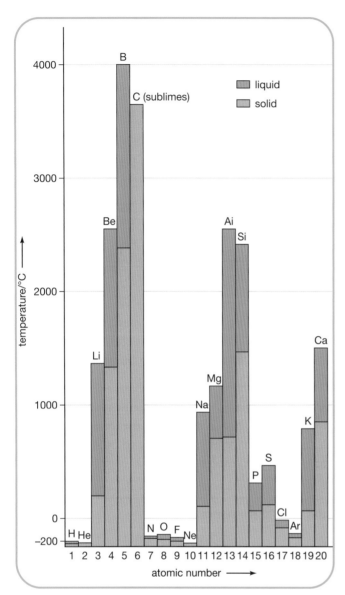

Figure 4.14 Melting and boiling points of elements 1–20

Hardness

As mentioned in the previous sections, hardness is related to bonding. In general, because giant covalent molecules have all their atoms interlinked by directional bonds, these substances, such as diamond, are very hard, but as the bonds will break on impact, the substances are brittle.

Small covalent molecules, like S_8 and P_4, are only held together by van der Waals' forces. The molecules can be forced apart easily and the substances are soft.

Although metallic bonds are strong, they are not directional like covalent bonds. Because of this metals can be distorted by impact or pressure, but they do not fall apart since the forces of attraction still exist between the positive ions and the negative delocalised electrons.

Study Questions

In questions 1–4 choose the correct word(s) from the following list to complete the sentence.

> covalent bonds   first   molecular   neutrons
> network   protons   second   van der Waals' forces

1 A fullerene is a form of carbon which has a covalent _____ structure.

2 The covalent radii of elements decrease from left to right across a period of the Periodic Table due to an increasing number of _____ in the nucleus.

3 The low melting point of sulphur is due to _____ between molecules.

4 The _____ ionisation energy of sodium is very high since it relates to the removal of an electron from an inner energy level.

Study Questions (continued)

5 Which element has discrete covalent molecules?

A Fluorine   B Silicon   C Lithium   D Boron

6 When liquefied chlorine evaporates

A   covalent bonds are broken

B   covalent bonds are formed

C   van der Waals' forces are broken

D   van der Waals' forces are formed.

7 A certain element contains only one type of bond, namely van der Waals' forces.

This element must be

A an alkali metal   B a noble gas
C a transition metal   D a halogen.

8 Which of the following statements is **not** true?

A   Diamond is very hard due to its covalent network structure.

B   Aluminium conducts electricity as it has delocalised electrons.

C   Nitrogen has weak covalent bonds since it is a gas at room temperature.

D   Phosphorus has a low melting point due to van der Waals' forces.

9 The second period of the Periodic Table consists of the elements lithium to neon.

a) Present the following information in the form of a table.

List the following elements: lithium, boron, nitrogen.

For each of these elements write down the melting point from your Data Book and indicate the type of bonding which is broken when each element melts.

b) The first ionisation energy of oxygen is 1320 kJ mol^{-1}.

Write the equation for this change.

c) What is the general trend in the first ionisation energies between lithium and neon? Explain this trend.

10* Diamond and graphite are well known forms of the element carbon. New forms of pure carbon have recently been made. They exist as individual molecules of different sizes and are called fullerenes. The main fullerene has the formula C_{60}.

a) How does the structure of a fullerene differ from that of diamond?

b) Fullerenes were first made by passing a high current of electricity through a graphite rod in an atmosphere of helium. This caused the graphite to vaporise. Suggest why helium gas was used.

c) Fullerenes can be made into hydrocarbons. One such hydrocarbon has the formula $C_{60}H_{36}$. Describe a chemical test which could be carried out on a solution of $C_{60}H_{36}$ to show whether the hydrocarbon is saturated or unsaturated.

11* The first 20 elements of the Periodic Table can be categorised according to their bonding and structure:

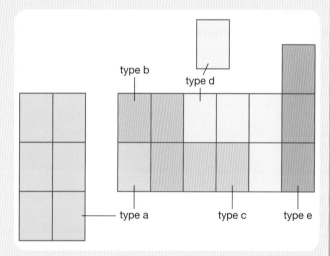

a) Copy and complete the following table by adding the appropriate letter for each type of element:

Type	Bonding and structure at normal room temperature and pressure
	Monatomic gases
	Covalent network solids
	Diatomic covalent gases
	Discrete covalent molecular solids
	Metallic lattice solids

b) Which of the types contains the element with the greatest attraction for the electrons in a covalent bond?

5 Bonding, Structure and Properties of Compounds

From previous work you should know and understand the following:

★ Ionic bonding

★ Covalent bonding.

Types of bonding in compounds

In this chapter you will see that the properties and structures of compounds also relate to the types of bonding present.

Ionic bonding

Ionic bonding is an electrostatic attraction between the positive ions of one element and the negative ions of another element.

Polar covalent bonding

Polar covalent bonding results from the same electron-sharing mechanism described in the previous chapter. In this chapter we shall see that, where atoms being joined are not of the same element, electrons are not always shared equally, so producing a *polar* bond. Covalent and polar covalent bonds are intramolecular bonds, i.e. bonds *within* molecules.

Intermolecular bonds

Intermolecular bonds are bonds *between* molecules. Van der Waals' forces mentioned in the last chapter are one type of intermolecular bond. As we shall see in this chapter other types of these bonds include permanent dipole–permanent dipole interactions and hydrogen bonds.

Ionic bonding

As described in Chapter 3, different elements have different attractions for bonding electrons, i.e. different electronegativities. Electronegativities are related to nuclear charge and the distance of outermost electrons from the nucleus. Two distinct trends are recognisable for the main group elements, i.e. excluding the transition elements:

1 electronegativity *increases* along a period from *left to right* up to group VII

2 electronegativity *decreases down* a group (see Table 3.3 on page 34).

It follows that fluorine has the greatest attraction for bonding electrons and, in practical terms, caesium has the least attraction.

The greater the difference in electronegativity between two elements, the less likely they are to share electrons, i.e. form covalent bonds. The element with the greater value of electronegativity is more likely to gain electrons to form a negative ion and the element with the smaller value is more likely to lose electrons to form a positive ion. This means electrostatic attraction results in ionic bonding, rather than covalent bonding. Elements far apart in the Periodic Table are more likely to form ionic bonds than elements close together. In practice, this means ionic compounds result from metals combining with non-metals. Typical ionic compounds are sodium fluoride and magnesium oxide. Caesium fluoride is the compound with the greatest degree of ionic bonding.

Questions

1 Calculate the difference in electronegativity for the elements in the following compounds and indicate which compound will be the most ionic and which the least ionic: **a)** KBr, **b)** NaI, **c)** CsCl.

2 The electronegativities of potassium and hydrogen are 0.8 and 2.2, respectively. Suggest what kind of bonding potassium hydride might have.

Structure of ionic compounds

Ionic compounds do *not* form molecules. Instead the positive and negative ions come together into various three dimensional structures called lattices. Electrostatic attraction holds the oppositely-charged ions together in appropriate numbers so that the total charge is zero. For example, in sodium chloride (NaCl), there are equal numbers of Na^+ and Cl^- ions, but in calcium fluoride (CaF_2) there are twice as many F^- ions as Ca^{2+}. When the lattice forms, energy is released: the lattice energy or enthalpy.

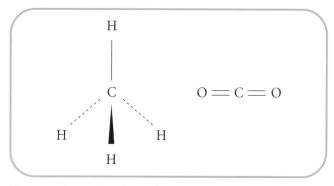

Figure 5.2 Bonding in methane and carbon dioxide

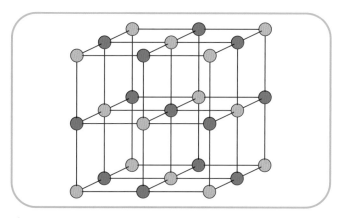

Figure 5.1 Sodium chloride lattice

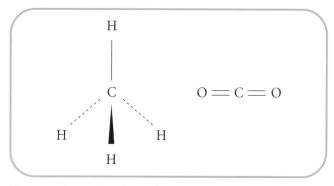

silicon atom

oxygen atom

Figure 5.3 Simplified structure of silicon dioxide

Question

3 Which of the following ionic compounds would you expect to have a sodium chloride type lattice: Na_2O, KBr, MgF_2, MgO?

Covalent compounds

Most covalent compounds are formed by combinations of non-metallic elements, and those encountered previously have mostly been molecular compounds such as methane and carbon dioxide (Figure 5.2).

A number of covalent compounds occur as **network structures**, of which silicon carbide and silicon dioxide are examples. Their structures are complex and can vary. Silicon carbide, SiC, has a structure similar to that of diamond while silicon dioxide, SiO_2, is a hexagonal crystal formed by the bonding of silicon atoms with four oxygen atoms to give SiO_4 tetrahedra. These tetrahedra then link by sharing each of their oxygen

atoms between two silicon atoms. A helical structure builds up with an overall ratio of two oxygen atoms for each silicon atom (Figure 5.3).

These network solids have very different properties from ordinary molecular compounds. Both silicon carbide and silicon dioxide, or quartz, have very high melting points since melting requires the breaking of strong covalent bonds. They are very hard, although brittle, because of these same strong directional bonds. Silicon carbide, with a network structure similar to that of diamond, is so hard that it is used as an abrasive in the cutting and grinding surfaces of tools, when it is known as 'carborundum'.

Figure 5.4 Silicon carbide grinding wheels

Molecular compounds such as CH_4, CO_2 and SiH_4 (silane) are gaseous at room temperature since there are only weak forces between the molecules – intermolecular forces. Melting and boiling of these compounds require only the overcoming of these weak intermolecular forces, whilst the covalent intramolecular forces remain intact.

Polar covalent bonding

Most compounds which are covalent are formed by elements with different electronegativities, although not so different as those forming ionic bonds. In these compounds, the bonding electrons are not shared equally. The atom with the greater share of electrons will end up with a slight negative charge by comparison with the other atom. For example in hydrogen chloride, the chlorine is more electronegative than the hydrogen and in water the oxygen is more electronegative than the hydrogen. Hydrogen chloride and water can be represented as in Figure 5.5. The symbols δ^+ and δ^- mean 'slightly positive' and 'slightly negative'. Covalent bonds with unequal electron sharing are called polar covalent bonds.

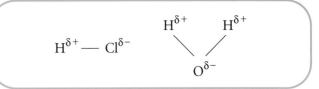

Figure 5.5 Bonding in hydrogen chloride and water

Some molecules containing such polar bonds end up with an overall polarity because the bonds are not arranged symmetrically in the molecule. HCl and H_2O are good illustrations of these, as is ammonia (Figure 5.6). Such molecules are said to have a permanent dipole.

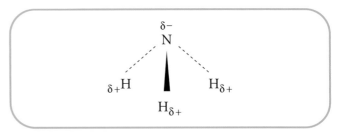

Figure 5.6 Bonding in ammonia

Other molecules have a symmetrical arrangement of polar bonds and the polarity cancels out over the molecule as a whole. This is the case in carbon dioxide and tetrachloromethane shown in Figure 5.7.

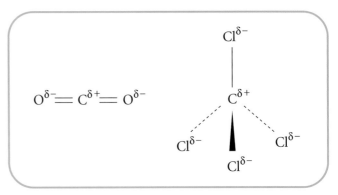

Figure 5.7 Bonding in carbon dioxide and tetrachloromethane

Behaviour in electric fields

One consequence of these two possibilities is the behaviour of the molecules in an electrical field as illustrated in Figure 5.8. Liquids with molecules which are non-polar behave as in Figure 5.8(a). Liquids with polar bonds behave as in Figure 5.8(b), unless the polarity cancels out in which case they behave as in Figure 5.8(a).

Question

4 Use Table 3.3 on page 34 to draw the structures of hydrogen sulphide, phosphorus hydride, hydrogen bromide and iodine monochloride showing polarities, where appropriate.

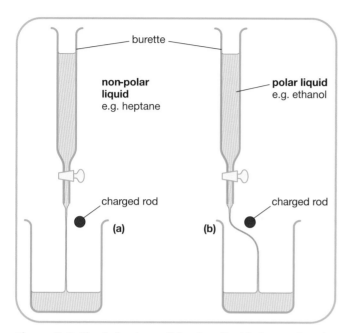

Figure 5.8 The behaviour of flowing liquids in an electric field

Heptane has almost non-polar bonds and the molecule is non-polar as a whole. Ethanol has one very polar O—H bond giving the molecule an overall polarity. Chloroform, unlike CCl_4, is polar overall because of the unsymmetrical arrangement of the C—Cl bonds. These structures are shown in Figure 5.9.

Heptane and CCl_4 behave as in Figure 5.8(a), ethanol and chloroform as in Figure 5.8(b).

Question

5 In which way would you expect the following liquefied gases to behave in an electric field: HCl, O_2, NH_3, CH_4?

Boiling points

Another consequence of molecules having an overall polarity is that their boiling points are higher than those of non-polar molecules of a similar molecular mass. The intermolecular forces are increased by the mutual attraction of the permanent dipoles on neighbouring molecules. Propanone and butane shown in Figure 5.10 are good examples of this behaviour.

Whether polar or non-polar, all covalent molecular compounds interact by van der Waals' bonding. This is because all molecules possess temporary or instantaneous dipoles caused by the constant

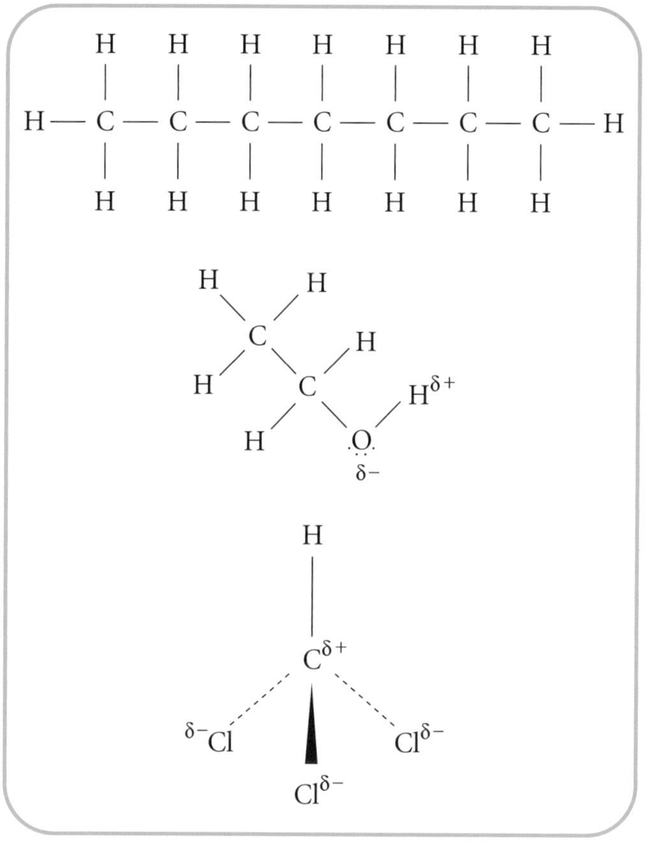

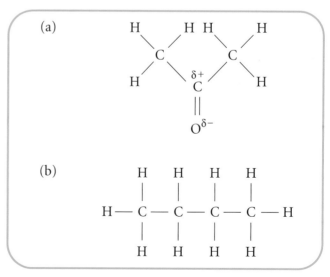

Figure 5.9 Bonding in heptane, ethanol, trichloromethane (chloroform)

movement of electrons within their molecules. Permanent dipole–permanent dipole attractions are stronger than van der Waals' bonds for molecules of equivalent size. Since propanone is polar, its boiling point is higher than that of butane which is non-polar.

Figure 5.10 (a) Propanone: Formula mass 58, Boiling point 56°C (b) Butane: Formula mass 58, Boiling point 0°C

Solvent action

Because of its polar nature, water is capable of dissolving other polar and ionic substances. It is a general rule that polar and ionic substances are more likely to dissolve in polar solvents, and non-polar substances are more likely to dissolve in non-polar solvents. For example, salt dissolves in water but not in heptane whereas wax dissolves in heptane but not in water. Hydrogen chloride is very soluble in water. See Figure 5.11 for the mechanism by which it dissolves.

The dissolving of the hydrogen chloride results in the uneven breaking of the H—Cl bond giving a strongly acidic solution – hydrochloric acid. Similarly, ionisation occurs when the other hydrogen halides, HBr and HI,

are dissolved in water and when pure sulphuric acid, also polar covalent, is dissolved in water. All give rise to strongly acidic solutions, i.e. fully ionised solutions, despite their original structures being polar covalent. If any of the hydrogen halides are dissolved in non-polar toluene, ionisation does not occur and the solutions are not acidic.

Ionic substances dissolve in a similar way, but of course the ions already exist in the initial lattice. The attraction drags ions out of the lattice and they go into solution surrounded by water molecules. The process is illustrated by Figures 5.12 and 5.13. Ions surrounded by a layer of water molecules, held by electrostatic attraction, are said to be hydrated.

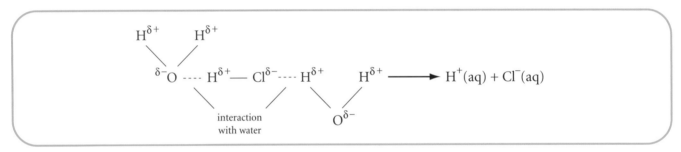

Figure 5.11 Hydrogen chloride dissolving in water

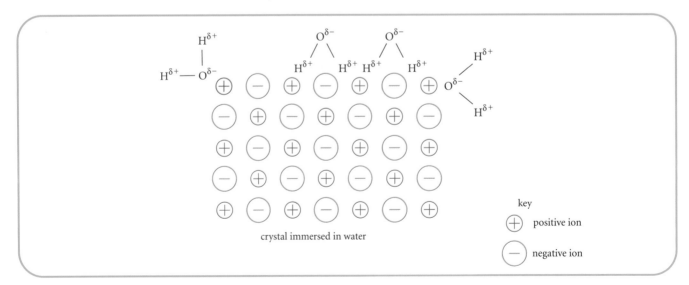

Figure 5.12 An ionic crystal dissolving in water

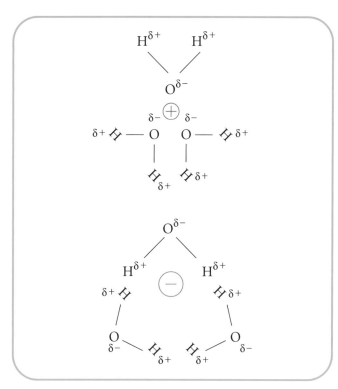

Figure 5.13 Hydrated ions

Anomalous physical properties of some hydrides

Boiling points

The accompanying graph, Figure 5.14, shows the expected increase in boiling point with molecular mass for the group IV hydrides. In the other three groups, however, the values of boiling points for NH_3, H_2O and HF (and HCl to some extent) are higher than would be expected for their molecular mass.

These anomalous properties would appear to indicate stronger bonding between the molecules than the expected van der Waals' bonding and simple permanent dipole–permanent dipole attraction.

The compounds showing these properties all contain bonds which are very polar, i.e. O—H, N—H, F—H (and Cl—H), as shown by the electronegativity differences in Table 5.1 below. These molecules can therefore interact in the fashion shown in Figure 5.15.

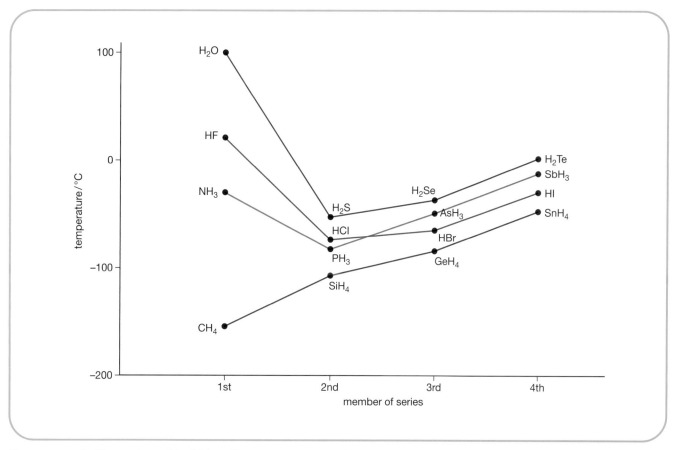

Figure 5.14 Boiling points of hydrides of groups IV, V, VI and VII

H—C	H—N	H—O	H—F
0.3	0.8	1.3	1.8
	H—P	H—S	H—Cl
	0.0	0.3	0.8
		H—Se	H—Br
		0.2	0.6
		H—Te	H—I
		0.1	0.4

Table 5.1 Electronegativity differences

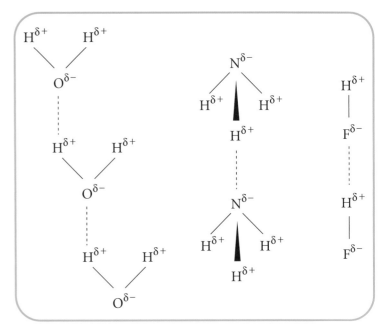

Figure 5.15 Hydrogen bonding in water, ammonia and hydrogen fluoride

This interaction is called **hydrogen bonding** since it occurs only for compounds containing a strongly electronegative element linked to hydrogen. The pull of electrons away from the hydrogen results in a positive charge located on a small atom, and hence a high positive charge density capable of interacting with the negative end of other molecules. Hydrogen bonding is stronger than van der Waals' bonding and stronger than ordinary permanent dipole–permanent dipole attractions but weaker than covalent bonds.

Melting points

Melting points show similar peculiarities to those of boiling points (Figure 5.16). The first members of the group V, VI and VII series have much higher

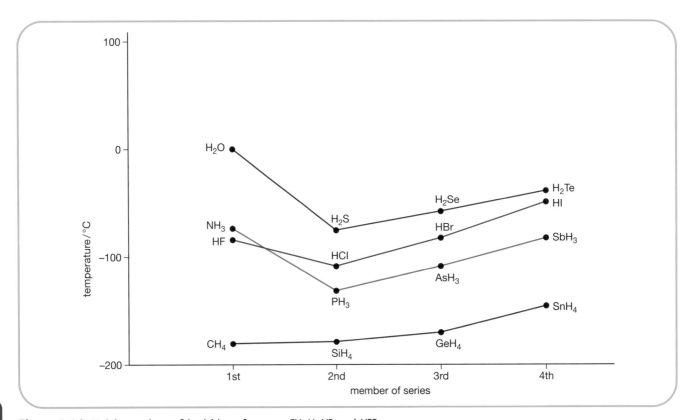

Figure 5.16 Melting points of hydrides of groups IV, V, VI and VII

melting points than CH_4, the first member of the group IV series, which does not have hydrogen bonding.

Viscosity

Comparison of the liquid fractions obtained from the distillation of crude oil shows that viscosity normally increases with molecular mass. The simple experiment illustrated in Figure 5.17 allows viscosities to be compared in a quick and simple way. If the tubes are inverted at the same time, the air bubble reaches the top fastest in the liquid of lowest viscosity. The results are as indicated, showing that the viscosity is related to molecular mass and to the number of —OH groups present. The —OH groups allow hydrogen bonding between the molecules and this increases the viscosity.

Miscibility

Miscible liquids mix thoroughly without any visible boundary between them, e.g. ethanol and water. On the other hand, water and hexane are immiscible, with the hexane forming a visible upper layer. Opportunities for hydrogen bonding caused by the presence of ——OH groups in the molecules of two liquids aid miscibility, but it should be noted that very strongly polar liquids like propanone, although without ——OH groups, are frequently miscible with water.

Density of water

An important consequence of hydrogen bonding is the unusual way in which water freezes. As with all liquids, water contracts on cooling, but when it reaches 4°C it begins to expand again, and at its freezing point it is less dense than the water which is about to freeze (see

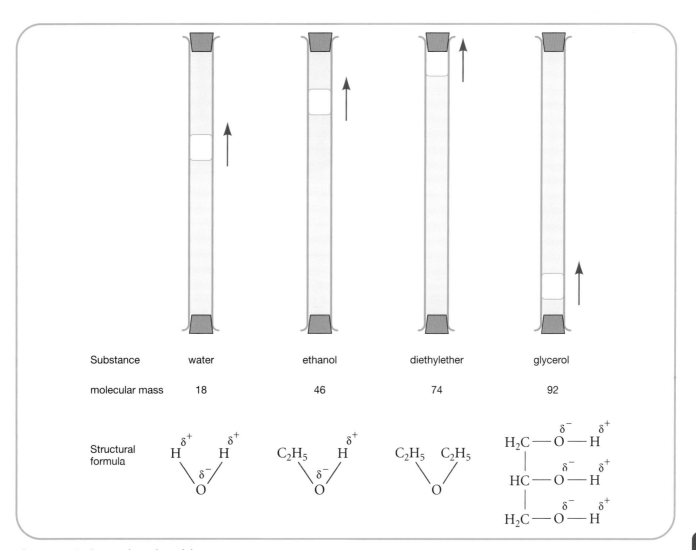

Figure 5.17 Comparing viscosities

Figure 5.18). The reason for this is the ordering of molecules into an open lattice, shown in Figure 5.19, as the hydrogen bonds are able to overcome the decreasing thermal motion of the molecules. As a result, ice floats on water, seas freeze from the top downwards – allowing fish to survive in the unfrozen water beneath – and, of course, pipes burst when water freezes inside them.

In a biochemical context, hydrogen bonds are responsible for the binding together of amino acid chains into the complex structures of proteins (see Chapter 12).

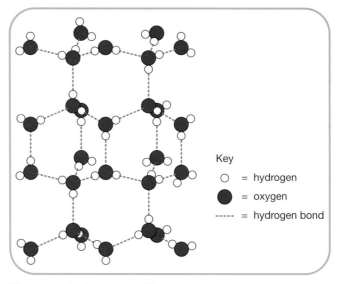

Figure 5.19 Structure of ice

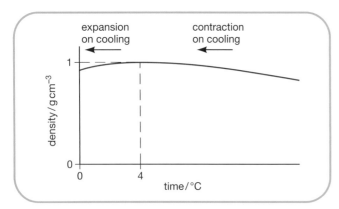

Figure 5.18 Water – change in density on cooling

Figure 5.20 Ice floats higher in salt water

Study Questions

In questions 1–4 choose the correct word(s) from the following list to complete the sentence.

| hydrogen bonding | large | molecular | non-polar |
| network | polar | small | van der Waals' forces |

1 Silicon carbide is a very hard substance due to its covalent _____ structure.

2 A liquid can be shown to have _____ molecules if it is attracted to a charged rod when it flows out of a burette.

3 When there is a _____ difference in electronegativity between two elements they are likely to combine by ionic bonding.

4 Ice has a relatively high melting point due to _____ between molecules.

5 Substance X melts below 2000°C and conducts electricity when solid and when liquid. Substance X could be

A copper

B graphite

C silicon dioxide

D calcium chloride.

6 Which of the following molecules would you expect to be the most polar?

A Hydrogen chloride

B Carbon monoxide

C Hydrogen iodide

D Nitrogen monoxide

Study Questions (continued)

7 Which type of bond forms when water vapour condenses?

    A Covalent

    B Metallic

    C Hydrogen

    D Ionic

8 Which of the following compounds has a covalent network structure?

    A $SiO_2(s)$

    B $HF(l)$

    C $MgO(s)$

    D $CS_2(l)$

9 Molecules of iodine monochloride, ICl, have a permanent dipole. The attraction between molecules of ICl will be

    A weaker than van der Waals' forces but stronger than hydrogen bonds

    B stronger than both van der Waals' forces and hydrogen bonds

    C stronger than van der Waals' forces but weaker than hydrogen bonds

    D weaker than both van der Waals' forces and hydrogen bonds.

10 Hydrogen bonds are present in

    A $H_2(g)$

    B $CH_4(g)$

    C $HI(l)$

    D $NH_3(l)$.

11* a) Explain the change in atomic (covalent) radius of the elements

    i)   across the Periodic Table from lithium to fluorine

    ii) down group I from lithium to caesium.

  b) Which two elements, of all those considered in **a)**, form the compound with most ionic character.

12* a) Which type of bonding exists in

    i)   sulphur dioxide

    ii) silicon dioxide?

  b) Use the Data Book to find the boiling points of these compounds.

  c) Explain why the boiling point of sulphur dioxide is low.

13* State whether you would expect each of the following to conduct electricity appreciably when connected to a low voltage source:

  a) solid rubidium chloride

  b) liquid gallium (element 31)

  c) liquid nitrogen.

Give your reasons briefly in terms of the type of bonding present.

14*

Compound	Formula	Molecular mass	Boiling point/°C
Ethane	CH_3CH_3	30	−89
Methanol	CH_3OH	32	64
Hydrazine	NH_2NH_2	32	113
Silane	SiH_4	32	−112

  a) From the information given, which of the compounds in the table above contain hydrogen bonding in the liquid state?

  b) Why does hydrogen bonding affect the boiling point of a substance?

  c) In the table we have compared substances of similar molecular mass. Why is molecular mass significant in this case?

15* The American scientist Linus Pauling devised a scale to compare the attraction of atoms for bonded electrons. This scale is called the electronegativity scale. Some electronegativity values are shown in the Data Book.

  a) Which group of the Periodic Table contains elements with no quoted values for electronegativity?

  b) Use the electronegativity values to explain why carbon disulphide contains pure covalent bonds.

  c) Explain the trend in the electronegativity values of the group VII elements.

Study Questions (continued)

16* The graph shows the boiling points of the group VI hydrides.

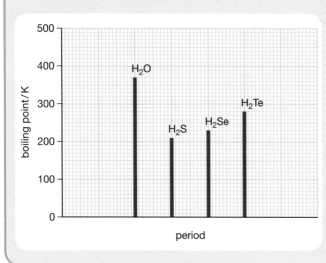

a) Explain why the boiling points increase from H_2S to H_2Te.

b) Why does H_2O have an unusually high boiling point compared to the other group VI hydrides?

6   The Mole

The Avogadro Constant

The mole is very important in chemistry since it enables us to compare quantities of different substances taking part in a chemical reaction. For example, magnesium and sulphur react together according to the following simple equation.

$$Mg + S \rightarrow MgS$$

The equation shows that this reaction needs equal numbers of magnesium and sulphur atoms. Since the elements differ in atomic mass, it will not be equal masses of each element that will combine.

Let us consider the three elements listed in Table 6.1.

	Magnesium	Sulphur	Iron
Average mass of 1 atom (amu)	24.3	32.1	55.8
Mass of n atoms (amu)	24.3 n	32.1 n	55.8 n
Ratio of mass of equal numbers of atoms	24.3 :	32.1 :	55.8
Mass of 1 mole	24.3 g	32.1 g	55.8 g

Table 6.1

As Table 6.1 shows, one mole of an element differs in mass from that of another element. However, since these quantities are in the same proportion as their relative atomic masses, they contain the same number of atoms. This number is known as the Avogadro Constant and is given the symbol L, after Loschmidt who was the first person to calculate its numerical value. It is a very large number which can be

determined by a variety of experimental methods, one of which will be referred to in Chapter 17.

> **AVOGADRO CONSTANT, L = 6.02×10^{23}**
> **One mole of any substance contains L,**
> **i.e. 6.02×10^{23}, formula units.**

The term 'formula unit' relates to the type of particle present in the substance. This is illustrated in sections a) to c) which follow.

a) **Metals and monatomic species,** e.g. noble gases
   Here the formula unit is an atom.

   Thus, 4 g of helium
   40 g of calcium } each contain L
   197 g of gold   (6.02×10^{23}) atoms

b) **Covalent substances**
   Here the formula unit is a molecule.

   The total number of atoms in a mole of molecules can be found by multiplying L by the number of atoms in each molecule, as shown in Table 6.2.

Quantity of substance	Number of molecules	Number of atoms per molecule	Total number of atoms
2 g hydrogen, H_2	L	2	2 L
18 g water, H_2O	L	3	3 L
30 g ethane, C_2H_6	L	8	8 L (2 L C atoms, 6 L H atoms)

Table 6.2

c) Ionic compounds

Here the formula unit consists of the ratio of ions expressed by the ionic formula of the compound. The total number of ions depends on the number of each kind of ion in the formula. Table 6.3 illustrates this with three examples.

The following equation describes a chemical reaction which involves different types of elements and compounds containing different types of particle. Figure 6.1 below then relates the number and type of particle to the stoichiometry of the reaction, i.e. the mole relationship of reactants and products as shown in the balanced equation.

$$Ca + 2H_2O \rightarrow H_2 + Ca(OH)_2$$
1 mol   2 mol   1 mol   1 mol

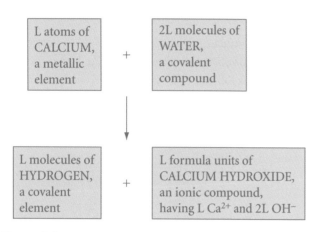

Figure 6.1

Since one mole of any substance contains the same number of formula units, it follows that equimolar quantities of different substances contain equal numbers of formula units. The following examples

Quantity of substance	Number of formula units	Number of positive and negative ions	Total number of ions
58.5 g Na$^+$Cl$^-$ (sodium chloride)	L	L Na$^+$ and L Cl$^-$	2 L
74 g Ca^{2+}(OH$^-$)$_2$ (calcium hydroxide)	L	L Ca^{2+} and 2 L OH$^-$	3 L
342.3 g (Al^{3+})$_2$ (SO$_4$$^{2-}$)$_3$ (aluminium sulphate)	L	2 L Al^{3+} and 3 L SO$_4$$^{2-}$	5 L

Table 6.3

illustrate this important point. Each of the quantities represents 0.2 moles of the substances and as a result contain 0.2 L formula units.

- 4.86 g of magnesium, Mg, contain 0.2 **L atoms**.
- 6.4 g of oxygen gas, O$_2$, contain 0.2 **L molecules**.
- 9.2 g of ethanol, C$_2$H$_5$OH, contain 0.2 **L molecules**.
- 12.4 g of sodium oxide, Na$_2$O, contain 0.2 L formula units consisting of 0.4 L Na$^+$ **ions** and 0.2 L O^{2-} **ions**.

Worked Example 6.1

10 g HF   0.2 mol CH$_4$ 17 g NH$_3$ 9 g H$_2$O

Which of the quantities given above contain(s)

a) 3.01 × 10^{23} molecules   b) 6.02 × 10^{23} atoms?

a) 6.02 × 10^{23} = **L**, hence, 3.01 × 10^{23} = 0.5 **L**

Gram formula masses: HF: 20 g, NH$_3$: 17 g, H$_2$O: 18 g

Hence, 10 g HF = 0.5 mol and contains 0.5 **L** molecules.

     0.2 mol CH$_4$ contains 0.2 **L** molecules.

Worked Example 6.1 (continued)

$$17 \, g \, NH_3 = 1 \, mol \text{ and contains } L \text{ molecules.}$$

$$9 \, g \, H_2O = 0.5 \, mol \text{ and contains } 0.5 \, L$$
molecules.

Answer: 10 g HF and 9 g H_2O each contain 3.01×10^{23} molecules.

b) Working for this part is shown in the following table.

Quantity per molecule	Number of molecules	Number of atoms per molecule	Total number of atoms
10 g HF	0.5 L	2	L
0.2 mol CH_4	0.2 L	5	L
17 g NH_3	L	4	4 L
9 g H_2O	0.5 L	3	1.5 L

The total number of atoms in a molecular substance can be found by multiplying the number of molecules by the number of atoms in each molecule.

Answer: 10 g HF and 0.2 mol CH_4 each contain 6.02×10^{23} atoms.

Worked Example 6.2

Which of the following quantities contains the Avogadro Constant of

   a) positive ions   b) negative ions?

80 g of NaOH   0.5 mol of $Fe_2(SO_4)_3$
250 cm^3 of 2 mol l^{-1} $CuCl_2$
1 mol NaOH = 40 g. Hence, 80 g NaOH = 2 mol, and contains 2 L formula units of NaOH made up of 2 L sodium ions and 2 L hydroxide ions.

1 mol of iron(III) sulphate contains L formula units made up of 2 L Fe^{3+} and 3 L SO_4^{2-}.
Hence, 0.5 mol of $Fe_2(SO_4)_3$ contains L iron(III) ions and 1.5 L sulphate ions.

Worked Example 6.2 (continued)

1 mol of $CuCl_2$ contains L formula units made up of L Cu^{2+} and 2 L Cl$^-$.
Number of moles of $CuCl_2$, $n = C \times V$

$$= 2 \times \frac{250}{1000}$$

$$= 0.5$$

0.5 mol of $CuCl_2$ contains 0.5 L copper ions and L chloride ions.

Answers:

   a) 0.5 mol of $Fe_2(SO_4)_3$ contains the Avogadro Constant of positive ions.

   b) 250 cm^3 of 2 mol l^{-1} $CuCl_2$ contains the Avogadro Constant of negative ions.

Questions

1 a) Which of the following quantities contains the Avogadro Constant of **molecules**?

   16 g O_2 60 g C_2H_6 96.15 g SO_2 159.8 g Br_2

   b) Which of the following quantities contains the Avogadro Constant of **atoms**?

   2 g H_2 8 g He   13.8 g Li   16 g O_2

2 Which of the following statements are true?

   A 40 g of argon contain L atoms.

   B 40 g of hydrogen fluoride contain 2 L molecules.

   C 14 g of nitrogen gas contain L molecules.

   D 14 g of carbon monoxide contain L atoms.

   E 1 litre of 1 mol l^{-1} sodium carbonate contains L sodium ions.

   F 93.45 g of magnesium fluoride contain 3 L fluoride ions.

3 Calculate the mass of

   a) Na_2S that contains L sodium ions

   b) Al_2O_3 that contains L oxide ions

   c) K_2SO_4 that contains L ions in total.

Molar volume

You will be familiar with problems in which a mole of a substance is related to mass. When dealing with a gas, however, it is usually more appropriate to measure its volume rather than its mass. In Table 6.4 several gases are compared. The molar volume of each gas at 0°C and one atmosphere pressure is calculated by dividing its molar mass by its density at that temperature and pressure.

$$\text{Density} = \frac{\text{mass}}{\text{volume}} \text{ and hence volume} = \frac{\text{mass}}{\text{density}}$$

From Table 6.4 it can be seen that the densities of gases increase in proportion to their gram formula masses. Consequently when the molar volumes of different gases are calculated, we observe that the value, within certain limits, is the same for all gases at the same temperature and pressure. The average molar volume for the gases listed in Table 6.4 measured at 0°C and one atmosphere pressure is 22.4 litres and all the values are within ±0.2 litres of the average.

The molar volume at *room temperature and pressure* can be tested experimentally in a variety of ways.

1   A round-bottomed flask can be evacuated and then weighed. It is reweighed after being filled with a gas as shown in Figure 6.2. This can be repeated using other gases. The volume of the flask can be found by filling it with water and emptying it into a measuring cylinder. The molar volume can then be

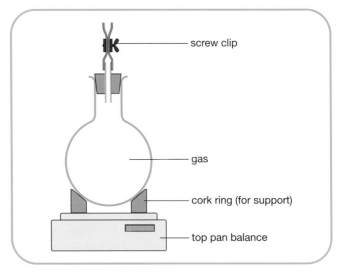

Figure 6.2 Molar volume experiment (safety screen not shown)

calculated from the data obtained as illustrated in Worked Example 6.3.

2   Instead of evacuating the flask it can be weighed with air in it. The density of air under experimental conditions of temperature and pressure can be obtained from a suitable data book. Once the volume of the flask is known, the mass of air in it can be calculated and hence the mass of the empty flask can be found.

3   Instead of a flask a gas syringe or a plastic syringe (preferably with a volume of at least 50 cm^3) may be used.

Gas	H_2	CH_4	N_2	O_2	Ar	CO_2
Gram formula mass/g	2	16	28	32	40	44
Density at 0°C and 1 atm pr/g l^{-1}	0.09	0.71	1.25	1.43	1.78	1.98
Molar volume at 0°C and 1 atm pr/l	22.2	22.5	22.4	22.4	22.5	22.2

Table 6.4

Worked Example 6.3

Calculate the molar volume of carbon dioxide from the following data obtained at 20°C and 1 atmosphere pressure.

Mass of empty flask	= 107.49 g
Mass of flask + carbon dioxide	= 108.37 g
Volume of flask	= 480 cm³

Calculation:

Mass of carbon dioxide
$$= 108.37 - 107.49$$
$$= 0.88\ g$$

0.88 g of carbon dioxide occupies a volume of 480 cm³.

Gram formula mass of CO_2 = 44 g

44 g of CO_2 occupies a volume of $\dfrac{480 \times 44}{0.88}$ cm³ = 24 000 cm³

Hence, the molar volume of CO_2 is 24.0 litres at 20°C and 1 atmosphere pressure.

Question

4 A 100 cm³ gas syringe was weighed empty and again when filled with ethane, C_2H_6, at 20°C and 1 atmosphere pressure. Calculate the molar volume of ethane from the following data.

Mass of syringe	= 151.51 g
Mass of syringe + ethane	= 151.63 g

Question

5 Calculate the density in g l⁻¹ of the following gases at room temperature and pressure. The molar volume under these conditions is 24 litres mol⁻¹.

a) neon   b) ammonia, NH_3 c) propane, C_3H_8

Since the volume of a gas changes if the temperature and/or the pressure changes, it is important to specify the temperature and pressure at which a volume is being measured. At room temperature and pressure, i.e. 20°C and 1 atmosphere pressure, the molar volume of any gas is approximately 24 litres mol⁻¹.

It may seem surprising at first that the molar volume is the same for all gases, even at the same temperature and pressure. This is certainly not true for either solids or liquids. However, in a gas the molecules have much greater kinetic energy and are relatively far apart so that the volume of the gas does not depend on the size of the actual molecules. It is possible to calculate that in a gas at room temperature and pressure the molecules themselves only occupy about 0.1% of the volume of the gas. The rest is empty space!

Provided the molar volume of a gas is known, the volume of the gas can be calculated from the number of moles of gas. Alternatively, the number of moles of a gas can be calculated from its volume. The relationship between volume of a gas, number of moles and molar volume can be expressed as follows.

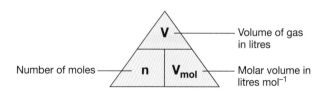

Hence, volume of gas, $V = n \times V_{mol}$

and number of moles, $n = \dfrac{V}{V_{mol}}$

Worked Example 6.4

The molar volume at 0°C and 1 atmosphere pressure is 22.4 litres mol^{-1}. Calculate **a)** the volume of 0.025 mol of oxygen, and **b)** the number of moles of nitrogen in 4.48 litres under these conditions.

a) Volume of oxygen, $V = n \times V_{mol}$

$$= 0.025 \times 22.4$$

$$= 0.56 \text{ litres} = 560 \text{ cm}^3$$

b) Number of moles of nitrogen, $n = \dfrac{V}{V_{mol}}$

$$= \dfrac{4.48}{22.4} = 0.2$$

Question

6 At room temperature and pressure the molar volume is 24 litres mol^{-1}.

a) Calculate the volume of **i)** 0.04 mol of CO_2 **ii)** 5 mol of CH_4.

b) Calculate the number of moles in **i)** 72 litres of H_2 **ii)** 360 cm^3 of Ar.

Reacting volumes

In this part of the chapter we shall apply the idea of molar volume to chemical reactions in which at least one of the reactants or products is a gas. There are two main types of calculation involved. The first type is often used in a reaction where only one of the reactants or products is a gas. The second type of calculation deals with reactions in which at least two gases take part and the volumes of the gases are then compared with each other. The difference between the two types of calculation should become apparent in the next two sections.

Calculations involving mass and volume

From previous work on the mole you should know how to use a balanced equation to calculate the mass of a reactant or product in a reaction. Revision examples are given in Chapter 1, page 11. This type of calculation can now be modified to include volume in a reaction where a gas is involved. This is illustrated in Worked Example 6.5.

Worked Example 6.5

Calculate the volume of carbon dioxide released when 0.4 g of calcium carbonate is dissolved in excess hydrochloric acid. The gas is collected at room temperature and pressure. The molar volume is 24 litres mol^{-1}. The equation for the reaction is

$$CaCO_3(s) + 2HCl(aq) \rightarrow CaCl_2(aq) + CO_2(g) + H_2O(l)$$

1 mol		1 mol
100 g		24 litres

Hence, 0.4 g of $CaCO_3$ gives $\dfrac{0.4}{100} \times 24$ litres of CO_2

$$= 0.096 \text{ litres (or 96 cm}^3\text{)}$$
$$\text{of } CO_2$$

This result could be tested experimentally using apparatus such as is shown in Figure 6.3.

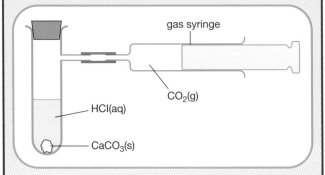

Figure 6.3

Alternatively, the gas could be collected over water in a measuring cylinder.

Questions

7 $Zn(s) + 2HCl(aq) \rightarrow ZnCl_2(aq) + H_2(g)$

Use the equation given above to calculate the volume of hydrogen produced when the following reactions go to completion. V_{mol} = 24 litres mol^{-1}.

a) 13.1 g of zinc are added to excess dilute hydrochloric acid.

b) Excess zinc is added to 100 cm^3 of 2 mol l^{-1} hydrochloric acid.

8 $2NH_3(g) + H_2SO_4(aq) \rightarrow (NH_4)_2SO_4(aq)$

Calculate the volume of ammonia, measured at room temperature and pressure, needed to neutralise 2.5 litres of 5 mol l^{-1} sulphuric acid. V_{mol} = 24 litres mol^{-1}.

Comparing volumes of gases

Let us consider the reaction in which carbon monoxide burns to form carbon dioxide.

$$2CO(g) + O_2(g) \rightarrow 2CO_2(g)$$

According to the equation, 2 moles of carbon monoxide combine with one mole of oxygen to form 2 moles of carbon dioxide.

Provided the volumes are measured at the same temperature and pressure, any volume of carbon monoxide will produce the same volume of carbon dioxide by reacting with half that volume of oxygen, e.g.

2 litres CO + 1 litre O$_2$ produces 2 litres CO$_2$

and

100 cm^3 CO + 50 cm^3 O$_2$ produces 100 cm^3 CO$_2$.

Worked Example 6.6

Calculate i) the volume of oxygen required for the complete combustion of 100 cm^3 of ethane and ii) the volume of each product. All volumes are measured at 150°C and 1 atmosphere pressure.

	$2C_2H_6(g) + 7O_2(g) \rightarrow 4CO_2(g) + 6H_2O(g)$			
Mole ratio of reactants and products	2	7	4	6
Volume ratio of reactants and products (at same T and P)	2	7	4	6
Simplified volume ratio	1	3.5	2	3

Hence 100 cm^3 of ethane i) requires 350 cm^3 of oxygen for complete combustion and ii) produces 200 cm^3 of CO$_2$(g) and 300 cm^3 of H$_2$O(g).

In many reactions which involve gases, one or more of the reactants or products may be a liquid or a solid. The volume of liquid or solid is so small compared with that of the gas or gases that it can be regarded as negligible. Thus if the reaction in the previous Worked Example is carried out when all the volumes are measured at room temperature and pressure, the products include water. Since water is a liquid its volume can be ignored.

$2C_2H_6(g) + 7O_2(g) \rightarrow 4CO_2(g) + 6H_2O(l)$				
Mole ratio:	2	7	4	6
Volume ratio:	2	7	4	negligible
i.e.	1	3.5	2	–

Questions

9 Nitrogen monoxide gas, NO, combines with oxygen to form brown fumes of nitrogen dioxide according to the following equation.

$$2NO(g) + O_2(g) \rightarrow 2NO_2(g)$$

Calculate the volume of oxygen needed to react completely with 60 cm³ of nitrogen monoxide, and calculate the volume of nitrogen dioxide formed.

Note: Balanced equations are required in the next two questions.

10 Calculate the volume of oxygen required for the complete combustion of 20 cm³ of propane, C_3H_8, and calculate the volume of each product. All volumes are measured at 130°C and 1 atmosphere pressure.

11 For each of the following gases, calculate **i)** the volume of oxygen required for the complete combustion and **ii)** the volume of carbon dioxide produced. All gases are measured under the same conditions of room temperature and pressure.

a) 200 cm³ of methane, CH_4

b) 5 litres of butane, C_4H_{10}

Thus 100 cm³ of ethane requires 350 cm³ of oxygen for complete combustion and produces 200 cm³ of carbon dioxide when the volumes are measured at room temperature and pressure. Incidentally, the volume of water formed would be about 0.2 cm³, less than 0.1% of the volume calculated in Worked Example 6.6.

Gas volume calculations involving excess reactant

In Chapter 1 we introduced the idea of excess reactant and saw how this affected mole calculations. At this point it is appropriate to apply this idea to calculations involving gas volumes. Worked Examples 6.7 and 6.8 illustrate the application of excess reactant to the types of calculation covered in the previous two sections. In each example it is necessary to find out which reactant is in excess before calculating the volume of gaseous product.

Questions on page 64 will give you some practice in relating the idea of excess to gas volume calculations.

Worked Example 6.7

Calculate the volume of carbon dioxide produced at room temperature and pressure when 0.4 g of calcium carbonate is added to 12 cm³ of 0.5 mol l⁻¹ hydrochloric acid. The molar volume is 24 litres mol⁻¹.

$$CaCO_3(g) + 2HCl(aq) \rightarrow CaCl_2(aq) + CO_2(g) + H_2O(l)$$

1 mol	2 mol		1 mol	
100 g			24 litres	

Number of moles of calcium carbonate, $n = \dfrac{m}{gfm}$

$$= \dfrac{0.4}{100} = 4 \times 10^{-3}$$

Number of moles of hydrochloric acid, $n = C \times V$

$$= 0.5 \times \dfrac{12}{1000} = 6 \times 10^{-3}$$

According to the equation, 1 mol of $CaCO_3$ requires 2 mol of HCl, and consequently 4×10^{-3} mol $CaCO_3$ require 8×10^{-3} mol HCl.

The number of moles of acid present is less than this, so the $CaCO_3$ is in excess and as a result the volume of CO_2 produced will depend on the number of moles of acid.

According to the equation, 2 mol of HCl produces 1 mol of CO_2, and consequently, 6×10^{-3} mol HCl produces 3×10^{-3} mol CO_2.

Hence, the volume of CO_2 produced $= 3 \times 10^{-3} \times 24$ litres

$$= 0.072 \text{ litres (or 72 cm}^3\text{)}$$

This could be verified by modifying the experiment referred to on page 60.

Worked Example 6.8

A mixture of 20 cm³ of propane and 130 cm³ of oxygen was ignited and allowed to cool. Calculate the volume and composition of the resulting gaseous mixture. All volumes are measured under the same conditions of room temperature and pressure.

$C_3H_8(g) + 5O_2(g) \rightarrow 3CO_2(g) + 4H_2O(l)$			
Mole ratio: 1	5	3	4
Volume ratio: 1	5	3	negligible

According to the equation,

20 cm³ of propane requires 5×20 cm³ of oxygen, i.e. 100 cm³.

Hence, oxygen is present in excess since its initial volume is 130 cm³.

Volume of excess oxygen $= (130 - 100) = 30$ cm³

Volume of carbon dioxide formed $= (3 \times 20) = 60$ cm³

Therefore the resulting gas mixture consists of 30 cm³ of O_2 and 60 cm³ of CO_2.

Questions

12 A piece of aluminium foil weighing 2.7 g was added to 100 cm^3 of 2 mol l^{-1} hydrochloric acid. The balanced equation for this reaction is

$$2Al + 6HCl \rightarrow 2AlCl_3 + 3H_2$$

a) Show by calculation which reactant is present in excess.

b) Calculate the volume of hydrogen produced (V_{mol} = 24 litres mol^{-1}).

13 In each of the following examples,

i) write the balanced equation for complete combustion and show by calculation which of the reactants is present in excess;

ii) calculate the volume and composition of the resulting gas mixture, assuming that all volumes are measured at the same room temperature and pressure.

a) A mixture of 10 cm^3 methane, CH_4, and 25 cm^3 oxygen was burned.

b) A mixture of 10 cm^3 propane, C_3H_8, and 25 cm^3 oxygen was burned.

Study Questions

Section A: The Avogadro Constant
[L = 6.02 × 10^{23} mol^{-1}]

Note: In the following questions you are advised to write the formula of any compound for which the formula is not provided.

1 What mass of water contains 6.02 × 10^{23} atoms?

　A 3 g　B 6 g　C 18 g　D 54 g

2 Which statement is true about methane?

　A　16 g of methane contain 6.02 × 10^{23} atoms.

　B　A molecule of methane is 16 times heavier than a molecule of hydrogen.

　C　16 g of methane contain the same number of atoms as 20 g of helium.

　D　A molecule of methane has a mass of 16 g.

3 Which of the following quantities of gases contains **L** molecules?

　A　16 g of oxygen

　B　16 g of methane

　C　16 g of sulphur dioxide

　D　16 g of hydrogen sulphide

4 The Avogadro constant is the same as the number of

　A　molecules in 19 g of fluorine gas

　B　positive ions in 62 g of sodium oxide

　C　atoms in 10 g of neon

　D　negative ions in 28 g of aluminium fluoride.

5 How many moles of ions are there in 3 moles of calcium chloride?

　A 3　B 6　C 9　D 12

6 Which of the following quantities contains 6.02 × 10^{23} positive ions?

　A　1 mol of Na_2SO_4

　B　2 mol of $MgCl_2$

　C　0.5 mol of $Fe_2(SO_4)_3$

　D　1.5 mol of $AlCl_3$

7 Calculate the mass of glucose, $C_6H_{12}O_6$, that contains

a) 6.02 × 10^{23} molecules

b) 6.02 × 10^{23} atoms.

8 Calculate the mass of calcium phosphate, $Ca_3(PO_4)_2$, that contains

a) L positive ions　**b) L** negative ions.

9 How many moles of chloride ions are there in a mixture which contains 0.1 moles of aluminium chloride and 0.2 moles of sodium chloride?

10 Calculate the number of calcium ions in a mixture of calcium fluoride and calcium carbonate which contains **L** fluoride ions and 1.5 **L** carbonate ions.

11
$$Zn + 2HCl \rightarrow ZnCl_2 + H_2$$

In an experiment 65.4 g of zinc were added to 250 cm^3 of 4 mol l^{-1} hydrochloric acid.

Study Questions (continued)

a) Show by calculation that the zinc is present in excess.

b) Which of the following statements are true about this experiment?

A   2 **L** zinc ions are produced.

B   **L** chloride ions are present.

C   0.5 **L** hydrogen molecules are produced.

D   **L** zinc atoms are oxidised.

E   2 **L** hydrogen ions are reduced.

Section B: Gas volumes

1 If the molar volume is 24.0 litres at room temperature and pressure, which of the following gases has a volume of 6.0 litres under these conditions?

A   4 g of oxygen

B   4 g of methane

C   4 g of helium

D   4 g of ammonia

2 $C_2H_4(g) + 3O_2(g) \rightarrow 2CO_2(g) + 2H_2O(l)$

When a mixture of 10 cm³ of ethene and 50 cm³ of oxygen was sparked, what was the final volume of gases? All volumes were measured at room temperature and pressure.

A 20 cm³   B 40 cm³   C 60 cm³   D 80 cm³

3 Mercury(II) nitrate decomposes when heated according to the following equation.

$$Hg(NO_3)_2(s) \rightarrow Hg(l) + 2NO_2(g) + O_2(g)$$

If the molar volume is 24.0 litres at room temperature and pressure, the total volume of gas produced when 0.01 mol of mercury(II) nitrate is completely decomposed by heat is

A 360 cm³   B 480 cm³   C 600 cm³   D 720 cm³.

4 Silver(I) oxide decomposes when heated as shown in the equation below.

$$2Ag_2O \rightarrow 4Ag + O_2$$

Calculate the mass of silver(I) oxide that would release 120 cm³ of oxygen, measured at room temperature and pressure. $V_{mol} = 24$ litres mol⁻¹.

5 Hydrogen sulphide, H_2S, is a gas which can be made by adding an acid to a metal sulphide. The equation

for the reaction between iron(II) sulphide and hydrochloric acid is shown below.

$$FeS + 2HCl \rightarrow FeCl_2 + H_2S$$

100 cm³ of 5.0 mol l⁻¹ hydrochloric acid was added to 30 g of iron(II) sulphide.

a) Show by calculation that excess metal sulphide was used.

b) Calculate the volume of gas produced ($V_{mol} = 24$ litres mol⁻¹).

6* The following apparatus can be used to determine the relative formula masses of liquids which are easily evaporated.

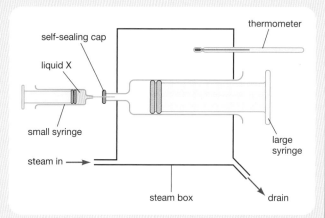

Some of liquid X is injected into the large syringe and it evaporates. The following results were obtained:

Mass of small syringe before injection = 5.774 g

Mass of small syringe after injection   = 5.648 g

Large syringe reading before injection = 5 cm³

Large syringe reading after injection   = 89 cm³

a) Calculate the relative formula mass of liquid X.

(Take the molar volume of a gas to be 30.6 litre mol⁻¹.)

b) Suggest why the above apparatus could *not* be used to determine the relative formula masses of liquids with boiling points above 100°C.

7* A mixture of 80 cm³ CO and 150 cm³ O_2 was exploded.

a) Write a balanced equation for the reaction.

After cooling the residual gas was shaken with sodium hydroxide solution.

Study Questions (continued)

b) Which gas would be absorbed by the sodium hydroxide?

c) What would be the reduction in volume of residual gas on shaking with the sodium hydroxide?

d) What volume of gas would remain? (Assume all volumes are measured at the same temperature and pressure.)

8* a) Write a balanced equation for the complete combustion of ethyne (C_2H_2).

b) If 50 cm^3 of ethyne are burned completely in 220 cm^3 of oxygen, what will be the volume and composition of the resulting gas mixture? (All volumes are measured under the same conditions of temperature and pressure.)

9* Hydrogen peroxide solution decomposes to produce oxygen gas.

$$2H_2O_2 \rightarrow 2H_2O + O_2$$

Solutions of hydrogen peroxide are marketed with a 'volume strength'. This relates to the volume of oxygen which can be produced.

Volume of oxygen produced = Volume strength $\times$ Volume of hydrogen peroxide solution

A student carried out an experiment.

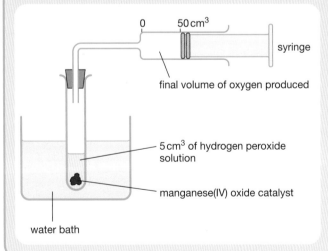

0   50 cm^3

syringe

final volume of oxygen produced

5 cm^3 of hydrogen peroxide solution

manganese(IV) oxide catalyst

water bath

a) Why is the catalyst more effective if used in powdered form?

b) Calculate the volume strength of the hydrogen peroxide solution used in the student's experiment.

c) The experiment was repeated with the temperature in the water bath increased by 20°C. State how this change would affect

i) the volume of oxygen produced in the first 20 s of the reaction

ii) the total volume of oxygen produced.

d) Calculate the total volume of oxygen produced by the decomposition of a solution containing 3.4 g of hydrogen peroxide. (Take the molar volume of oxygen as 24.0 litres.)

Unit 2

The World of Carbon

Carbon compounds play a vital role in our daily lives and your appreciation of this should increase through the study of topics such as fuels and polymers, as well as fats, oils and proteins. Embedded within this unit you will find a more systematic study of hydrocarbons and oxygen-containing carbon compounds with regard to nomenclature, structural formulae, reactions and uses.

7 Fuels

From previous work you should know and understand the following:

★ Fossil fuels   ★ Hydrocarbons

★ Crude oil   ★ Alkanes and cycloalkanes

★ Fractional distillation   ★ Isomerism

★ Cracking   ★ Greenhouse effect.

It has been necessary to introduce in this chapter some terms which are explained in detail later in the text. There is sufficient explanation of the terms in this chapter to put them in context. Terms in blue type appear in the Chemical Dictionary.

Petrol

Figure 7.1 A fractional distillation plant

The first stage in the refining of oil is often called primary distillation, in which the components, or fractions, of crude oil are separated out by their differing boiling points, as shown in Figure 7.2. Further separation of the residue is achieved under reduced pressure by a process called vacuum distillation.

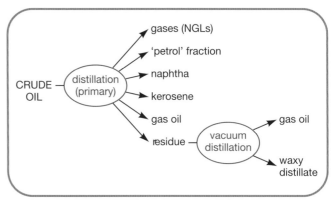

Figure 7.2 Fractions from crude oil

The petrol fraction so obtained is not ready to be used as fuel but requires further processing. Straight-chain hydrocarbons do not perform well as fuel in a petrol-burning engine. Better performance occurs if **aromatic** hydrocarbons, containing rings of six carbon atoms, or **branched-chain** hydrocarbons, are incorporated in the petrol.

The naphtha fraction is an important **feedstock** for the production of the group of compounds known as aromatic hydrocarbons. Steam cracking of naphtha, described in Chapter 11, yields a small proportion of aromatic compounds.

Cyclohexane, which is present in naphtha, can undergo **dehydrogenation**, or removal of hydrogen, to produce benzene, which is the simplest aromatic hydrocarbon.

$$C_6H_{12} \rightarrow C_6H_6 + 3H_2$$
cyclohexane   benzene

This process is an example of **reforming** and involves passing the naphtha fraction over a catalyst, platinum or molybdenum (VI) oxide, at about 500°C. Straight-chain alkanes may also be converted to aromatic hydrocarbons, e.g. heptane undergoes **cyclisation**, or ring formation, as well as dehydrogenation to form methylbenzene or toluene.

$$C_7H_{16} \rightarrow C_7H_8 + 4H_2$$
$$\text{heptane} \quad \text{methylbenzene}$$

Benzene has a ring structure which is usually drawn as shown below. Each corner of the hexagon represents a carbon atom with a hydrogen atom attached. Methylbenzene (toluene), $C_6H_5CH_3$, has one of these hydrogen atoms replaced by a methyl group as illustrated. Aromatic compounds, including benzene and toluene, are dealt with in detail in Chapter 8.

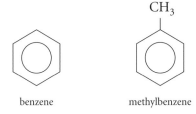

benzene   methylbenzene

Reforming also includes processes in which molecular rearrangement or isomerisation occurs, i.e. where the number of atoms per molecule do not change. In particular, straight-chain hydrocarbons undergo rearrangement to form branched-chain hydrocarbons, as illustrated by the following example.

$$CH_3CH_2CH_2CH_2CH_2CH_2CH_2CH_3 \rightarrow \underset{\underset{CH_3}{|}}{\overset{\overset{CH_3 \quad CH_3}{|\quad\;\;|}}{CH_3CCH_2CHCH_3}}$$

octane   2, 2, 4-trimethylpentane

Question

1 Write equations for the reforming of **a)** hexane to benzene **b)** heptane to an isomer with two methyl branches.

The products obtained on reforming naphtha, as well as those from the catalytic cracking of heavier fractions, are blended with the petrol fraction to give a more efficient fuel. Butane is added in the winter to make the

fuel more volatile – it dissolves in the petrol. The flow diagram, Figure 7.3, summarises how petrol is made. The end-product contains straight-chain alkanes, branched-chain alkanes, cycloalkanes and aromatic hydrocarbons. The blend's components, which are of different volatilities, are adjusted to take account of prevailing temperatures.

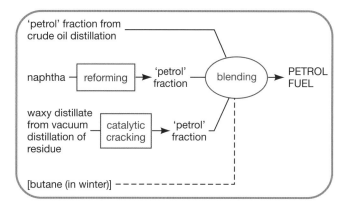

Figure 7.3 Making petrol

In a petrol engine, a mixture of petrol vapour and air is drawn into the cylinder – or air is drawn in and petrol injected into the air to form a mixture – compressed and ignited by a high voltage electric spark, as shown in Figure 7.4.

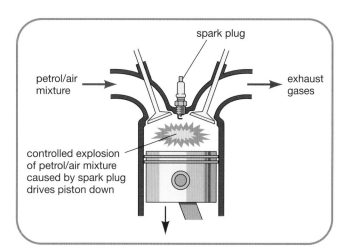

Figure 7.4 Petrol engine

Unfortunately, the fuel–air mixture can auto-ignite or pre-ignite when the engine is hot. This is heard as 'knocking' or 'pinking' of the engine and is inefficient and potentially harmful to the engine. Leaded petrol contained the anti-knocking agent, lead tetra-ethyl, $Pb(C_2H_5)_4$. Dibromoethane was also added to act as a scavenger for lead by producing volatile lead(II) bromide which escaped in the exhaust fumes. The harmful effects of lead compounds in the environment led to the introduction of unleaded petrol. Unleaded petrol requires a greater proportion of branched and aromatic hydrocarbons to increase the efficiency of burning in the absence of the lead compounds.

Concern was expressed at the presence of benzene, a known carcinogen (cancer-causing substance) in petrol. In 1998 the average benzene contents were:

Petrol	Average benzene content/%
Leaded 4*	1.9
Unleaded	2.4
Super unleaded	2.9

Table 7.1

From the year 2000 the EU limit for benzene in all petrols has been 1%. Leaded petrol was phased out at about the same time. To produce fuels of the required octane ratings, 'oxygenates' – compounds containing oxygen – are increasingly being added. Methyl tertiary butyl ether, MTBE, or 2-methoxy, 2-methyl propane is one such compound.

$$CH_3 - \underset{\underset{CH_3}{|}}{\overset{\overset{CH_3}{|}}{C}} - OCH_3$$

MTBE

Alternative fuels

There are many alternative fuels to petrol for vehicle propulsion. Currently diesel and liquified petroleum gas, LPG, are being strongly advocated. Both have advantages and disadvantages compared with petrol, but both share petrol's major disadvantages, i.e. since all are derived from the fossil fuel crude oil, a finite resource, they are going to run out and, since they contain carbon, their combustion is going to produce vast quantities of carbon dioxide which will increase global warming by the 'greenhouse effect'.

Ethanol

One of the major disadvantages of fossil fuels, such as oil, coal and natural gas, is that they are not renewable. Ethanol, which can be produced by fermentation from renewable resouruces, is therefore being used increasingly as a blending agent in motor fuel. For example, 'Gasohol' is a lead-free petrol containing between 10% and 20% ethanol. Since ethanol is an 'oxygenate' it gives a high octane rating.

Figure 7.5 The increase in types of fuel could cause problems!

The use of pure ethanol as a fuel is being encouraged in certain countries, such as Brazil, where it is economic to produce it by fermentation using surplus sucrose from sugar cane. Thus sugar cane can be regarded as a renewable source of engine fuel. Since the cane took in carbon dioxide during its growth, the combustion of the ethanol to produce the same amount of carbon dioxide is not increasing the overall concentration of carbon dioxide in the atmosphere.

$$C_2H_5OH + 3O_2 \rightarrow 2CO_2 + 3H_2O$$

Ethanol is therefore neutral in its contribution to 'greenhouse' gases.

Methanol

The manufacture of methanol from synthesis gas is described in Chapter 11. It is used as a fuel in racing cars and has been suggested as a fuel for general use in cars or, like ethanol, as part of a blend with petrol. Some of its advantages and disadvantages are listed in Table 7.2.

Advantages
Virtually complete combustion, less carbon monoxide than from petrol.
Contains no aromatic carcinogens (cancer-causing agents).
Cheaper to produce than primary fractionation gasoline.
Less volatile, and explosive, than petrol.
The car engine requires little modification.

Disadvantages
Difficulty in mixing methanol and petrol without an extra co-solvent.
Methanol absorbs water, tending to form immiscible layers, and becomes corrosive to car engines.
Methanol is toxic.
Less energy produced volume for volume than from petrol.
Bigger car fuel tanks needed.
Methanol is made from synthesis gas made from fossil methane. Increases 'greenhouse' gases, unless 'biogas' is used to make synthesis gas.

Table 7.2 Advantages and disadvantages of methanol as a fuel

Methane

When vegetable and animal matter decays in the absence of air, anaerobic respiration occurs with the formation of a gaseous mixture known as biogas. This consists mainly of methane (about 60%) and carbon dioxide. It is a useful fuel especially in rural areas. On a farm, manure and straw can be fed into a large tank called a biogas digester (Figure 7.6).

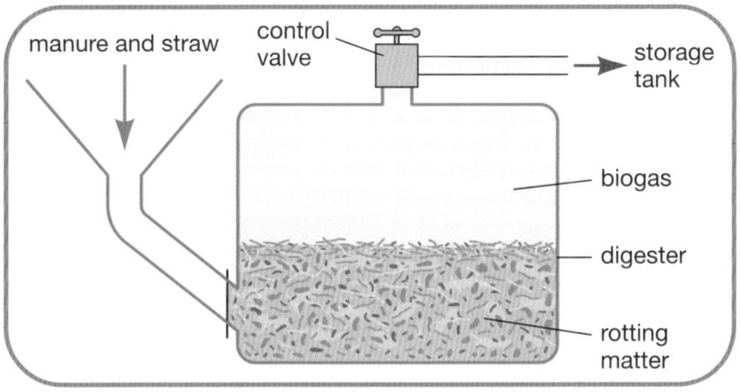

Figure 7.6 A biogas digester

Biogas is produced by the action of bacteria on the decaying matter. The bacteria function best in warm conditions, about 35°C. Hence in colder countries some of the biogas is burned to keep the digester sufficiently warm.

In China there are over 7 million biogas digesters providing energy for about five times that many people. In the West, biogas is beginning to be taken seriously as an energy source. Intensively reared animals produce so much manure that disposal becomes a problem.

Increasingly the manure is being used to produce biogas which can keep the animal sheds heated and can be used locally. Human waste has similar potential. Sewage works can now be powered by the biogas generated in the plant and the gases emitted from decaying matter in landfill sites are used in district heating schemes. Motor vehicles have already been successfully developed with engines modified to run on liquefied or compressed natural gas, methane, so there is scope to develop the use of biogas in the same way.

Figure 7.7 A car that can run on natural gas

Since biogas is produced by the decay of recently grown material which was photosynthesised from atmospheric CO_2, the combustion of biogas does not contribute to the greenhouse effect. Methane from coal mines and natural gas, however, does contribute.

Hydrogen

Hydrogen is increasingly being seen as the key portable fuel of the future when fossil fuels run out. Since its combustion produces only water, it is also seen as a way of reducing greenhouse emissions at an earlier date. The spectacle of the launch of the liquid hydrogen-fuelled space shuttle is evidence of hydrogen's potential to produce energy. Figure 7.8 gives comparisons with other fuels. (Although there are other methods, it is presumed here that hydrogen would be stored and carried as a liquid.)

Figure 7.9 A hydrogen-powered bus

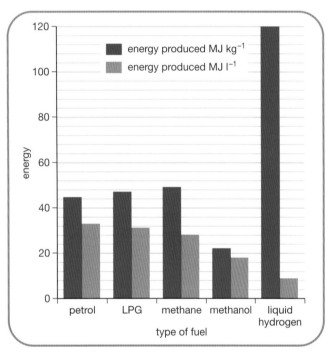

Figure 7.8 Approximate amount of energy available by mass (MJ kg^{-1}) and volume (MJ l^{-1}) from various fuels

It is more correct to regard hydrogen as an energy carrier rather than as a primary fuel. This is because hydrogen does not occur in significant quantities uncombined but has to be produced by splitting up hydrogen compounds. The energy used to do this will, because of losses, always exceed the energy from the combustion of the hydrogen obtained. To provide a portable fuel this could be regarded as acceptable. If the energy used to produce the hydrogen comes from fossil fuels they contribute to the greenhouse effect and there is no real benefit from using hydrogen. Nuclear power sources could be used, and, after a number of years when their expansion was not possible politically, are now being regarded more favourably in some quarters.

Figure 7.10 Shuttle launch. The main hydrogen-fuelled engines are assisted by solid fuel boosters at lift-off

Renewable energy sources like wind and wave power – but especially solar power – are regarded as the best ways to provide the energy to split up water by electrolysis. Solar cells are becoming more efficient and even in temperate regions enough of them could, linked to electrolysis cells, produce significant quantities of hydrogen.

Currently some buses in the UK are fuelled, experimentally, by hydrogen, and an Anglo–Belgian company has been set up to produce a taxi using hydrogen.

The major problem with hydrogen as a fuel is the difficulty in a small vehicle of carrying enough of it as liquid which must be kept below −250°C in well-insulated tanks. The weight of the storage system is considerable. Although the energy output of hydrogen per kg is the highest of all the fuels compared, the low density of liquid hydrogen results in the lowest energy output per litre. This means that the volume of liquid hydrogen which would have to be carried to provide a vehicle with a useful operating range would often be unmanageable.

Studies have shown, however, that it is possible to produce hydrogen-fuelled airliners where, because of

the large quantities involved, there are actually weight savings for the storage system and fuel together compared with conventionally-fuelled craft.

Fuel cells have been used for many years on board spacecraft. The fuel cell uses hydrogen and air in the reverse of electrolysis to form water and electrical energy.

Question

2   Use the enthalpies of combustion quoted in your Data Book to calculate the theoretical amount of energy (in MJ) available on complete combustion of 1 kg of each of the following compounds:
a) ethanol **b)** benzene.

Study Questions

In questions 1–4 choose the correct word from the following list to complete the sentence.

absence	decreases	ethanol	increases
less	methanol	more	presence

1   Reforming of naphtha _____ the proportion of branched chain hydrocarbons.

2   Biogas is produced by the action of bacteria on organic waste in the _____ of air.

3   In winter butane may be added to petrol to make it _____ volatile.

4   _____ is an alternative liquid fuel which can be made by fermentation.

5   The following processes are involved in the use or production of petrol.

Which process is **not** a chemical change?

A Blending   B Combustion
C Reforming   D Cracking

6   Which of the following is a renewable source of fuel?

A Natural gas

B Oil

C Sugar cane

D Coal

7   Which equation shows the process most likely to occur during reforming of naphtha?

A $C_7H_{16} \rightarrow C_2H_4 + C_5H_{12}$

B $C_6H_{14} + \frac{13}{2}O_2 \rightarrow 6CO + 7H_2O$

C $C_6H_{14} \rightarrow C_6H_6 + 4H_2$

D $C_7H_{16} + 11O_2 \rightarrow 7CO_2 + 8H_2O$

8* In some countries, cow dung is fermented and the mixture of gases produced, known as biogas, is used as a fuel. The mixture contains a small amount of carbon dioxide.

a) Name the main component of the biogas mixture.

b) The percentage of carbon dioxide in a biogas sample can be estimated by experiment. Part of the apparatus is shown in the diagram.

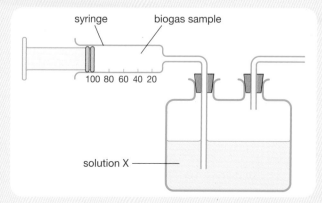

Study Questions (continued)

i) Solution X is used to absorb the carbon dioxide. Give a name for solution X.

ii) Complete the diagram to show all of the apparatus which could be used to carry out the experiment.

c) Name *one* other fuel which can be made by fermentation.

9* Methanol and ethanol can be used as alternative fuels in car engines.

a) Methanol

Methanol can be made as follows.

$$X + H_2O \xrightarrow{\text{steam reforming}} \underset{\text{synthesis gas}}{CO + 3H_2} \rightarrow CH_3OH$$

i) Identify X.

ii) Methanol is less volatile than petrol and less likely to explode in a car accident.
Explain why methanol is less volatile than petrol.

b) Ethanol

In some countries, ethanol for fuel is made by fermentation. Why is ethanol considered to be a 'renewable' fuel?

10 Hydrogen is used to fuel large rockets and has been suggested as a fuel for airliners.

a) Compared with a kerosine-fuelled airliner, what would be the advantage in terms of air pollution when the hydrogen-fuelled airliner is actually in flight?

b) Why would there be an overall disadvantage in terms of air pollution if the hydrogen was produced by electrolysing water using energy from kerosine-fuelled gas turbines?

11 Which of the following sources of energy make an overall contribution to global warming?

A Burning alcohol fermented from waste sugar cane.

B Burning alcohol produced from ethene made from natural gas.

C Burning methane from abandoned coal mines.

D Burning methane from digestion of sewage sludge.

Explain your choice of answer(s).

Hydrocarbons

From previous work you should know and understand the following:

★ How fossil fuels – coal, oil and natural gas – were formed

★ How and why crude oil is separated into fractions by fractional distillation

★ That molecular structure and physical properties of hydrocarbons are related

★ The names and molecular and structural formulae of alkanes (C_1–C_8), alkenes (C_2–C_6) and cycloalkanes (C_3–C_6)

★ How to identify isomers and draw their structural formulae

★ What is meant by saturated and unsaturated carbon compounds and how they can be distinguished

★ What happens during cracking and addition reactions.

Homologous series

From previous work you will be familiar with the idea of different series of hydrocarbons such as alkanes, alkenes and cycloalkanes. Alkanes and alkenes have carbon atoms linked in chains, while cycloalkanes have a ring structure.

Alkanes and cycloalkanes are saturated as all of their carbon–carbon bonds are single bonds. Alkenes are unsaturated since they contain carbon–carbon double bonds. The carbon–carbon double bond is an example of what is called a functional group since it has a major influence on the chemical behaviour of the compound.

Whether a hydrocarbon is saturated or unsaturated can be shown by testing it with bromine water. Alkenes rapidly decolourise the bromine water, while alkanes and cycloalkanes do not do so.

Alkenes are not the only type of unsaturated hydrocarbon. Cycloalkenes, e.g. cyclohexene, are unsaturated, as are compounds which have more than one double bond per molecule. Buta-1,3-diene, an example of a compound with more than one double bond, is important in the manufacture of synthetic rubber.

It is also possible to have a carbon–carbon triple bond in a molecule. The alkynes are a series of unsaturated hydrocarbons which possess this functional group. The first member of this series is called **ethyne**, commonly known as acetylene.

cyclohexene

buta–1, 3–diene

Members of the three homologous series – alkanes, alkenes and alkynes – are listed in Table 8.1. Names and molecular formulae of compounds with up to eight carbon atoms per molecule are shown along with structural formulae of the first few members.

Each of these series is an example of an homologous series. Members of a given series, known as **homologues,** have the following characteristics:

◆ Physical properties show a gradual change from one member to the next.

◆ Chemical properties and methods of preparation are very similar.

◆ Successive members differ in formula by —CH_2— and their molecular masses differ by 14.

◆ They can be represented by a **general formula**.

Alkanes	Alkenes	Alkynes
Methane CH_4 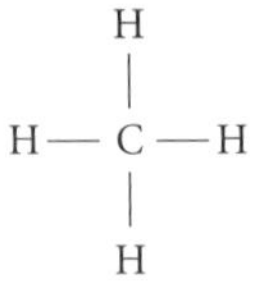 ![H—C—H with H above and below]		
Ethane C_2H_6 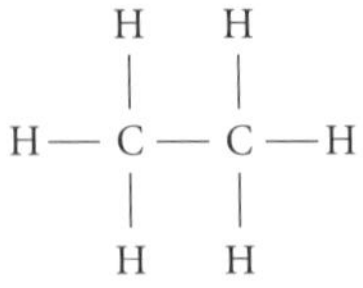 [H—C—C—H with H's]	Ethene C_2H_4 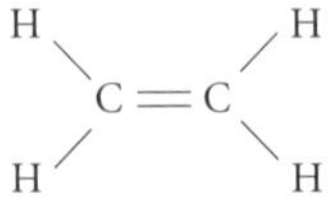 [C=C structure]	Ethyne C_2H_2 $H—C\equiv C—H$
Propane C_3H_8 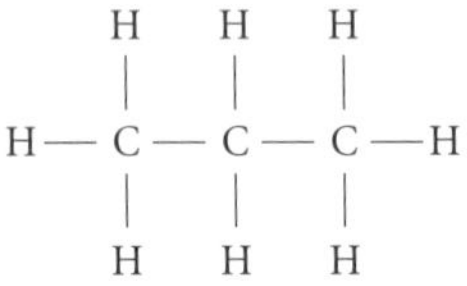 [H—C—C—C—H]	Propene C_3H_6 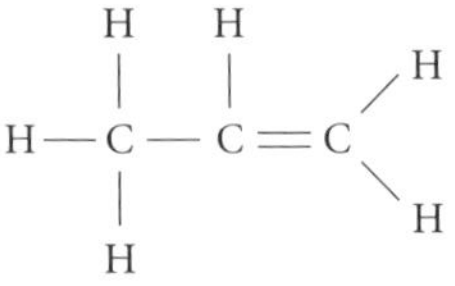 [H—C—C=C structure]	Propyne C_3H_4 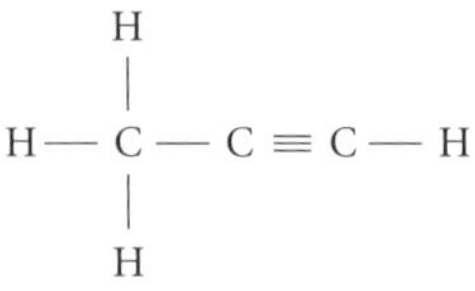 [H—C—C≡C—H]
Butane C_4H_{10}	Butene C_4H_8	Butyne C_4H_6
Pentane C_5H_{12}	Pentene C_5H_{10}	Pentyne C_5H_8
Hexane C_6H_{14}	Hexene C_6H_{12}	Hexyne C_6H_{10}
Heptane C_7H_{16}	Heptene C_7H_{14}	Heptyne C_7H_{12}
Octane C_8H_{18}	Octene C_8H_{16}	Octyne C_8H_{14}
General formula: C_nH_{2n+2}	General formula: C_nH_{2n}	General formula: C_nH_{2n-2}

Table 8.1

♦ Members of a series possess the same **functional group**, i.e. a certain group of atoms or type of bond which is mainly responsible for the characteristic chemical properties of that homologous series.

Naming isomers of alkanes

As you already know, isomers are compounds which have the same numbers of atoms of each kind but differ in the way in which these atoms are arranged, i.e. isomers have the same molecular formula but different structural formulae. At this stage we will expand this topic to include the systematic naming of isomers.

The first three members of the series – methane, ethane and propane – do not have isomers since in each case there is only one way of arranging the atoms in the molecule. Their structural formulae are shown in Table 8.1.

Butane, C_4H_{10}, has two isomers. Their full and shortened structural formulae are shown here.

a)

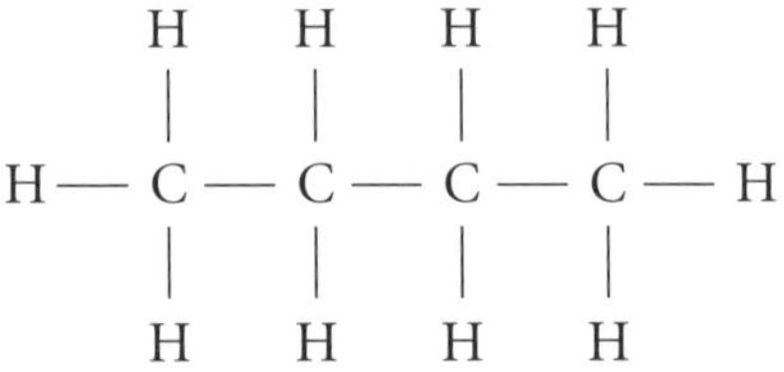

[H—C—C—C—C—H structure with H atoms]

$CH_3CH_2CH_2CH_3$

b)

[branched C structure]

CH_3
|
CH_3CHCH_3

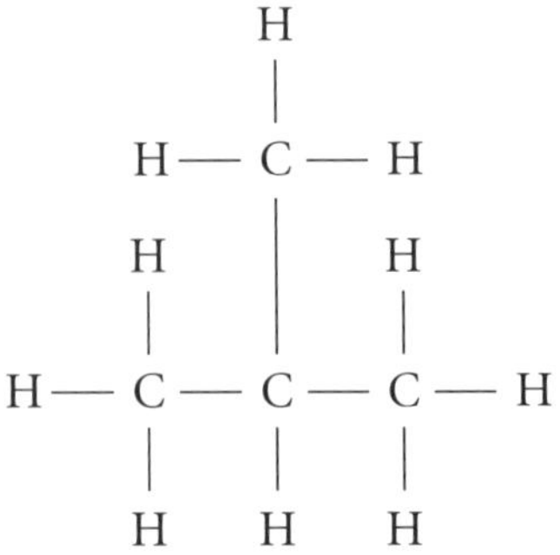

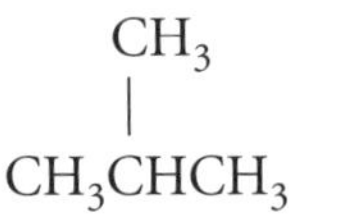

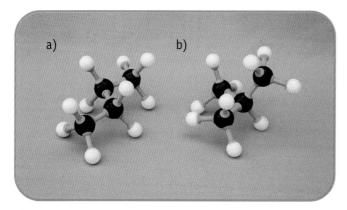

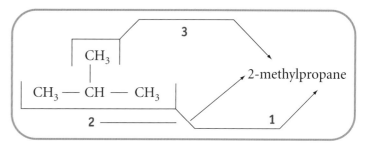

Figure 8.2 Naming a branched alkane

Figure 8.1 Isomers of butane: a) butane
b) 2-methylpropane

Compound a) is called butane and is an example of a straight-chain hydrocarbon. Compound b) is a branched hydrocarbon and its name is 2-methylpropane. These compounds have similar chemical behaviour but differ in physical properties, e.g. boiling point. Butane boils at 0°C, while 2-methylpropane boils at −10°C.

The method of naming isomers follows a system laid down by the International Union of Pure and Applied Chemistry (IUPAC) and operates as follows:

1 Select the longest chain of carbon atoms in the molecule and name this chain after the appropriate compound.

2 Find out which atom the branch is attached to by giving each carbon in the longest chain a number. Begin numbering from the end which is nearer the branch.

3 Identify the branch, i.e. whether it is a methyl group (CH$_3$—), an ethyl group (C$_2$H$_5$—) or some other alkyl group.

Figure 8.2 illustrates how this system applies to 2-methylpropane.

Pentane, C$_5$H$_{12}$, has three isomers as shown below Figure 8.2 (boiling points given in brackets).

There is a tendency for the boiling point to decrease as the amount of branching increases, since this causes the molecule to become more compact and have a smaller surface area. The weak forces of attraction between the molecules, i.e. van der Waals' forces, are even weaker as a result.

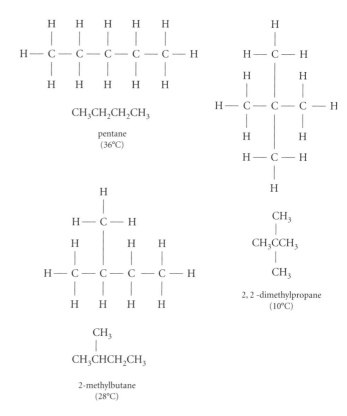

The following examples illustrate some of the problems which arise when dealing with structural formulae. Four other ways of representing the same isomer, i.e. 2-methylbutane, different from the structural formula given above, are shown on the following page. A new isomer is *not* obtained:

◆ by drawing the branch below instead of above the carbon atom chain, nor

◆ by drawing the branch nearer the right-hand end of the molecule instead of the left-hand end, nor

◆ by drawing a bend in the longest chain.

It is important to be able to recognise the isomer whichever way it is drawn.

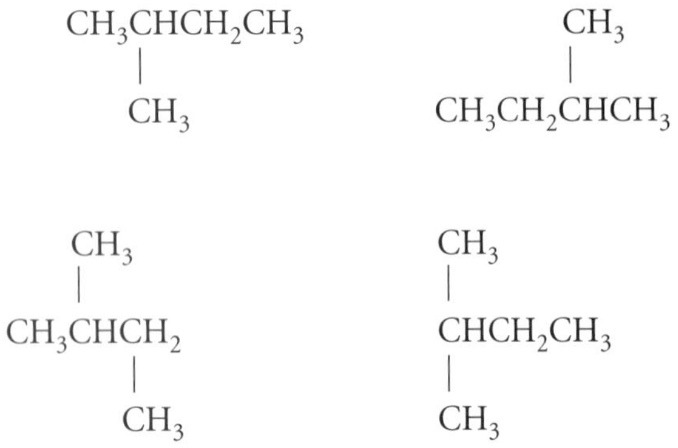

$$CH_3CHCH_2CH_3$$
$$|$$
$$CH_3$$

$$CH_3$$
$$|$$
$$CH_3CH_2CHCH_3$$

$$CH_3$$
$$|$$
$$CH_3CHCH_2$$
$$|$$
$$CH_3$$

$$CH_3$$
$$|$$
$$CHCH_2CH_3$$
$$|$$
$$CH_3$$

Many compounds were given a common or trivial name before the need for systematic naming was appreciated. For example, the branched isomer of butane was commonly known as iso-butane. In the following sections of this unit, trivial or traditional names will frequently be given in brackets after the systematic name.

The system of naming can be applied to more complicated molecules as shown by the following two examples.

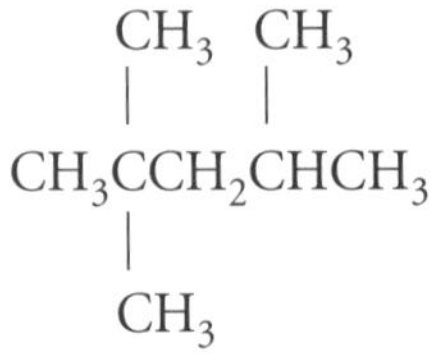

$$CH_3 \quad CH_3$$
$$| \quad |$$
$$CH_3CCH_2CHCH_3$$
$$|$$
$$CH_3$$

2, 2, 4-trimethylpentane

$$C_2H_5$$
$$|$$
$$CH_3CHCHCHCH_2CH_3$$
$$| \quad |$$
$$CH_3 \quad CH_3$$

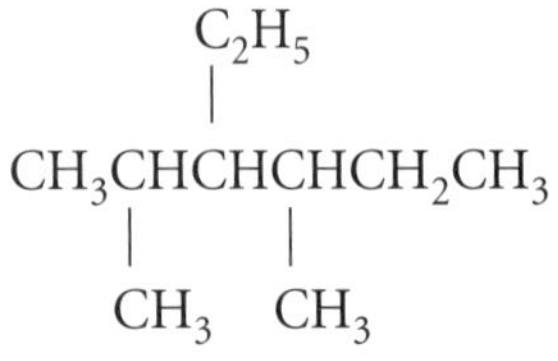

3-ethyl-2, 4-dimethylhexane
(an isomer of decane, $C_{10}H_{22}$)

2, 2, 4 – trimethylpentane is an isomer of octane and is the standard by which petrol is given an octane rating.

Questions

1 Draw the full structural formulae of all five isomers of hexane and work out their systematic names.

2 Draw the shortened structural formulae of
 a) 2,4-dimethylhexane
 b) 4-ethyl-3-methyloctane
 c) 2,2,4,4-tetramethylpentane.

3 Work out the systematic names of the following compounds.

 a)
$$CH_3$$
$$|$$
$$CH_3CH_2CHCHCH_3$$
$$|$$
$$CH_3$$

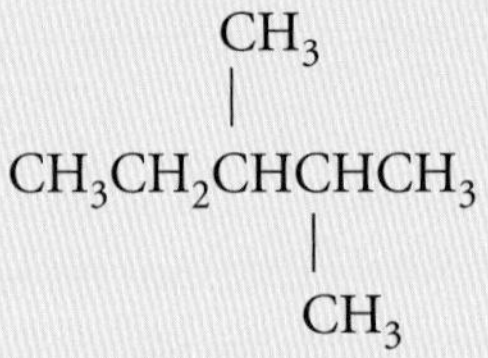

 b)
$$C_2H_5$$
$$|$$
$$CH_3CH_2CCH_2CH_2CH_3$$
$$|$$
$$C_2H_5$$

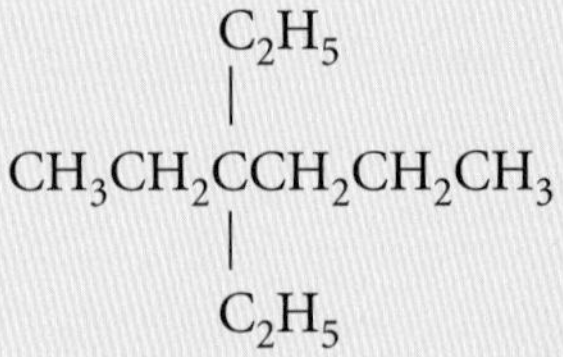

 c)
$$CH_3 \qquad CH_3$$
$$| \qquad |$$
$$CH_3CHCH_2CH_2CH_2CCH_3$$
$$|$$
$$CH_3$$

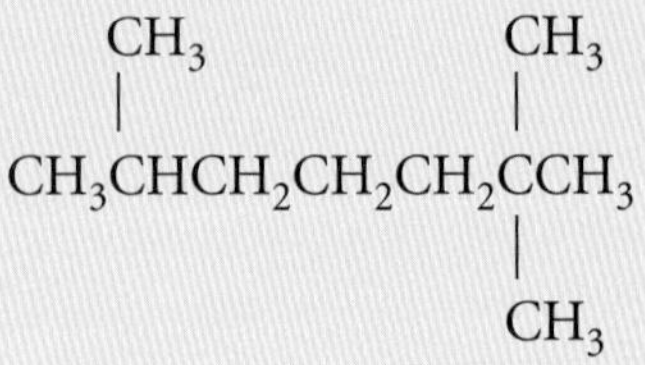

Naming isomers of alkenes and alkynes

Ethene has no isomers. Propene has no isomers which are alkenes but it is isomeric with cyclopropane.

Thereafter isomers of alkenes can arise for two reasons, since

◆ the position of the double bond in the chain can vary.

◆ the chain can be straight or branched.

This is illustrated by the following examples which are isomers of butene, C_4H_8:

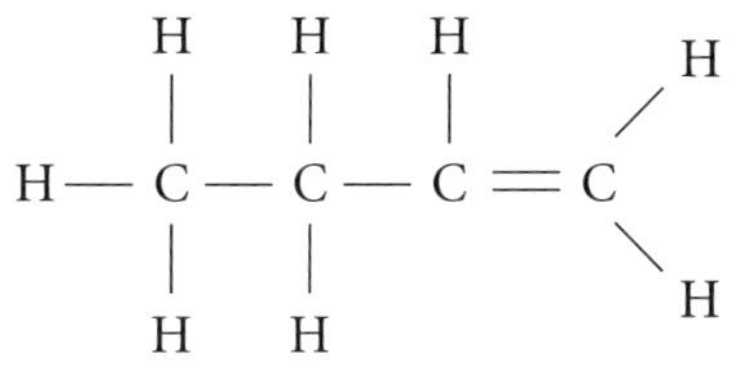

$CH_3CH_2CH=CH_2$

but-1-ene

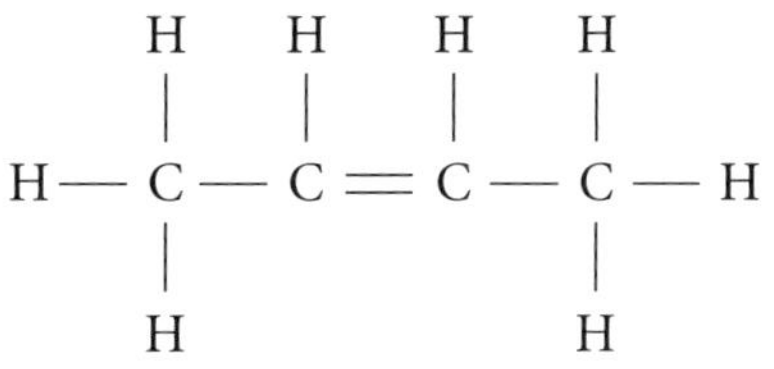

$CH_3CH=CHCH_3$

but-2-ene

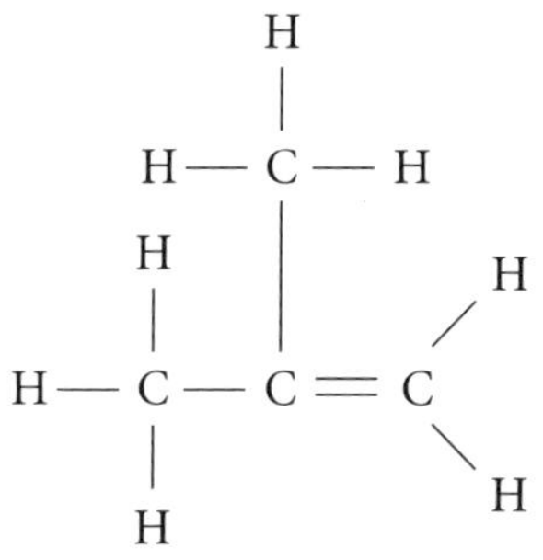

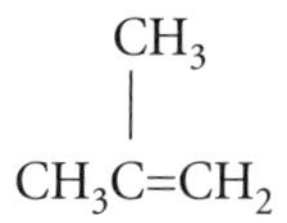

CH_3
|
$CH_3C=CH_2$

2-methylpropene

Where necessary, the name shows the position of the double bond. Thus but-2-ene has the double bond between the second and third carbon atoms in the chain.

When naming a branched alkene the position of the double bond is more important than the position of the branch. The name of the alkene whose structure is shown below is 3-methylbut-1-ene (and *not* 2-methylbut-3-ene). Numbering the carbon atoms from the right-hand end of the chain gives the lowest number, in this case 1, to indicate the position of the double bond and consequently the methyl branch is attached to the third carbon atom.

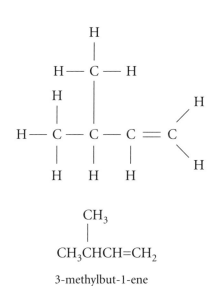

CH_3
|
$CH_3CHCH=CH_2$

3-methylbut-1-ene

Questions

4   Name the following alkenes.

a) $CH_3CH_2CH=CHCH_3$

b) CH_3
                           |
      $CH_2=CHCH_2CHCHCH_3$
                              |
                           CH_3

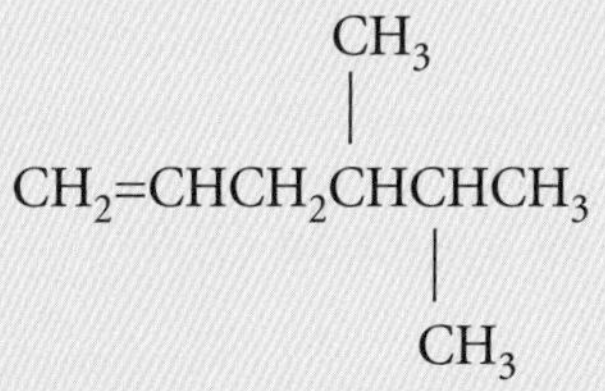

c) CH_3
                           |
         $CH_3CH=CCH_3$

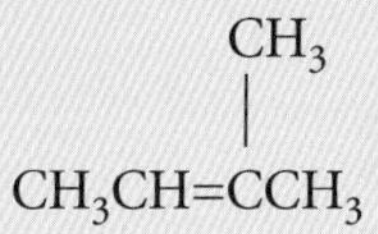

5   Draw the full structural formulae of the following compounds.

a) oct-4-ene

b) 3-ethylpent-1-ene

c) 4,4-dimethylhex-2-ene

Each of the first two members of the alkynes, i.e. ethyne and propyne, has only one possible structural formula as shown in Table 8.1.

Butyne, C_4H_6, has two isomers which are shown below.

$CH_3CH_2C\equiv CH$ $CH_3C\equiv CCH_3$

but-1-yne   but-2-yne

When naming a branched alkyne the position of the triple bond takes precedence over the position of the branch. Thus the name of the alkyne shown here is 4-methylpent-1-yne (and *not* 2-methylpent-4-yne).

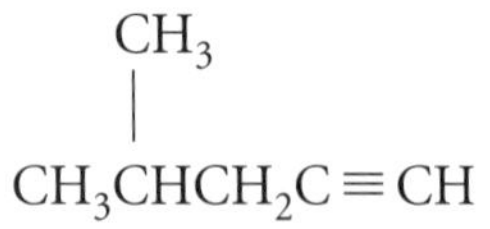

$$CH_3$$
$$|$$
$$CH_3CHCH_2C \equiv CH$$

Question

6 Draw the full structural formulae of the three isomers of pentyne and work out their systematic names.

Addition reactions of alkenes

You are already familiar with the idea that alkenes can undergo addition reactions because they contain C=C bonds. Alkanes are unable to react by addition since they contain only single bonds. We shall consider four different examples of addition reactions in this section.

Addition of hydrogen (H₂)

In general:

> **An ALKENE reacts with HYDROGEN to form an ALKANE.**
>
> e.g. $CH_3CH_2CH=CH_2 + H_2 \rightarrow CH_3CH_2CH_2CH_3$
> but-1-ene   butane

This example of an addition reaction is also known as **hydrogenation**. The reaction is carried out at 200°C in the presence of a nickel catalyst. Hydrogenation of vegetable oils is an important application of this type of reaction which we will refer to again in Chapter 12.

Addition of halogens (Br₂, Cl₂)

In general:

> **An ALKENE reacts with BROMINE to form a DIBROMOALKANE.**

Similarly, an **ALKENE reacts with CHLORINE to form a DICHLOROALKANE.**

e.g. $CH_3CH_2CH=CH_2 + Cl_2 \rightarrow CH_3CH_2CHClCH_2Cl$
but-1-ene   1,2-dichlorobutane

Note that the halogen atoms form bonds with the carbon atoms which were originally held together by the double bond.

Addition of hydrogen halides (HX, where X is F, Cl, Br or I)

In general:

> **An ALKENE reacts with a HYDROGEN HALIDE to form a HALOGENOALKANE.**
>
> e.g. $CH_2=CH_2 + HBr \rightarrow CH_3CH_2Br$
> ethene   bromoethane

Addition of water (H₂0)

In general:

> **An ALKENE reacts with WATER (steam) to form an ALKANOL.**
>
> e.g. $CH_2=CH_2 + H_2O \rightarrow CH_3CH_2OH$
> ethene   ethanol

This example of addition is also known as **hydration**. A mixture of ethene and steam is pressurised to about 65 atmospheres and passed over a catalyst (phosphoric acid) at 300°C. Only about 5% conversion of ethene is achieved under these conditions. However, unreacted ethene is separated from the liquid products and recycled to give a 95% yield of ethanol.

This method can be applied to the production of other alkanols from alkenes with more carbon atoms, e.g. the catalytic hydration of propene produces propanol.

The reverse process, i.e. **dehydration** of an alkanol, can easily be demonstrated in the laboratory using the apparatus illustrated in Figure 8.3. When ethanol

e.g.

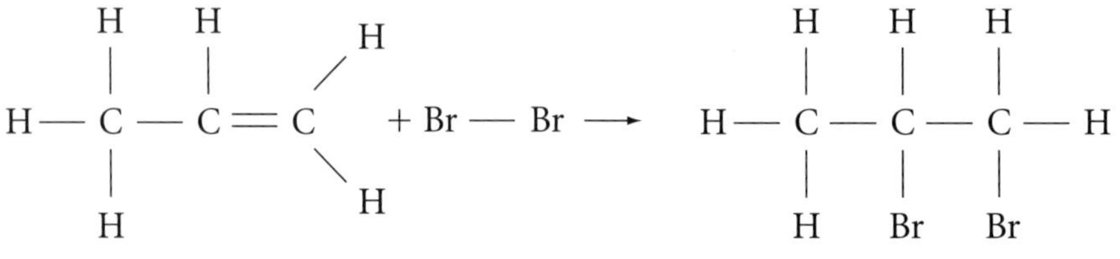

$$
\begin{array}{c}
\quad H \quad\quad H \\
\quad | \quad\quad | \\
H - C - C = C \overset{\displaystyle H}{\underset{\displaystyle H}{\diagup}} \quad + Br - Br \longrightarrow \\
\quad | \\
\quad H
\end{array}
\qquad
\begin{array}{c}
H \quad\quad H \quad\quad H \\
| \quad\quad | \quad\quad | \\
H - C - C - C - H \\
| \quad\quad | \quad\quad | \\
H \quad\quad Br \quad\quad Br
\end{array}
$$

propene   1,2-dibromopropane

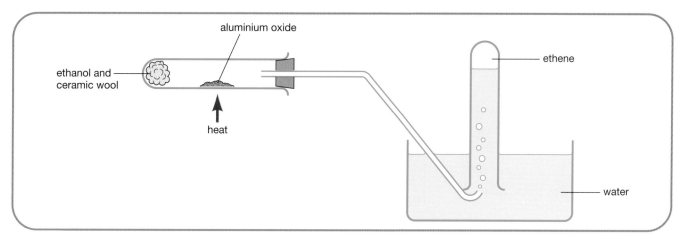

Figure 8.3 Dehydration of ethanol

vapour is passed over heated aluminium oxide, which acts as a catalyst, dehydration occurs forming ethene.

$$C_2H_5OH \rightarrow CH_2{=}CH_2 + H_2O$$

Concentrated sulphuric acid has a strong attraction for water and can also be used to dehydrate alkanols.

This process is not confined to the production of ethene. Other alkenes can be produced by dehydration of alkanols, e.g. dehydration of propanol produces propene. When an alkanol is dehydrated, the —OH group is removed along with a hydrogen atom from an adjacent carbon atom. In some cases this will result in the formation of two different alkenes, e.g. dehydration of butan-2-ol can produce either but-1-ene or but-2-ene as shown below.

Note that in each of the addition reactions described above, the product is saturated. The atoms of the substance being added form bonds with the carbon atoms which were originally held together by the double bond. This point is illustrated in Figure 8.4, which shows the full structural formulae of ethene and each of the products of the four types of addition reaction.

$$
\begin{array}{ccccc}
\text{H} & \text{H} & \text{H} & \text{H} \\
| & | & | & | \\
\text{H}-\text{C}-\text{C}-\text{C}-\text{C}-\text{H} \\
| & | & | & | \\
\text{H} & \text{H} & \text{OH} & \text{H}
\end{array}
$$

or

$CH_3CH_2CH{=}CH_2 + H_2O$

but-1-ene

$CH_3CH{=}CHCH_3 + H_2O$

but-2-ene

butan-2-ol

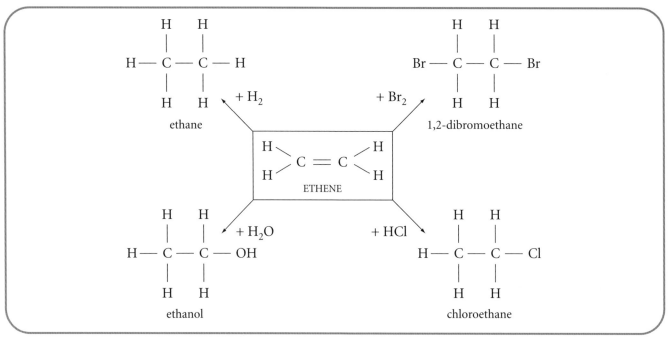

Figure 8.4 Addition reactions of ethene

Questions

7 Draw the full structural formula of the product of each of the following addition reactions. Give the name of each product.

a) Pent-2-ene + chlorine

b) ethene + hydrogen iodide

c) 3-methylbut-1-ene + hydrogen (think carefully when naming this product!)

8 Hydration of propene produces two different alkanols. Draw their full structural formulae.

9 Draw the full structural formulae of the alkenes which produce the following compounds by addition reactions. Name these alkenes.

a)

$$H-\underset{\underset{H}{|}}{\overset{\overset{H}{|}}{C}}-\underset{\underset{F}{|}}{\overset{\overset{H}{|}}{C}}-\underset{\underset{H}{|}}{\overset{\overset{H}{|}}{C}}-H$$

b)

$$H-\underset{\underset{H}{|}}{\overset{\overset{H}{|}}{C}}-\underset{\underset{H}{|}}{\overset{\overset{CH_3}{|}}{C}}-\underset{\underset{H}{|}}{\overset{\overset{H}{|}}{C}}-\underset{\underset{H}{|}}{\overset{\overset{H}{|}}{C}}-OH$$

c)

$$H-\underset{\underset{H}{|}}{\overset{\overset{H}{|}}{C}}-\underset{\underset{Br}{|}}{\overset{\overset{CH_3}{|}}{C}}-\underset{\underset{Br}{|}}{\overset{\overset{H}{|}}{C}}-H$$

Uses of halogenoalkanes

Halogenoalkanes have a wide variety of applications and only a few of the most important uses are mentioned in this section.

Trichloromethane, $CHCl_3$, is more commonly known as chloroform. In 1847 James Simpson, an Edinburgh obstetrician, was the first to use its powerful anaesthetic properties to reduce the pain of childbirth. However, chloroform is toxic and can damage the liver. During the 1950s, research by ICI for a suitable alternative led to the development of 2-bromo-2-chloro-1,1,1-trifluoroethane, referred to as 'halothane'.

<pre>
        F   Br
        | |
   F — C — C — H   halothane
        | |
        F   Cl
</pre>

Chloroethane, C_2H_5Cl, is useful as a local anaesthetic but its most important use has been in the manufacture of lead tetraethyl, $Pb(C_2H_5)_4$, the 'anti-knocking' agent present in leaded petrol. Another halogenoalkane, 1,2-dibromoethane $(CH_2Br)_2$, was also included in leaded petrol so that lead could be removed from the engine as volatile lead(II) bromide.

Chlorofluorocarbons, or CFCs, were developed for a number of uses where a non-flammable, volatile liquid was required. Dichlorodifluoromethane was important as a refrigerant and as an aerosol propellent. Other CFCs found uses as degreasing solvents and 'blowing agents' for making plastic foam, e.g. expanded polystyrene.

<pre>
        F
        |
   F — C — Cl   dichlorodifluoromethane
        |
        Cl
</pre>

During the 1980s, however, there was growing concern over the depletion of the 'ozone layer' and conviction that CFCs were chiefly responsible. In the upper atmosphere ozone has an important role in absorbing ultra-violet (UV) radiation from the sun. CFCs are chemically unreactive but in the upper atmosphere UV radiation provides enough energy to break C—Cl bonds producing chlorine atoms, i.e. Cl•, which convert ozone molecules, O_3, into O_2.

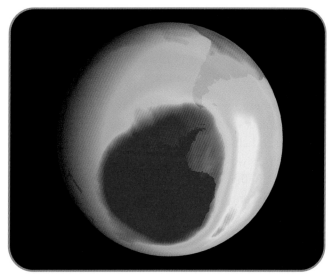

Figure 8.5 Satellite image showing the 'hole' in the ozone layer

<pre>
        F   H
        | |
   F — C — C — F   1,1,1,2–tetrafluoroethane
        | |
        F   H
</pre>

HFAs, i.e. hydrofluoroalkanes, are being developed as alternatives to CFCs, e.g. 1,1,1,2-tetrafluoroethane is a refrigerant. HFAs do not contain chlorine and, being relatively more reactive than CFCs, are less likely to persist in the atmosphere.

Halogenoalkanes also play an important part in the chemistry of carbon compounds. Carbon–halogen bonds are polar which makes halogenoalkanes more prone than hydrocarbons to attack by ionic reagents, e.g. an alkali can react with bromoethane to form ethanol.

$$C_2H_5Br + OH^- \rightarrow C_2H_5OH + Br^-$$

Addition reactions of ethyne

Ethyne is formed when water is added to calcium carbide, CaC_2.

$$CaC_2 + 2H_2O \rightarrow C_2H_2 + Ca(OH)_2$$

Figure 8.6 shows a suitable method for preparing ethyne in the laboratory and testing the gas for unsaturation. Alternatively the gas can be collected over water and then tested with bromine water.

Ethyne decolourises bromine water which shows that ethyne is unsaturated. It burns with a very sooty flame due to its high carbon content.

Ethyne undergoes similar addition reactions to ethene when it reacts with hydrogen, hydrogen halides, bromine and chlorine. However, in this case the addition is a two-stage process. In the first stage one molecule of ethyne combines with one molecule of the addition reagent to form a compound with a C=C bond. In the second stage another molecule of the

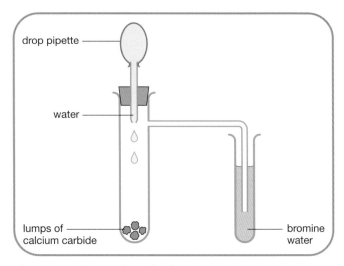

Figure 8.6 Preparation of ethyne

reagent is needed to form a saturated product. These points are illustrated in Figure 8.7 which summarises the reactions of ethyne with hydrogen, hydrogen chloride and bromine.

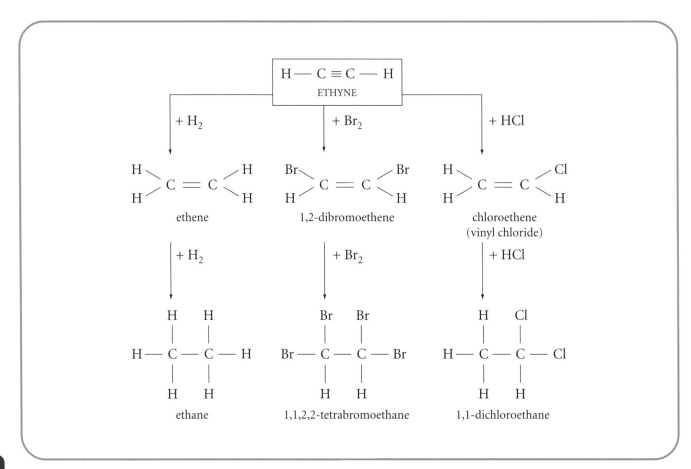

Figure 8.7 Addition reactions of ethyne

Questions

10 Draw the full structural formula of the product of each of the following reactions. Name each product.

a) One mole of ethyne reacts with one mole of hydrogen bromide.

b) One mole of propyne reacts with two moles of chlorine gas.

c) One mole of but-2-yne reacts with one mole of hydrogen gas.

11 Describe the molecular shapes of ethane, ethene and ethyne (see Figure 8.8).

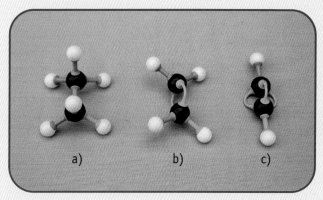

a)   b)   c)

Figure 8.8 a) Ethane, b) ethene, c) ethyne

Aromatic hydrocarbons

The carbon compounds we have referred to so far in this chapter can be classified as either

1 **aliphatic** in which carbon atoms are linked together to form *chains* (e.g. butane, propene, ethanol) or

2 **alicyclic** in which carbon atoms are linked together to form *rings* (e.g. cyclopentane, cyclohexene).

There is another class of carbon compounds which was referred to in Chapter 7. **Aromatic compounds** contain a distinctive ring structure of six carbon atoms. The simplest aromatic compound is the hydrocarbon called benzene, molecular formula: C_6H_6.

Aliphatic and alicyclic compounds tend to have similar chemical behaviour, e.g. cyclohexene undergoes similar addition reactions to alkenes. Aromatic compounds, on the other hand, often behave in a markedly different way due to their characteristic chemical structure.

Structure of benzene

The structure of benzene is quite complicated and it is useful to begin by comparing the experimental results obtained on testing various hydrocarbons which contain six carbon atoms per molecule. Table 8.2 summarises these results and conclusions.

From its formula, C_6H_6, we might have expected benzene to be unsaturated. Indeed, a possible structural formula frequently seen in organic chemistry textbooks shows an arrangement of alternate double and single

Name	Formula	Effect on bromine water	Saturated/ unsaturated
Hexane	C_6H_{14}	No change	Saturated
Hexene	C_6H_{12}	Decolourised	Unsaturated
Cyclohexane	C_6H_{12}	No change	Saturated
Cyclohexene	C_6H_{10}	Decolourised	Unsaturated
Hexyne	C_6H_{10}	Decolourised	Unsaturated
Benzene	C_6H_6	No change	?

Table 8.2

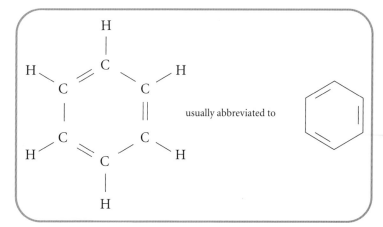

Figure 8.9 Traditional formula of benzene

bonds (see Figure 8.9). However, the bromine water test shows that C=C bonds are not present in benzene.

There is also experimental evidence that benzene has a planar molecule, i.e. all of its atoms lie in the same plane, and that all of its carbon–carbon bonds are of the same length. The carbon–carbon bond length in benzene is greater than the C=C bond length but is less than that of the C—C bond as can be seen from the data given in Table 8.3.

Bond type	Bond length/pm = 10^{-12} m
C⋯C (aromatic)	139
C=C	134
C—C	154

Table 8.3

Each carbon atom has four outer electrons and in benzene three of these are used in forming bonds with a hydrogen atom and the two adjacent carbon atoms. The six remaining electrons, one from each carbon atom, occupy electron clouds which are not confined or localised between any one pair of carbon atoms. These electrons are said to be delocalised. Figure 8.10 illustrates this description of the structure of benzene.

Consequently, it is appropriate to represent the structure of benzene as a regular hexagon suggesting a single bond 'framework'. The circle inside the hexagon indicates the additional bonding due to the delocalised electrons. Figure 8.11 shows this usual way of

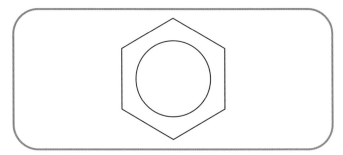

Figure 8.11 Structural formula of benzene

representing the structural formula of benzene. Each corner of the hexagon represents a carbon atom with one hydrogen atom attached.

All of this may suggest that benzene is incapable of reacting by addition. This is not true. Hydrogenation of benzene produces cyclohexane and benzene reacts with chlorine under certain conditions to form BHC (benzene hexachloride, $C_6H_6Cl_6$), an important insecticide.

Aromatic compounds

Benzene can be regarded as the basis to which all other aromatic compounds relate. In the most important reactions of benzene at least one of the hydrogen atoms on the benzene ring is replaced by another atom or group of atoms. For example, if one of these hydrogen atoms is replaced or substituted by a methyl group, a compound called methylbenzene (toluene), molecular formula $C_6H_5CH_3$, is produced.

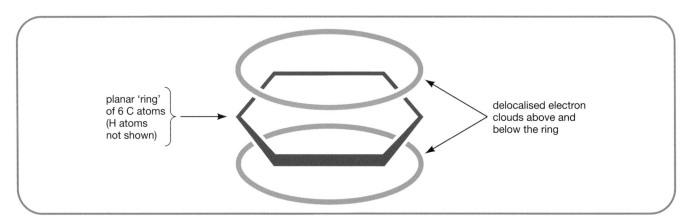

planar 'ring' of 6 C atoms (H atoms not shown)

delocalised electron clouds above and below the ring

Figure 8.10 Planar shape of benzene showing delocalised electrons

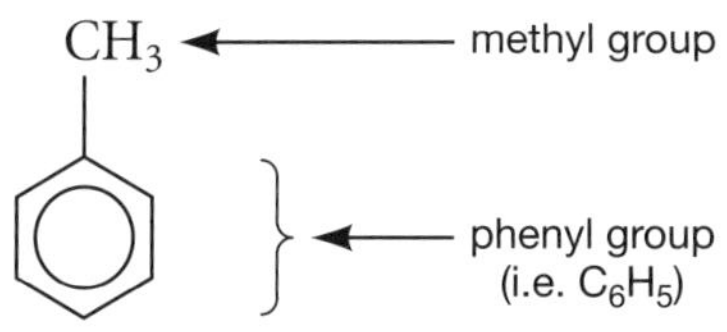

methylbenzene

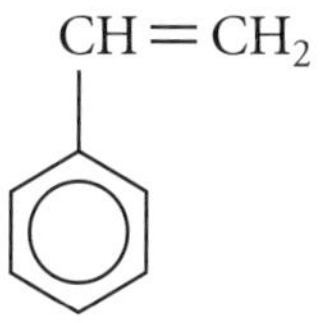

phenylethene

This formula also shows what is meant by the term 'phenyl group' which is present in many aromatic compounds. It consists of a benzene ring minus one hydrogen atom, i.e. C_6H_5-. The phenyl group does not exist on its own but must be attached to another atom or group of atoms. The prefix 'phenyl' is sometimes used in naming aromatic compounds. For example, phenylethene is the systematic name for styrene which is the monomer for producing poly(phenylethene) or polystyrene, as shown in Figure 8.12.

It is possible to have more than one of the hydrogen atoms in benzene replaced by methyl groups, e.g. 1,4-dimethylbenzene (para-xylene), $C_6H_4(CH_3)_2$, and 1,3,5-trimethylbenzene, $C_6H_3(CH_3)_3$.

$CH_3C_6H_4CH_3$

1, 4–dimethylbenzene

From these formulae you should realise that the number of hydrogen atoms still attached to the benzene ring will depend on how many have been substituted.

Some examples of important monosubstituted aromatic compounds are given below.

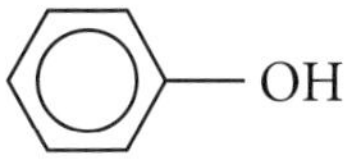

phenol

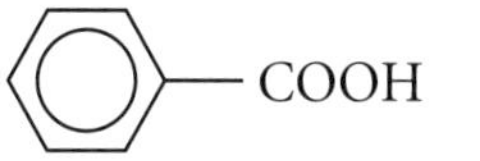

benzoic acid

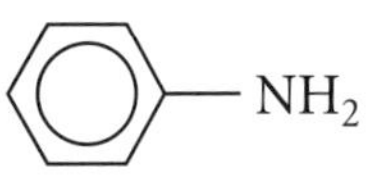

phenylamine (aniline)

Many aromatic compounds, including benzene itself, are important as feedstocks for the production of everyday materials. Some of these feedstocks and their products are illustrated in Figure 8.12 as a summary of synthetic routes emanating from benzene.

Questions

12 From the information given above, work out the molecular formula of

   a) phenol

   b) benzoic acid

   c) phenylethene.

13 a) Dimethylbenzene has three isomers, one of which is shown on the left. Draw the structural formulae of the other two isomers and name them.

   b) Draw the structural formulae of the isomers of dichlorobenzene and name them.

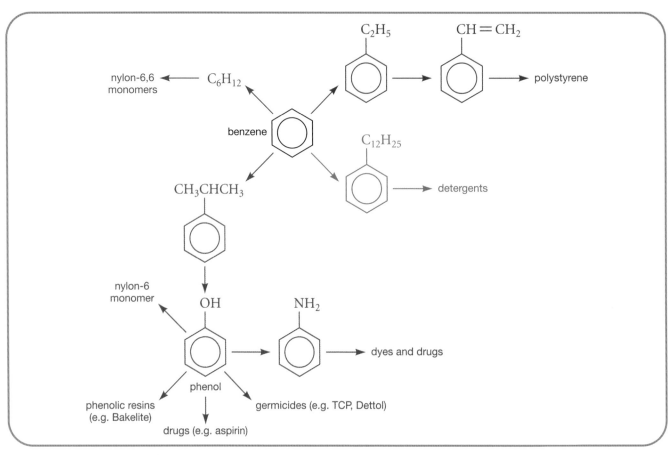

Figure 8.12 Benzene as a feedstock

Study Questions

In questions 1–4 choose the correct word from the following list to complete the sentence.

| benzene | branched | cyclohexene | double |
| hydration | hydrogenation | straight | triple |

1 2,2,4-trimethylpentane is an example of a _____ chain hydrocarbon.

2 The functional group in an alkyne is a carbon-carbon _____ bond.

3 _____ is the type of reaction occurring when ethanol is produced from ethene.

4 The simplest aromatic hydrocarbon is called _____ .

5 Which of the following could be the molecular formula of an alkyne?

A C_3H_8 B C_4H_6 C C_5H_{10} D C_6H_6

6 In which of the following pairs do **both** compounds rapidly decolourise bromine water?

A Cyclohexene and propyne

B Ethyne and cyclopentane

C Butane and butene

D Propene and benzene

7 Which of the following compounds is an isomer of cyclohexane?

A Benzene

B Cyclopentane

C 3-methylpent-1-yne

D 2,3-dimethylbut-2-ene

8 Hydrocarbons suitable for use in unleaded petrol are produced in oil refineries. An example of a typical reaction is shown at the top of page 89.

Study Questions (continued)

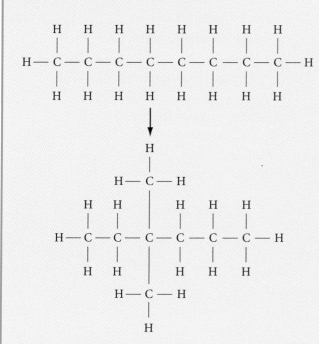

a) Name the industrial process in which reactions like the one above occur.

b) State the systematic name for the product shown above.

c) With reference to its structural formula indicate why the product is suitable for use in unleaded petrol.

9

$$C_2H_4 \xrightarrow[+Br_2]{(1)} X \xrightarrow{(2)} Ethyne \xrightarrow{(3)} Y$$

X and Y are saturated compounds which are also isomers.

a) Reactions (1) and (3) are examples of the same type of reaction. What type of reaction is this?

b) Draw the full structural formulae and give the systematic names of X and Y.

c) Reaction (2) is achieved by heating X in the presence of KOH solution.

  i) What substance is removed during reaction (2)?

  ii) Suggest why KOH is used to remove this substance.

10* A ratio line can be used to illustrate the carbon to hydrogen ratio in different compounds.

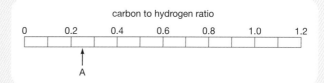

Methane would appear at point A.

a) At what value on the line would butane appear?

b) A hydrocarbon X with six carbon atoms per molecule has a carbon to hydrogen ratio of 0.5. X does *not* immediately decolourise bromine solution. Give a name for X.

c) Reforming is an industrial process which can convert alkanes into aromatic hydrocarbons.

  i) In relation to the alkane hydrocarbons, where on the line would the reformed hydrocarbons appear?

  ii) Name the aromatic hydrocarbon produced by reforming hexane.

11* Straight chain hydrocarbons, branched chain hydrocarbons, cyclic hydrocarbons and aromatic hydrocarbons are all obtained from petroleum oil.

a) State the systematic name for the following molecule:

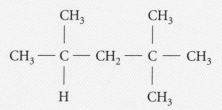

b) Draw the *full* structural formula for 1,2-dimethylcyclohexane.

c) Name a straight chain hydrocarbon which is an isomer of 1,2-dimethylcyclohexane.

d) Draw the *full* structural formula for an aromatic hydrocarbon with eight carbon atoms in the molecule.

12* Petrol is a mixture of chemicals.

a) The structural formula for a hydrocarbon found in petrol is

Study Questions (continued)

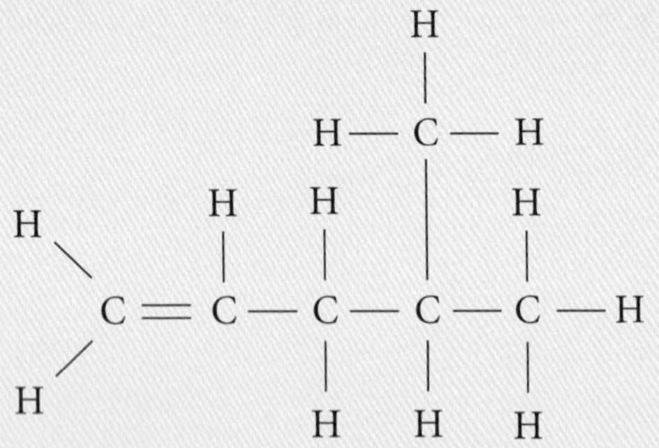

```
                          H
                          |
                  H — C — H
          H   H   |   H
  H       |   |   |   |
   \      |   |   |   |
    C = C — C — C — C — H
   /      |   |   |
  H       H   H   H
```

State the systematic name for this hydrocarbon.

b) A possible route to another compound found in petrol is:

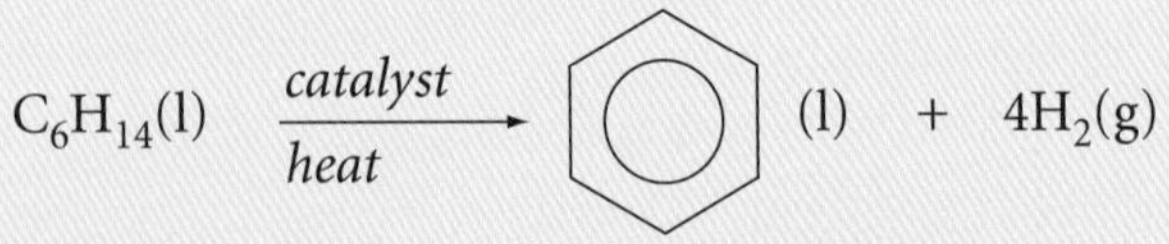

$$C_6H_{14}(l) \xrightarrow[\text{heat}]{\text{catalyst}} \bigcirc (l) \ + \ 4H_2(g)$$

product
X

i) Name product X.

ii) Name the process taking place.

c) Nitrogen oxides are polluting gases which are present in the exhaust mixture from petrol engines.

In a catalytic converter, nitrogen oxides can react with carbon monoxide to form two non-toxic gases.

Name the two gases.

Alcohols, Aldehydes and Ketones

Alcohols

Alcohols

In everyday language the word 'alcohol' is used to represent an alcoholic drink. From a chemical point of view an alcohol is a carbon compound which contains the hydroxyl functional group, —OH. As far as naming is concerned, an alcohol is characterised by the name-ending '-ol'.

Ethanol, or ethyl alcohol as it is traditionally called, is the most important alcohol and is the alcohol present in alcoholic drinks. However, there are many other alcohols, some of which are shown below.

CH_3CH_2OH

ethanol

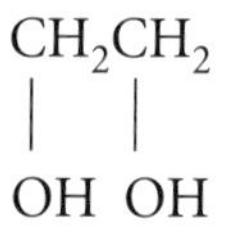

$\begin{array}{cc} CH_2 & CH_2 \\ | & | \\ OH & OH \end{array}$

ethane-1,2-diol

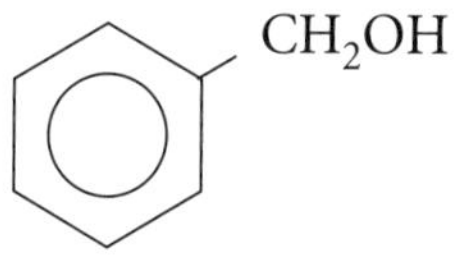

phenylmethanol

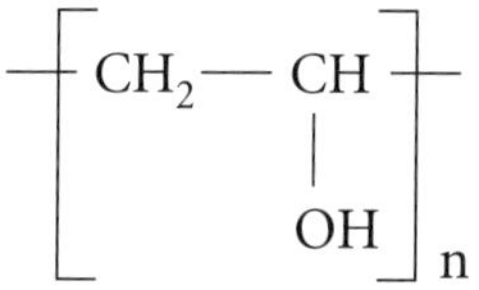

poly(ethenol)

Ethane-1,2-diol (ethylene glycol) is very important as an antifreeze and as one of the monomers used in making polyester fibres. Phenylmethanol (benzyl alcohol) is an aromatic alcohol. Poly(ethenol) has many uses, one of which is that it is the adhesive on the back of British postage stamps. You will come across this polymer again in Chapter 11.

Alkanols

Alkanols belong to a homologous series of alcohols which runs parallel to the alkane series of hydrocarbons. An alkanol can be regarded as a substituted alkane in which one of the hydrogen atoms has been replaced by the hydroxyl group. The name of an alkanol is obtained by replacing the final letter of the corresponding alkane by the name-ending '-ol'. These points are illustrated in Table 9.1.

Alkanes		Alkanols	
Methane	CH_4	Methanol	CH_3OH
Ethane	C_2H_6	Ethanol	C_2H_5OH
Propane	C_3H_8	Propanol	C_3H_7OH
Butane	C_4H_{10}	Butanol	C_4H_9OH
Pentane	C_5H_{12}	Pentanol	$C_5H_{11}OH$
Hexane	C_6H_{14}	Hexanol	$C_6H_{13}OH$
Heptane	C_7H_{16}	Heptanol	$C_7H_{15}OH$
Octane	C_8H_{18}	Octanol	$C_8H_{17}OH$
General formula: C_nH_{2n+2}		General formula: $C_nH_{2n+1}OH$ or $C_nH_{2n+2}O$	

Table 9.1

Structural formulae and isomers

Methanol has no isomers. Ethanol does have an isomer but it does not have an —OH group and is therefore not an alkanol.

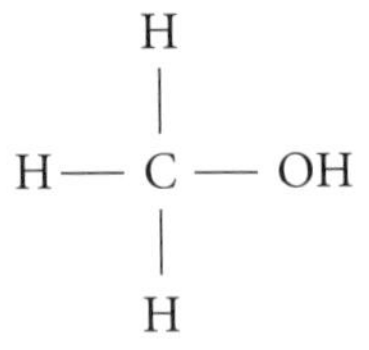

methanol

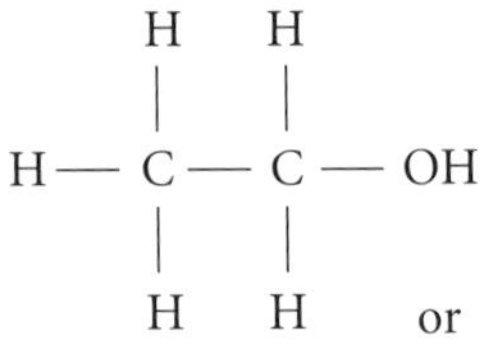

ethanol

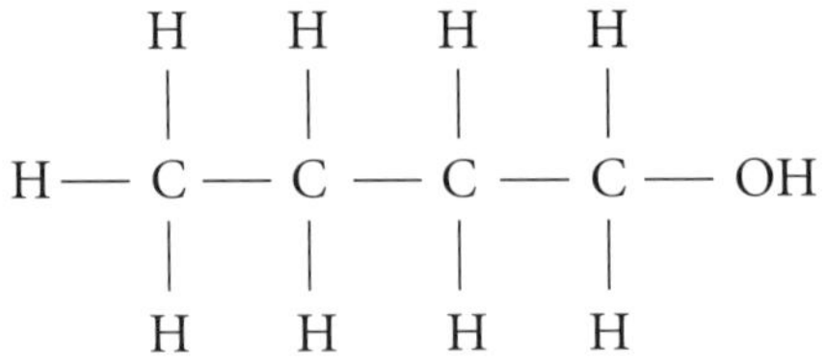

$CH_3CH_2CH_2CH_2OH$

butan-1-ol

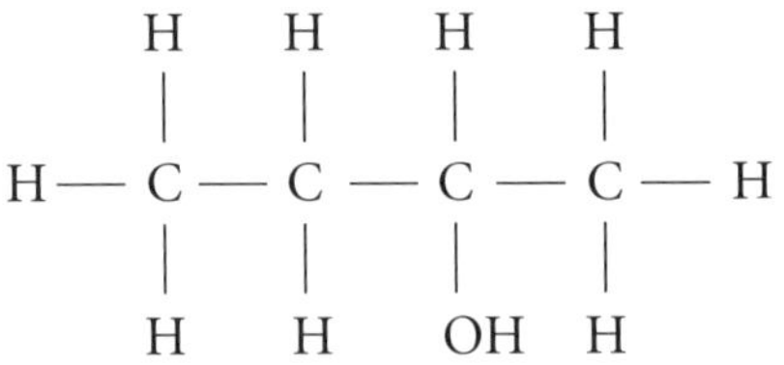

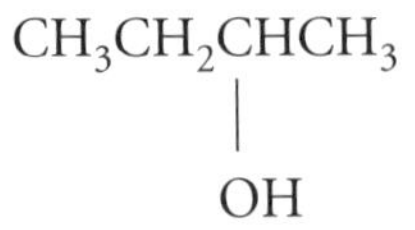

$CH_3CH_2CHCH_3$ or $CH_3CH_2CH(OH)CH_3$

OH   butan-2-ol

Propanol has two alkanol isomers since the hydroxyl group can be attached either to a carbon atom at the end of the chain or to the carbon atom in the middle of the chain. Their structural formulae – full and shortened – as well as their systematic names are given below.

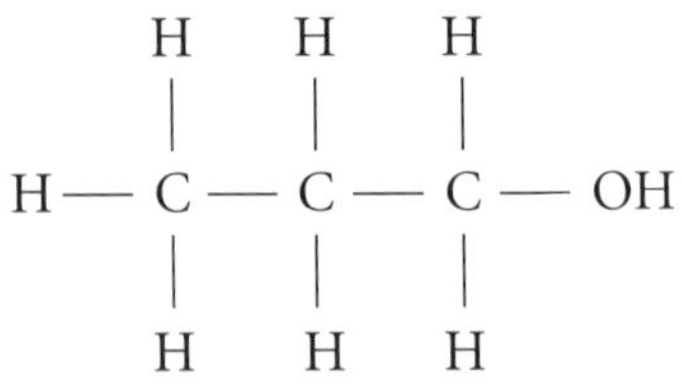

$CH_3CH_2CH_2OH$

propan-1-ol

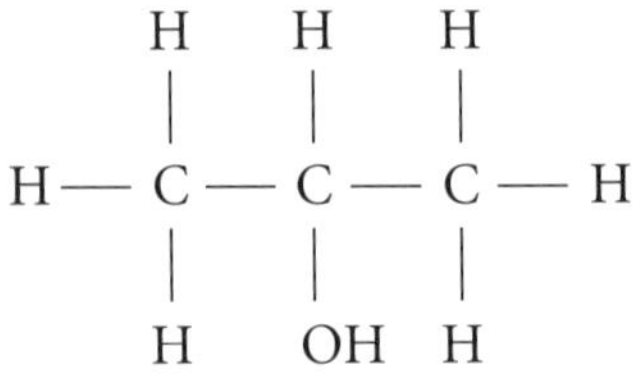

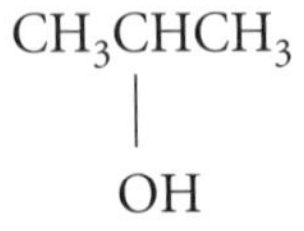

CH_3CHCH_3 or $CH_3CH(OH)CH_3$

OH

propan-2-ol

CH_3CCH_3 2-methylpropan-2-ol

OH

When naming alkanols which have branched chains, the position of the hydroxyl group takes precedence over the position of the branch. Thus the alkanol shown below is called 3-methylbutan-2-ol. In this example, the carbon atoms in the chain are numbered from the right-hand end so that the carbon which has the —OH group attached has the lowest possible number.

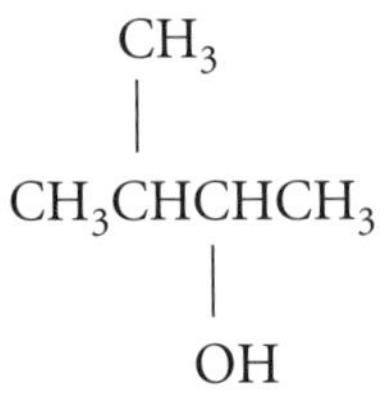

Butanol has several isomers. As with propanol, the position of the —OH group can vary but in addition the chain may be branched. Three isomers of butanol are shown here.

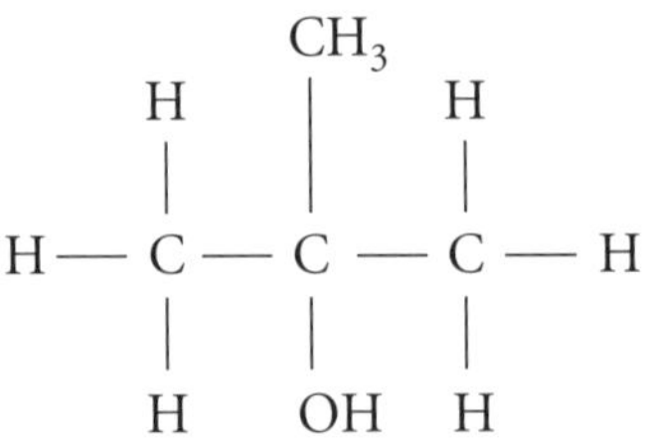

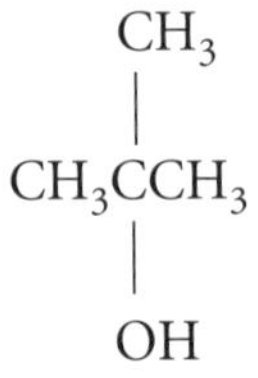

Naming alkanols

When an alkanol contains three or more carbon atoms per molecule it is necessary when naming it to specify the carbon atom to which the —OH group is attached. Figure 9.1 on page 93 explains how the name of the third isomer of butanol shown above is derived.

CH_3

$CH_3CHCHCH_3$ or $CH_3CHCH(OH)CH_3$

OH   3-methylbutan-2-ol

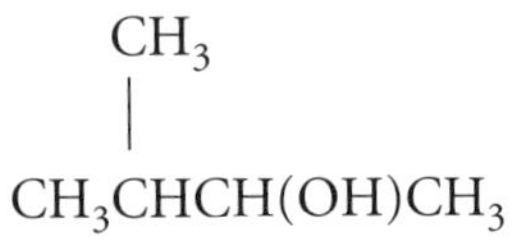

CH_3

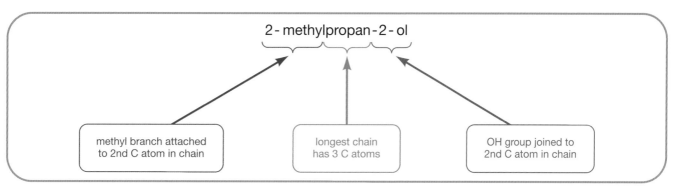

Figure 9.1 Naming a branched alkanol

Summary

ALCOHOLS are carbon compounds which contain the hydroxyl functional group, —OH.

Alcohols include the homologous series of ALKANOLS (general formula: $C_nH_{2n+1}OH$), e.g.

propan-1-ol, $CH_3CH_2CH_2OH$.

$$R\text{ —— }CH_2OH \qquad\qquad R\text{ —— }CH\text{ —— }R'$$
$$|$$
$$OH$$

PRIMARY   SECONDARY
e.g. propan-1-ol   e.g. propan-2-ol
butan-1-ol   butan-2-ol

TERTIARY
e.g.
2-methylpropan-2-ol

Types of alcohol

Alcohols can be subdivided into three different types depending on the position of the hydroxyl group. These are summarized in Table 9.2.

In a secondary or tertiary alkanol, the alkyl groups may be the same, as in propan-2-ol, or different, as in butan-2-ol. In 2-methylpropan-2-ol all three alkyl groups are the same, namely, methyl groups.

Type	Primary	Secondary	Tertiary
Position of –OH group	Joined to the *end* of the carbon chain	Joined to an *intermediate* carbon atom	Joined to an *intermediate* carbon atom which also has a *branch* attached
Characteristic group of atoms	— CH_2OH	— CH — | OH	| — C — | OH

Table 9.2

Primary, secondary and tertiary alkanols can be represented as follows, where the symbols R, R′ and R″ stand for alkyl groups, e.g. methyl, CH_3—, ethyl, C_2H_5—, etc.

Methanol and ethanol are also examples of primary alkanols.

Figure 9.2 Isomers of butanol

Questions

1 There is a fourth isomer of butanol which is not shown on page 92. Draw its full structural formula, give its systematic name and decide what type of alkanol it is.

2 Draw the shortened structural formula and name the type of each of the following alkanols.

   a) pentan-3-ol,

   b) 2-methylbutan-1-ol,

   c) 3-ethylpentan-3-ol.

3 Write down the systematic name and type of each of the following alkanols.

   a)

$$CH_3CH_2\overset{\displaystyle \overset{CH_3}{|}}{\underset{\displaystyle \underset{OH}{|}}{C}}CH_3$$

   b)

$$HOCH_2\overset{\displaystyle \overset{CH_3}{|}}{\underset{\displaystyle \underset{CH_3}{|}}{C}}CH_2CH_3$$

   c)

$$CH_3\overset{\displaystyle \overset{CH_3}{|}}{C}HCH_2\underset{\displaystyle \underset{OH}{|}}{C}HCH_3$$

Oxidation of alcohols

Oxidation occurs when a substance combines with oxygen. From previous work you know that metals combine with oxygen to form oxides and that metals lose electrons during oxidation. For example, when magnesium is oxidised its atoms lose their outer electrons to form magnesium ions. This can be expressed in the form of an ion-electron equation as follows:

$$Mg \rightarrow Mg^{2+} + 2e^-$$

When studying the oxidation of carbon compounds it is often more helpful to focus on the oxygen : hydrogen ratio in the compounds. This is dealt with in the summary at the end of this chapter.

Combustion of an alcohol is an example of oxidation, although somewhat extreme. If sufficient oxygen is available, an alcohol will burn completely to produce carbon dioxide and water, e.g.

$$C_2H_5OH(l) + 3O_2(g) \rightarrow 2CO_2(g) + 3H_2O(l)$$

In Chapter 2, page 21, we introduced the idea of enthalpy of combustion of a substance and you will have carried out an experiment to determine the enthalpy of combustion of an alkanol. During complete combustion of an alkanol all of the C—C and C—H bonds are broken.

In this section, however, we will mainly be concerned with oxidation reactions which do not involve such a drastic change to the structure of the alcohol. As we shall see, only a few specific bonds will be affected. Furthermore, whether the alcohol is a primary, secondary or tertiary alcohol has an important bearing on what happens.

Primary and secondary alcohols can be oxidised by various oxidising agents but tertiary alcohols do not readily undergo oxidation. Acidified potassium dichromate solution is a suitable oxidising agent. When mixed with a primary or secondary alcohol and heated in a water bath, the orange colour due to dichromate ions changes to a blue-green colour showing that chromium(III) ions have been formed. A different smell may be detected showing that the alkanol has changed.

The reduction of dichromate ions to chromium(III) ions can be expressed in the following ion–electron equation. The dichromate ions gain electrons from the alcohol which, therefore, has been oxidised.

$$Cr_2O_7^{2-} + 14H^+ + 6e^- \rightarrow 2Cr^{3+} + 7H_2O$$
dichromate ions

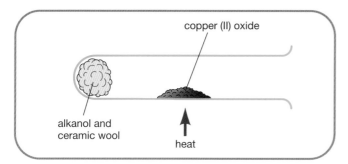

Figure 9.3 Oxidation of an alkanol

Oxidation can also be achieved by passing the alcohol vapour over heated copper(II) oxide as shown in Figure 9.3 During the reaction the oxide is reduced to copper. If ethanol is used, it is oxidised to form a compound called ethanal. The equation for the reaction is:

$$CH_3CH_2OH + CuO \rightarrow CH_3CHO + Cu + H_2O$$

The following section describes what happens to alcohols when they react with either of these oxidising agents, i.e. $Cr_2O_7^{2-}/H^+$ (aq) or CuO(s).

♦ When **primary alcohols** are oxidised they produce aldehydes.

e.g. ethanol   → ETHANAL

CH_3CH_2OH → CH_3CHO

ethanal

propan-1-ol   → PROPANAL

$CH_3CH_2CH_2OH$ → CH_3CH_2CHO

propanal

♦ When **secondary alcohols** are oxidised they produce ketones.

e.g. propan-2-ol   → PROPANONE

$CH_3CH(OH)CH_3$ → CH_3COCH_3

propanone

butan-2-ol   → BUTANONE

$CH_3CH_2CH(OH)CH_3$ → $CH_3CH_2COCH_3$

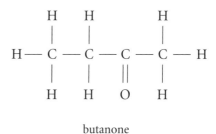

butanone

Aldehydes and ketones can be identified by the presence of a carbon-oxygen double bond, C=O, which is known as the carbonyl group. In an aldehyde, the carbonyl group is at the *end* of the carbon chain and has a hydrogen atom attached to it. In a ketone, the carbonyl group is joined to two other carbon atoms.

When primary alcohols are oxidised to aldehydes and secondary alcohols are oxidised to ketones, two hydrogen atoms are removed – namely the hydrogen atom of the —OH group and a hydrogen on the adjacent carbon atom. A tertiary alcohol cannot be similarly oxidised since it does not have a hydrogen atom attached to the carbon atom which is adjacent to the hydroxyl group. Figure 9.4 illustrates these points.

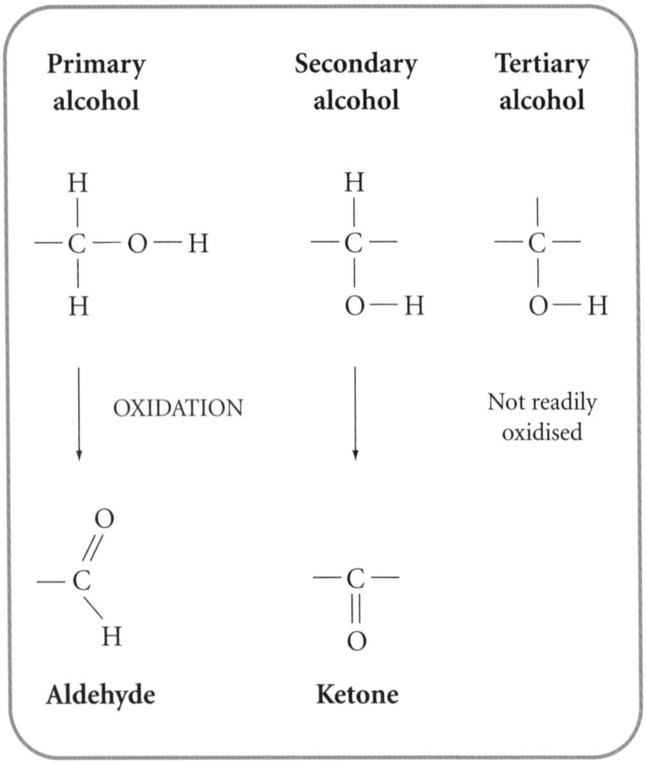

Alkanals		Alkanones
Methanal (formaldehyde) HCHO		
Ethanal (acetaldehyde) CH_3CHO		
Propanal CH_3CH_2CHO		Propanone (acetone) CH_3COCH_3
Butanal $CH_3CH_2CH_2CHO$		Butanone $CH_3CH_2COCH_3$
General formula: $C_nH_{2n}O$ [n ⩾ 1]		$C_nH_{2n}O$ [n ⩾ 3]
Functional group:		

Table 9.3

Figure 9.4 Oxidation of alcohols

At the beginning of this chapter it was mentioned that the alkanols form a homologous series of alcohols. In a similar way there is a homologous series of aldehydes called alkanals, and a homologous series of ketones called alkanones. Table 9.3 shows the functional group, general formula and the first few members of each series. Traditional names of some of these compounds are given in brackets.

Note that in the shortened structural formula of an alkanal, the functional group is written as —CHO and not —COH, which would imply that the hydrogen atom is joined to the oxygen atom.

Branched alkanals and alkanones do exist. When naming these the position of the functional group again takes precedence over the position of the branch, as shown in the following examples. Note that there is no need to indicate a number for the functional group when naming an alkanal as it must always be at the end of the chain. However, when naming an alkanone it is usually necessary to specify the number of the carbonyl group.

$$CH_3$$
$$CH_3CHCH_2CHO$$

3-methylbutanal

$$CH_3$$
$$CH_3COCHCH_2CH_3$$

3-methylpentan-2-one

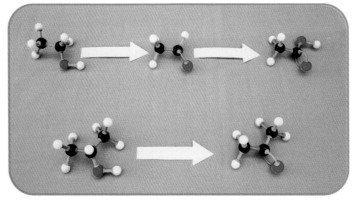

Figure 9.5 Oxidation of alkanols: ethanol → ethanal → ethanoic acid and propan-2-ol → propanone

Oxidation of aldehydes

So far we have tended to emphasise the similarity between aldehydes and ketones in that both types of compound contain the carbonyl group and are obtained by oxidation of alcohols. They do have several chemical properties which are similar as you will discover if you study chemistry beyond Higher level.

However, there is one important reaction which shows a difference between aldehydes and ketones, namely, that aldehydes are readily oxidised when ketones are not. Several oxidising agents can be used to distinguish aldehydes from ketones:

♦ Acidified potassium dichromate solution, $Cr_2O_7^{2-}(aq) + H^+(aq)$

♦ Benedict's solution or Fehling's solution, $Cu^{2+}(aq)$ in alkaline solution

♦ Tollen's reagent, i.e. ammoniacal silver(I) nitrate solution, $Ag^+(aq) + NH_3(aq)$.

Questions

4 For each of the following compounds **i)** draw its structural formula and **ii)** name the alkanol which produces it on oxidation.

  a) 4-methylpentanal,

  b) 3-methylbutan-2-one

5 Name the following compounds.

  a)

$$CH_3CH_2COCHCH_2CH_3$$

with C_2H_5 branch

  b)

$$CH_3CCH_2CHO$$

with CH_3 branches above and below

Summary

ALDEHYDES are carbon compounds which contain a carbonyl group, C=O, at the end of the chain of carbon atoms, i.e. they contain the —CHO functional group. Aldehydes include the homologous series of ALKANALS (general formula: $C_nH_{2n}O$), e.g. propanal, CH_3CH_2CHO. Alkanals are formed when primary alkanols, e.g. propan-1-ol, are oxidised.

Summary

KETONES are carbon compounds which contain a carbonyl group on a middle carbon atom, i.e. they contain the —CO— functional group. Ketones include the homologous series of ALKANONES (general formula: $C_nH_{2n}O$), e.g. propanone, CH_3COCH_3. Alkanones are formed when secondary alkanols, e.g. propan-2-ol, are oxidised.

Prescribed Practical Activity

In the experiment two samples of each of the oxidising agents listed above are poured into separate test tubes. A few drops of an aldehyde are added to one set of each oxidising agent and a few drops of a ketone are added to the other set. The test tubes are placed in a hot water bath for several minutes as shown in Figure 9.6.

Table 9.4 summarises what is observed when an aldehyde reacts with each oxidising agent and explains each observation with reference to what happens to the oxidising agent. A ketone does not react with any of these oxidising agents.

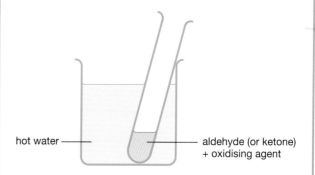

hot water — aldehyde (or ketone) + oxidising agent

Figure 9.6

Prescribed Practical Activity (continued)

Oxidising agent	Observations	Explanation
Acidified potassium dichromate solution	Orange solution → blue-green solution	$Cr_2O_7^{2-}$ (aq) reduced to Cr^{3+}(aq)
Benedict's solution or Fehling's solution	Blue solution → orange-red precipitate	Cu^{2+}(aq) reduced to Cu_2O(s), which is copper(I) oxide, i.e. $Cu^{2+} + e^- \rightarrow Cu^+$
Tollen's reagent	Colourless solution → silver mirror	Ag^+(aq) reduced to Ag(s), i.e. $Ag^+ + e^- \rightarrow Ag$

Teachers may wish to consult SSERC's *Science and Technology Bulletin*, Number 196 (Spring 1999) for a discussion on the use of Benedict's, Fehling's and other solutions. The article suggests that Fehling's solution gives a distinction between propanal and propanone, but that Benedict's solution does not do so.

Table 9.4

Acidified potassium dichromate solution oxidises an aldehyde to form a carboxylic acid. If the aldehyde is an alkanal, it is oxidised to form an alkanoic acid (see Chapter 10), e.g.

$$CH_3CH_2CH_2CHO \rightarrow CH_3CH_2CH_2COOH$$
butanal → butanoic acid

The other two oxidising agents are alkaline and as a result oxidise an alkanal to give the appropriate alkanoate ion, e.g.

$$CH_3CH_2CHO \rightarrow CH_3CH_2COO^-$$
propanal → propanoate ion

The change that occurs to the functional group when an aldehyde is oxidised is shown below.

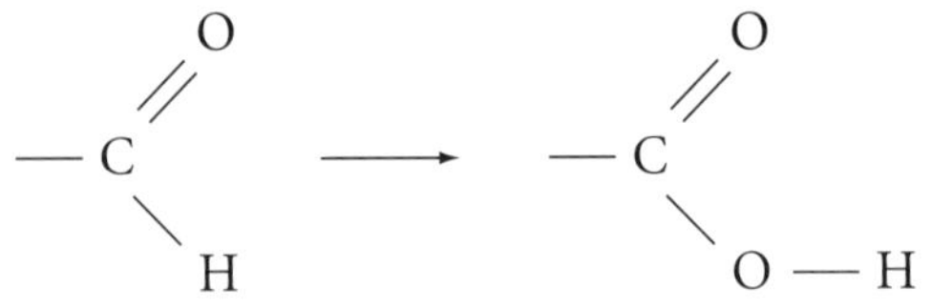

This shows that during the oxidation of an aldehyde, the C—H bond next to the carbonyl group is broken and the hydrogen atom is replaced by a hydroxyl group. This cannot happen with a ketone since it does not have a hydrogen atom attached to its carbonyl group.

Summary

The oxidation of alcohols can be summarised as follows:

1 Primary alcohol → Aldehyde → Carboxylic acid
2 Secondary alcohol → Ketone (Not readily oxidised)
3 Tertiary alcohol (Not readily oxidised)

Let us apply this specifically to alkanols. The ratio of oxygen atoms to hydrogen atoms is quoted for each compound. Note that this ratio changes when a compound is oxidised.

1   Primary alkanol   →   Alkanal   →   Alkanoic acid

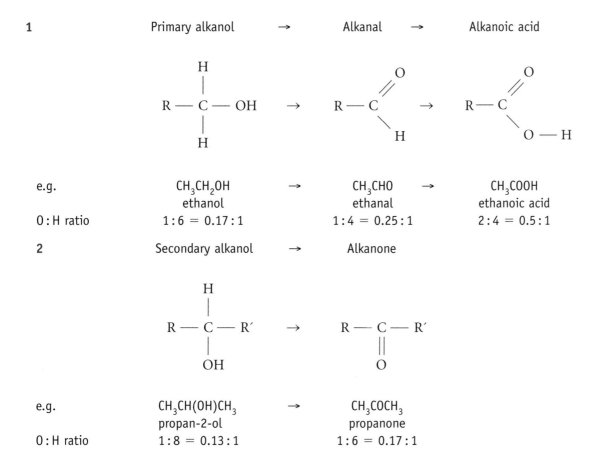

e.g. CH_3CH_2OH → CH_3CHO → CH_3COOH

                   ethanol   ethanal   ethanoic acid

O : H ratio $1:6 = 0.17:1$ $1:4 = 0.25:1$ $2:4 = 0.5:1$

2   Secondary alkanol   →   Alkanone

e.g. $CH_3CH(OH)CH_3$ → CH_3COCH_3

                   propan-2-ol   propanone

O : H ratio $1:8 = 0.13:1$ $1:6 = 0.17:1$

In each of the examples above we can see that oxidation of an alkanol or an alkanal has resulted in an increase in the oxygen : hydrogen ratio. The reverse of these reactions would be reductions and would involve a decrease in the oxygen : hydrogen ratio.

Although it is possible to write ion–electron equations to show that the above changes are oxidations, it is often considered more appropriate when dealing with carbon compounds to define oxidation and reduction in terms of the O : H ratio as follows:

◆ **oxidation** occurs when there is an *increase* in the oxygen to hydrogen ratio.

◆ **reduction** occurs when there is a *decrease* in the oxygen to hydrogen ratio.

Question

6   a) 'Oxidation occurs when methanol is converted to methanal and then to methanoic acid.' Justify this statement by working out the oxygen to hydrogen ratio in each compound shown in the following sequence.

$$CH_3OH \rightarrow HCHO \rightarrow HCOOH$$

   b) Write similar oxidation sequences for

     i) butan-1-ol and

     ii) pentan-2-ol.

     Show that the reactions involve oxidation by reference to O : H ratios.

Study Questions

In questions 1–4 choose the correct word from the following list to complete the sentence.

| blue-green | carbonyl | decreased | hydroxyl |
| increased | secondary | orange-red | tertiary |

1 Pentan-3-ol is an example of a _____ alcohol.

2 When an alcohol is oxidised the ratio of oxygen atoms to hydrogen atoms is _____ .

3 The functional group in a ketone is known as the _____ group.

4 When acidified potassium dichromate solution reacts with an aldehyde the colour of the solution changes to _____ .

5 Which of the following is an alkanone?

A
$$\begin{array}{ccccc} & H & H & H & \\ & | & | & | & \\ H- & C & - & C & - & C & -H \\ & | & | & | & \\ & H & OH & H & \end{array}$$

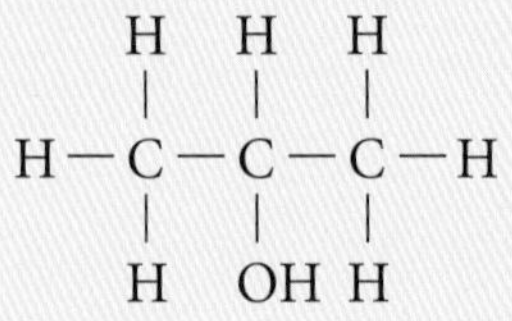

B
$$\begin{array}{c} H \\ | \\ H-C-C \\ | \\ H \end{array} \begin{array}{c} O \\ \diagup\diagup \\ \diagdown \\ OH \end{array}$$

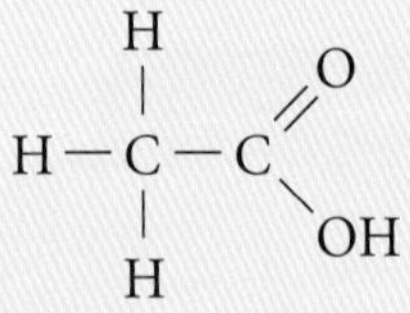

C
$$\begin{array}{ccccc} & H & & H & \\ & | & & | & \\ H- & C & - & C & - & C & -H \\ & | & || & | & \\ & H & O & H & \end{array}$$

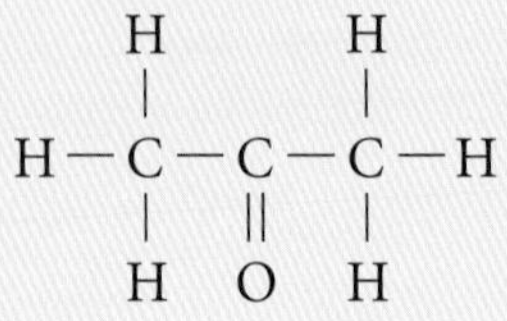

D
$$\begin{array}{c} H \\ | \\ H-C-C \\ | \\ H \end{array} \begin{array}{c} O \\ \diagup\diagup \\ \diagdown \\ H \end{array}$$

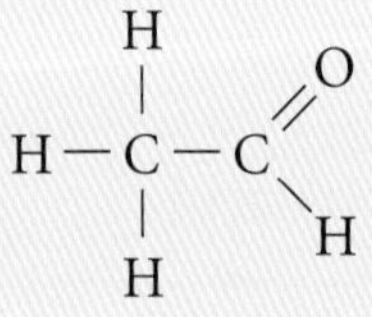

6 Which of the following is a pair of isomers?

A Ethanol and ethanal

B Butanoic acid and butanal

C Propan-2-ol and propanal

D Pentan-2-one and pentanal

7 Which of the following can form a silver mirror with Tollen's reagent?

A $CH_3CH_2CH(OH)CH_3$

B $CH_3CH_2CH_2CHO$

C $CH_3COCH_2CH_3$

D $CH_3CH_2CH_2CH_2OH$

8

$$\begin{array}{ccc} & 1 & 2 \\ HCHO & \rightarrow HCOOH & \rightarrow CH_3OH \end{array}$$

In the reaction sequence shown above

A both steps involve oxidation

B step 1 is reduction and step 2 is oxidation

C both steps involve reduction

D step 1 is oxidation and step 2 is reduction.

9* Copy the following carbon skeleton *three* times.

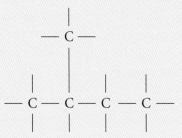

Add a hydroxyl group to each skeleton to make

a) a primary alkanol

b) a secondary alkanol

c) a tertiary alkanol.

10 Give the systematic name of each of the compounds for which you have drawn a skeleton structure in question 9.

Study Questions (continued)

11* There are four isomeric alcohols of molecular formula C_4H_9OH. Their structural formulae are as follows:

CH$_3$CH$_2$CH$_2$CH$_2$OH   **I**

CH$_3$CH$_2$CHCH$_3$ **II**
|
OH

CH$_3$
|
CH$_3$—C—CH$_3$ **III**
|
OH

CH$_3$
|
CH$_3$CHCH$_2$OH   **IV**

a) Give systematic names for I, II, III, and IV.

b) State which of the compounds I–IV are primary, which secondary, and which tertiary alcohols.

c) The four alcohols are contained, separately, in four bottles marked A, B, C and D. From the following information decide which bottle contains which alcohol. State your reasoning briefly at each stage.

   i) The contents of A, B and C can readily be oxidised by acid potassium dichromate solution, while those of D cannot.

   ii) A and B, on complete oxidation by the dichromate, give acids of formulae C_3H_7COOH. C does not give this acid.

   iii) All four substances can be dehydrated to give alkenes. A and D can both form the same alkene. B and C can both form the same alkene, which is an isomer of that formed by A and D.

12

CH$_3$CH=CHCH$_3$ → CH$_3$CH$_2$CHClCH$_3$ → CH$_3$CH$_2$CH(OH)CH$_3$

     A   B   C

a) Give the systematic name of each of the compounds A, B and C.

b) Name the chemical required to convert A into B.

c) C reacts with acidified potassium dichromate solution to form compound D.

   i) What colour change occurs in this reaction?

   ii) What type of reaction has occurred when C changes to D?

   iii) Draw the full structural formula of D and give its systematic name.

   iv) When D is tested with Tollen's reagent will it form a silver mirror? Explain your answer.

10 Carboxylic Acids and Esters

From previous work you should know and understand the following:

★ Alcohols and aldehydes

★ Oxidation of aldehydes

★ Condensation reaction.

Carboxylic acids

A carboxylic acid is characterised by its functional group, called the carboxyl group, and by its name ending, i.e. '-oic acid'.

Carboxyl functional group:

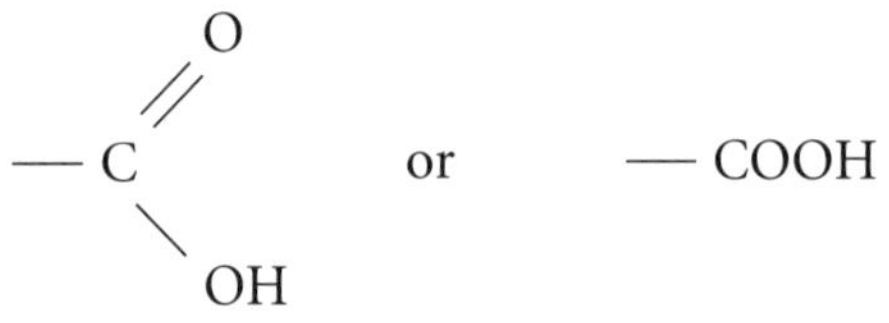

$$-C\overset{\displaystyle O}{\underset{\displaystyle OH}{\Big\langle}} \quad \text{or} \quad -COOH$$

Examples:

$$CH_3COOH$$
ethanoic acid

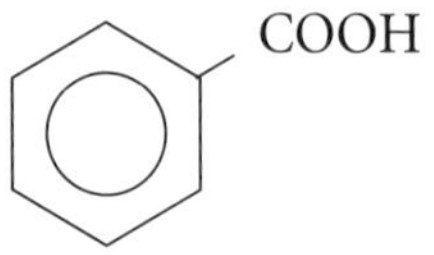

benzoic acid

Name of acid	Structural formulae
Methanoic acid (formic acid)	HCOOH 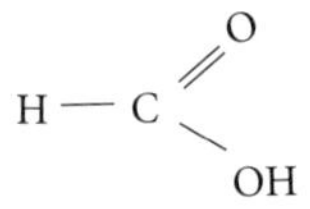
Ethanoic acid (acetic acid)	CH_3COOH 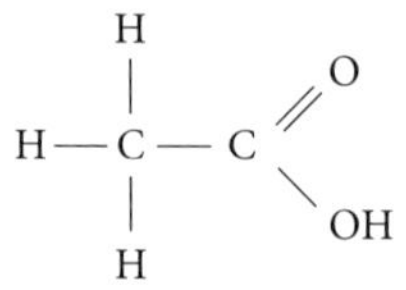
Propanoic acid	CH_3CH_2COOH 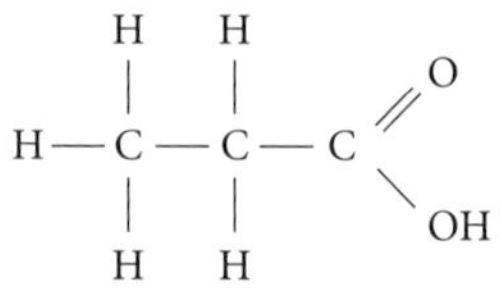
Butanoic acid	$CH_3CH_2CH_2COOH$
Pentanoic acid	$CH_3CH_2CH_2CH_2COOH$
Hexanoic acid	$CH_3CH_2CH_2CH_2CH_2COOH$
Heptanoic acid	$CH_3CH_2CH_2CH_2CH_2CH_2COOH$
Octanoic acid	$CH_3CH_2CH_2CH_2CH_2CH_2CH_2COOH$
General formula:	$C_nH_{2n}O_2$

Table 10.1

There is a homologous series of carboxylic acids called alkanoic acids which are based on the corresponding parent alkanes. The first eight members of this series are listed in Table 10.1 along with their structural formulae.

When studying fats and oils in Chapter 12 you will come across carboxylic acids with even longer chains of carbon atoms.

Branched-chain alkanoic acids can also be obtained and these are named as before with the functional group taking precedence. As with the naming of aldehydes there is no need to give a number for the functional group as it must be at the end of the chain,

i.e. the carbon atom of the —COOH group is the first carbon atom in the chain. Thus the compound shown here is called 3-methylbutanoic acid.

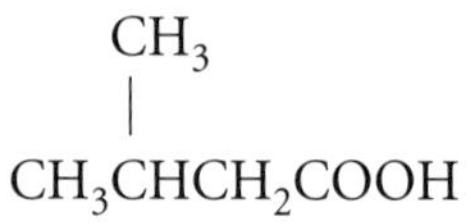

$$\begin{array}{c} CH_3 \\ | \\ CH_3CHCH_2COOH \end{array}$$

The fact that these compounds are acids, and consequently release H^+ ions when in aqueous solution, is very important and we will return to this aspect in Chapter 16.

Summary

CARBOXYLIC ACIDS are carbon compounds which contain the carboxyl functional group, —COOH. Carboxylic acids include the homologous series of ALKANOIC ACIDS (general formula: $C_nH_{2n}O_2$), e.g. propanoic acid, CH_3CH_2COOH.

Questions

1 Draw the shortened structural formula of

   a) 2,2-dimethylpropanoic acid

   b) 3-ethyl-5-methylhexanoic acid.

2 Name the following acid and draw the shortened structural formula of the alkanol from which it can be formed.

$$CH_3$$
$$|$$
$$CH_3CH_2CHCHCOOH$$
$$|$$
$$CH_3$$

Uses of carboxylic acids

Ethanoic acid is a feedstock for the production of several important materials. Its principal use is in the manufacture of ethenyl ethanoate (vinyl acetate) which is polymerised to give the plastic component of vinyl emulsion paints. Other important uses include the making of cellulose ethanoate, which is used to produce films and lacquers, and benzene-1,4-dicarboxylic acid, a diacid required for the manufacture of polyester. Another diacid, hexanedioic acid, is a monomer used in the production of nylon.

$$CH_3COO$$
$$|$$
$$CH = CH_2$$

ethenyl ethanoate

HOOC — ⬡ — COOH

benzene-1,4-dicarboxylic acid

$$HOOC(CH_2)_4COOH$$

hexanedioic acid

Benzoic acid, C_6H_5COOH, occurs naturally in cherry bark, raspberries and tea but it is synthesised for use as the food additive, E210. It acts as a preservative and antioxidant and is often used in fruit products and soft drinks. Its sodium salt, E211, has similar properties and is used in bottled sauces.

Making esters

When a carboxylic acid reacts with an alcohol a new type of carbon compound, called an ester, is formed. The functional groups present in the reactants interact to form an ester link and, at the same time, produce a water molecule. Therefore, the process of making an ester is an example of a condensation reaction. The interaction between the functional groups is illustrated in Figure 10.1 on the following page.

The reaction between ethanol and ethanoic acid produces the ester, ethyl ethanoate. The equation showing full structural formulae is given in Figure 10.1 also.

This equation can be rewritten using shortened structural formulae:

$$CH_3CH_2OH + HOOCCH_3 \rightleftharpoons CH_3CH_2OOCCH_3 + H_2O$$

In order to show the reaction between the functional groups it is advisable to reverse the formula of one of the reactants. In the example above the formula of the carboxylic acid has been reversed. The equation can, of course, be written with the formula of the carboxylic acid shown first and then the formula of the alcohol written in reverse. Applying this to the above example gives the following equation:

$$CH_3COOH + HOCH_2CH_3 \rightleftharpoons CH_3COOCH_2CH_3 + H_2O$$
ethanoic acid   ethanol   ethyl ethanoate

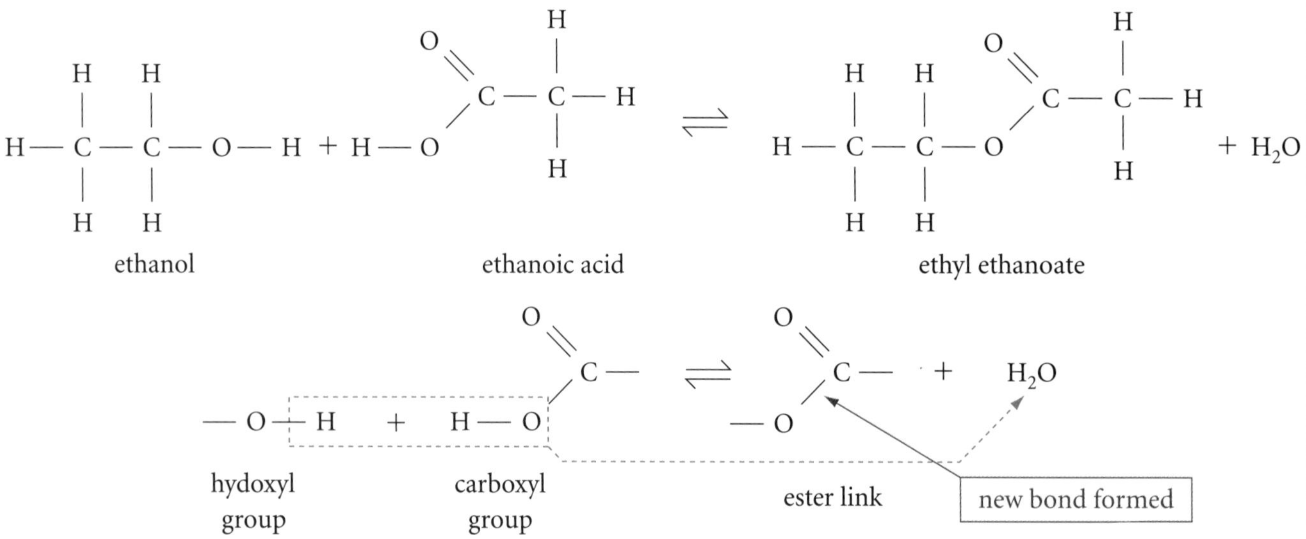

Figure 10.1 showing the formation of an ester:

ethanol + ethanoic acid ⇌ ethyl ethanoate + H_2O

hydoxyl group + carboxyl group ⇌ ester link + H_2O (new bond formed)

Figure 10.1 Formation of an ester, also showing how the functional groups interact

Prescribed Practical Activity

When a carboxylic acid reacts with an alcohol a new type of carbon compound, called an **ester**, is formed. A convenient method of preparing an ester is described below.

About 2 cm³ of an alcohol and an equal volume of a carboxylic acid are mixed together and a few drops of concentrated sulphuric acid are added. This mixture is heated, as shown in Figure 10.2, for several minutes.

The test tube is removed from the water bath and its contents added to about 20 cm³ of sodium hydrogencarbonate solution as shown in Figure 10.3. This solution neutralises the sulphuric acid, as well as any unreacted carboxylic acid, releasing carbon dioxide in the process. The ester appears as an immiscible layer on the surface of the solution. The ester can also be detected by its characteristic smell.

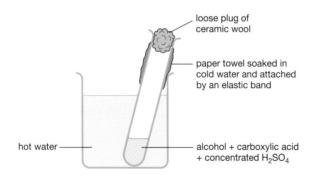

loose plug of ceramic wool

paper towel soaked in cold water and attached by an elastic band

hot water

alcohol + carboxylic acid + concentrated H_2SO_4

Figure 10.2

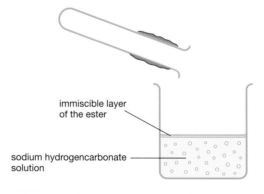

immiscible layer of the ester

sodium hydrogencarbonate solution

Figure 10.3

Concentrated sulphuric acid plays an important part in the reaction. It provides hydrogen ions which act as a catalyst. It also helps to increase the yield of ester by absorbing water, the other product. Why this happens should become apparent once you have studied the topic of chemical equilibrium in Chapter 15.

Referring back to Figure 10.3, it is worth noting that both reactants contain polar O—H bonds while the ester produced does not. Hydrogen bonding between molecules in alcohols, and also in carboxylic acids, means that these types of carbon compounds have relatively high boiling points. Furthermore, short-chain alcohols and carboxylic acids are miscible with water. Because of a lack of hydrogen bonding, esters, on the other hand, tend to be more volatile and less miscible with water. You may have observed this last property during the experiment on ester formation.

Naming esters

The name of an ester depends on which alcohol and which carboxylic acid have been used in preparing it. The first part of the name, which ends in '-yl', comes from the alcohol and the second part, which ends in '-oate', comes from the acid.

Two examples are illustrated in Figures 10.4 and 10.5 below. The first example shows full structural formulae, the second uses shortened structural formulae. In each case the formula of the ester is shown twice since there are two ways of drawing it as explained above.

When naming an ester formed from an alkanol, the first part of the name is derived by deleting the letters '-anol' from the name of the alkanol and replacing them with the ending '-yl'. Thus, **pentanol** produces **pentyl** esters, **octanol** produces **octyl** esters etc.

If an ester is prepared using an alkanoic acid, the second part of the ester's name ends in '-oate'. Thus, **butanoic acid** produces a **butanoate** ester, **hexanoic acid** produces a **hexanoate** ester, etc.

Esters prepared from alkanols and alkanoic acids which have the same total number of carbon atoms will, as a result, have the same molecular formula, e.g. ethyl ethanoate and methyl propanoate are isomers, molecular formula: $C_4H_8O_2$. Butanoic acid, C_3H_7COOH, is also isomeric with these esters although it belongs to a different homologous series.

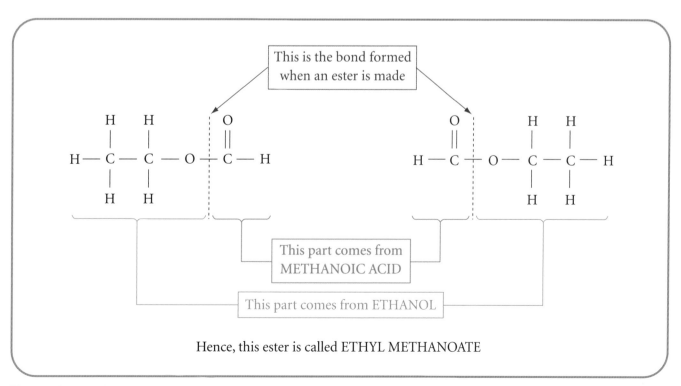

Figure 10.4 Naming an ester using full structural formulae

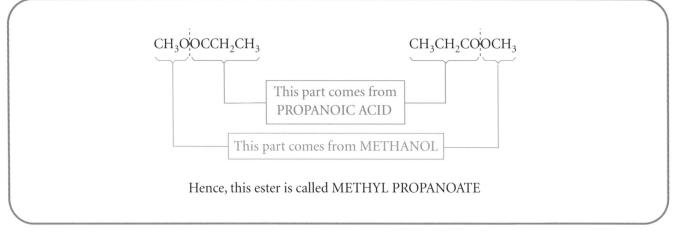

Figure 10.5 Naming an ester using shortened structural formulae

Questions

3 Draw the full structural formulae of methanol, ethanoic acid and the ester produced by reacting them together. Name the ester.

4 Draw the shortened structural formulae of butanoic acid, propan-1-ol and the ester produced by reacting them together. Name the ester.

Uses of esters

Esters have a variety of uses. You will be aware from having prepared different esters that they differ widely in smell. This property makes them useful in making perfumes and as artificial flavourings in the food industry. Table 10.2 lists several esters and the flavours associated with them.

Name	Structural formulae	Flavour
Pentyl ethanoate	$CH_3CH_2CH_2CH_2CH_2OOCCH_3$	Pear
1-methylbutyl ethanoate	CH_3 | $CH_3CH_2CH_2CHOOCCH_3$	Banana
Octyl ethanoate	$CH_3(CH_2)_7OOCCH_3$	Orange
Methyl butanoate	$CH_3OOCCH_2CH_2CH_3$	Pineapple
Pentyl butanoate	$CH_3(CH_2)_4OOCCH_2CH_2CH_3$	Apricot

Table 10.2

The actual flavour of a fruit may be due to a mixture of various esters and other substances. When an artificial flavouring is being produced several natural and synthetic materials are blended together. Similarly, perfumes are often mixtures containing various natural and synthetic components including esters.

In terms of quantity used, the principal use of esters is as solvents, e.g. in car body paints, radiator enamels, adhesives and cosmetic preparations. Rapid evaporation of the solvent is often important so esters with relatively few carbon atoms are used as they tend to be more volatile. Ethyl ethanoate is the most common ester to be used as a solvent.

Esters also have medicinal uses, e.g. methyl salicylate or 'oil of wintergreen' is used in analgesic rubs for strains and sprains. Aspirin is another ester derived from salicylic acid, which is an aromatic carboxylic acid.

Hydrolysis of esters

The equations for the formation of an ester on pages 103–104 show that the reaction is reversible. The reverse process, in which the ester is split by reaction with water to form an alcohol and a carboxylic acid, is an example of hydrolysis. Therefore, hydrolysis is the opposite of condensation which occurs when an ester is formed. The C—O bond formed when an ester is made is the bond which is broken when the ester is hydrolysed – look back at Figure 10.3 on page 104.

Figure 10.6 Esters are used as solvents in car body paints

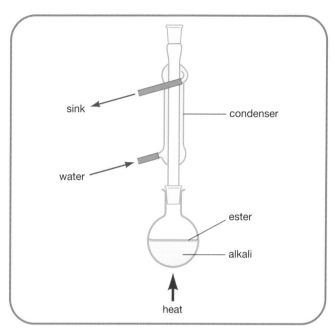

Figure 10.7 Heating under reflux

Although hydrolysis is the process by which water breaks down a carbon compound, in practice, water *on its own* is rarely successful in achieving this to any appreciable extent. Hydrolysis of an ester can be carried out by heating it in the presence of a dilute acid, e.g. HCl(aq) or H_2SO_4(aq), to provide hydrogen ions to catalyse the reaction.

The equation for the acid-catalysed hydrolysis of ethyl ethanoate is as follows:

$$CH_3CH_2OOCCH_3 + H_2O \rightleftharpoons CH_3CH_2OH + HOOCCH_3$$
ethyl ethanoate   ethanol   ethanoic acid

Acid-catalysed hydrolysis is reversible and as a consequence will be incomplete. Complete hydrolysis can, however, be achieved by adding the ester to a strong alkali, e.g. NaOH(aq) or KOH(aq), and heating under reflux for about 30 minutes as shown in Figure 10.7.

The products of the reaction are the alcohol and the salt of the carboxylic acid and these can be separated by distillation using the reassembled apparatus shown in Figure 10.8. The alcohol is distilled off and the salt of the carboxylic acid is left behind in the aqueous solution. When this solution is acidified with hydrochloric acid, the salt is converted to the carboxylic acid.

The equations for a) the hydrolysis of ethyl ethanoate by sodium hydroxide solution and b) subsequent acidification of the salt solution (after removal of ethanol) are as follows:

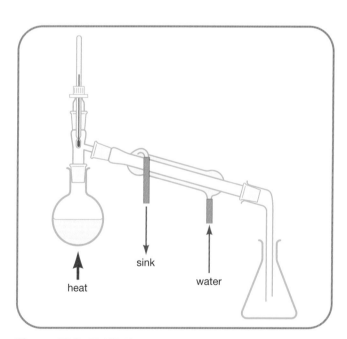

Figure 10.8 Distillation

a) $$CH_3COOCH_2CH_3 + Na^+ + OH^- \rightarrow$$
$$CH_3COO^- + Na^+ + HOCH_2CH_3$$
sodium ethanoate   ethanol

b) $$CH_3COO^- + Na^+ + H^+ + Cl^- \rightarrow$$
$$CH_3COOH + Na^+ + Cl^-$$
ethanoic acid

107

Alkaline hydrolysis of an ester has two main advantages over acid-catalysed hydrolysis.

1 Alkaline hydrolysis is not reversible and hence there should be a higher yield of products.

2 It is usually easier to separate the products when alkaline hydrolysis has been used.

Ester hydrolysis, especially by alkali, is an important process in the manufacture of soap from fats and oils. This will be dealt with in more detail in Chapter 12.

Questions

5 Name the following esters and draw the full structural formulae of the alcohols and carboxylic acids produced when they are hydrolysed.

a) $CH_3CH_2CH_2OOCH$

b) (refer to page 102 if necessary)

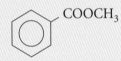

6 Write equations, giving shortened structural formulae for the carbon compounds involved, for the following reactions. Beneath each carbon compound write its name.

a) Hydrolysis of ethyl butanoate by KOH(aq).

b) Acid-catalysed hydrolysis of pentyl ethanoate.

Percentage yield

Balanced equations enable us to calculate the quantity of product which we would expect to obtain from a known mass of reactant, assuming that 100% conversion takes place. The quantity of product thus calculated is called the theoretical yield.

The actual yield, i.e. the quantity actually obtained, is usually less than the theoretical yield. This is particularly so when carrying out a reaction involving carbon compounds. The reaction may not go to completion, e.g. preparation of an ester, or other reactions may also occur which 'compete' with the main reaction. Separation of the desired product may be difficult or the product may be impure and some of it lost during purification. There is also the human factor – poor practical technique can have a drastic effect on the actual yield!

In industrial processes a high percentage yield as well as high purity of product is often desirable. Where this is not possible or economically feasible, unconverted reactants are frequently recycled for further reaction, e.g. making ammonia by the Haber Process and making ethanol by hydration of ethene.

The percentage yield of product can be calculated using the following relationship.

$$\text{Percentage yield} = \frac{\text{Actual yield}}{\text{Theoretical yield}} \times 100\%$$

Alternatively, if the percentage yield is known, the actual yield can be calculated as follows.

$$\text{Actual yield} = \text{Percentage yield} \times \text{Theoretical yield}$$

Worked Example 10.1

A sample of methyl ethanoate weighing 6.9 g was obtained from a reaction mixture containing 9.0 g of ethanoic acid, excess methanol and a small volume of concentrated sulphuric acid. Calculate the percentage yield of ester using the following equation.

$$CH_3OH + HOOCCH_3 \rightleftharpoons CH_3OOCCH_3 + H_2O$$

$$\begin{array}{ccc} & 1\ mol & 1\ mol \\ & 60\ g & 74\ g \end{array}$$

In theory, 60 g of ethanoic acid should yield 74 g of methyl ethanoate.
Hence, 9.0 g of ethanoic acid should yield

$$9.0 \times \frac{74}{60}\ g$$

of methyl methanoate = 11.1 g

i.e. the theoretical yield is 11.1 g, while the actual yield is 6.9 g.

$$\text{Percentage yield } \frac{6.9}{11.1} \times 100\% = 62.2\%$$

Questions

7 Excess ethyne was reacted with 0.1 mol of hydrogen chloride and 4.1 g of the product, 1,1-dichloroethane, were obtained. Calculate the percentage yield using the equation:

$$C_2H_2 + 2HCl \rightarrow C_2H_4Cl_2$$

8 A student hydrolyses 7.5 g of ethyl benzoate and obtains a 71% yield of benzoic acid. Calculate the actual yield of benzoic acid, given that, theoretically, one mole of the ester ($C_6H_5COOC_2H_5$) produces one mole of the acid (C_6H_5COOH).

Study Questions

1 Which compound is an ester?

A   COOCH$_3$

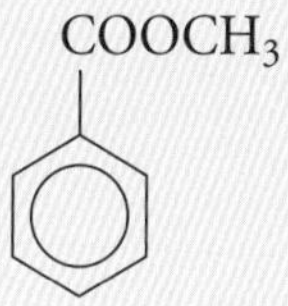

B   CH$_2$OH

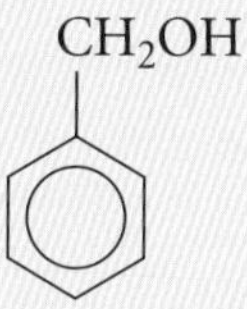

C   COOH

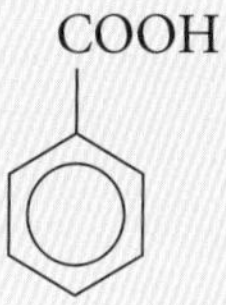

D   COCH$_3$

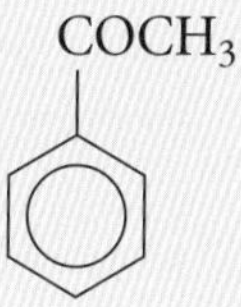

2 Which of the following is **not** an isomer of hexanoic acid?

A   2-methylpentanoic acid

B   Ethyl propanoate

C   2,2-dimethylbutanoic acid

D   Butyl ethanoate

3 $HCOOCH_3 + H_2O \rightarrow HCOOH + CH_3OH$

The type of reaction shown above is

A   hydration

B   condensation

C   addition

D   hydrolysis.

4 Banana flavouring contains the ester:

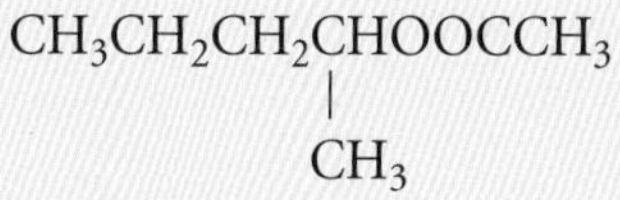

$$CH_3CH_2CH_2CHOOCCH_3$$
$$|$$
$$CH_3$$

This ester can be made from

A   pentan-1-ol and ethanoic acid

B   ethanol and pentanoic acid

C   pentan-2-ol and ethanoic acid

D   ethanol and 2-methylbutanoic acid

5* a) Esters are formed when carboxylic acids react with alcohols, e.g.

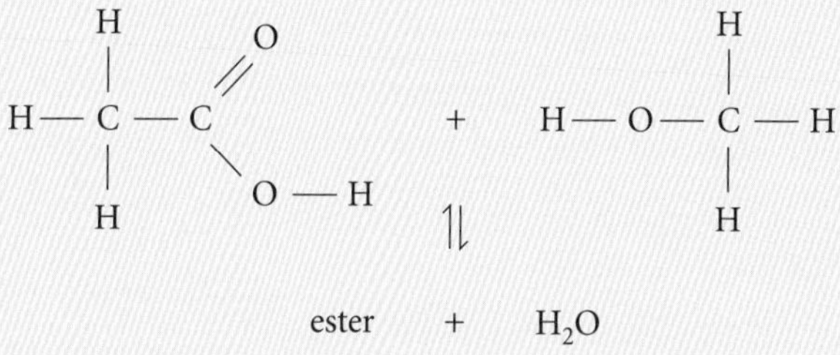

ester   + H_2O

i)   Name the ester formed in this reaction.

ii)   Name the type of reaction which takes place.

iii) To find out which atoms of the alcohol and carboxylic acid go to form the water molecule, the reaction was carried out using an alcohol in which the ^{16}O atom was substituted by the ^{18}O isotope. All of the ^{18}O was found in the ester and none in the water.

Study Questions (continued)

Copy the equation and circle the atoms in the acid and the alcohol which combine to form the water molecule.

b) Esters can be prepared in the laboratory by heating an alcohol and a carboxylic acid with a few drops of concentrated sulphuric acid in a water bath. After 10 minutes or so, the reaction mixture is poured into sodium hydrogencarbonate solution.

    i)   What evidence, apart from smell, shows that the ester has been formed?

    ii)   State *two* safety precautions that should be adopted when carrying out this experiment.

6 The following diagram shows three different reactions of pentan-2-ol.

$$pent\text{-}2\text{-}ene$$
$$\uparrow \quad I$$
$$an\ ester \xleftarrow{\quad II \quad} pentan\text{-}2\text{-}ol \xrightarrow[oxidation]{\quad III \quad} compound\ X$$

a) What type of reaction occurs in **i)** reaction I **ii)** reaction II?

b) Pent-2-ene is not the only alkene formed in reaction I. Name the other alkene which can be produced.

c) Name compound X and name a suitable oxidising agent for reaction III.

d) Draw the structural formula of the ester formed when pentan-2-ol reacts with methanoic acid.

7* Synthetic perfumes are cheaper and easier to produce than natural perfumes.

a) Cinnamyl alcohol smells pleasantly of hyacinths; it can be described as unsaturated.

$$\overset{\displaystyle H}{\underset{|}{C}}=\overset{\displaystyle H}{\underset{|}{C}}-\overset{\displaystyle H}{\underset{|}{\underset{H}{C}}}-OH$$

Give *two* other terms which could be used to describe this alcohol.

b) Phenylethanol has a smooth rose-like odour and is used in floral perfumes together with its propanoate ester.

$$CH_2CH_2OH \qquad + \qquad CH_3CH_2COOH$$

phenylethanol
Mass of one mole
= 122 g

propanoic acid
Mass of one mole
= 74 g

$$X \quad + \quad H_2O$$

propanoate ester
Mass of one mole
= 178 g

    i)   Draw the structural formula for ester X.

    ii)   3.05 tonnes of phenylethanol was refluxed with 1.48 tonnes of propanoic acid. Show, by calculation, that the phenylethanol is in excess.

         (One tonne = 1000 kg)

    iii) The formation of the propanoate ester gives a 70% yield after refluxing. Calculate the mass of ester obtained. (Show your working clearly.)

8 In theory, one mole of benzene, C_6H_6, can be converted to one mole of methylbenzene, $C_6H_5CH_3$, which in turn can yield one mole of benzoic acid, C_6H_5COOH. In an experiment, 0.1 mole of benzene produced 4.6 g of methylbenzene which was then converted to benzoic acid, the percentage yield for this reaction being 70%. Calculate:

a) the percentage yield for the first reaction

b) the mass of benzoic acid finally obtained

c) the overall percentage yield of benzoic acid based on the original quantity of benzene.

Study Questions (continued)

9* The flavour in plums comes from naturally-occurring esters. Part of a workcard outlining the laboratory preparation of an ester is shown below.

PREPARATION OF AN ESTER

1 Mix 1 cm³ of the alkanol with 1 cm³ of the alkanoic acid in a test tube.

2 Wrap a piece of paper soaked in cold water around the top of the test tube and hold in place as shown in the diagram.

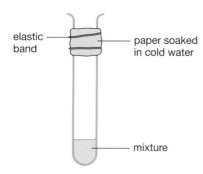

elastic band — paper soaked in cold water

mixture

3

4

5 After 20 minutes, pour the contents of the test tube into a beaker containing sodium hydrogen-carbonate solution.

---

a) Complete the workcard by adding appropriate instructions for steps 3 and 4.

b) Why is paper soaked in cold water put around the test tube?

c) The structural formula for a plum-flavoured ester is shown below.

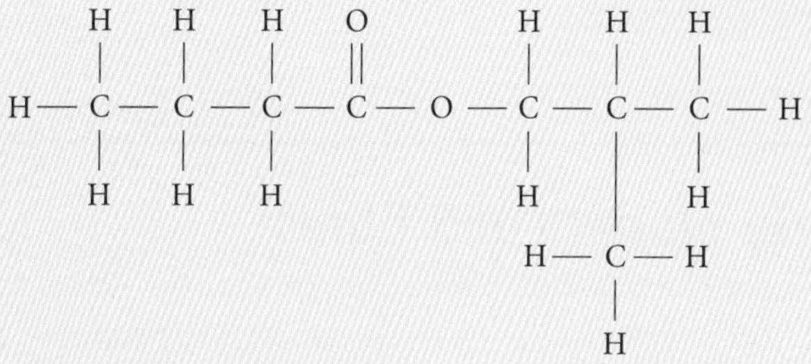

Give the systematic name of the alkanol used in making this ester.

11 Polymers

From previous work you should know and understand the following:

★ Alkenes

★ Addition reactions

★ Polymerisation

★ Monomer

★ Polymer

★ Thermoplastic

★ Thermosetting.

Early plastics and fibres

Successful large scale manufacture of artificial plastics and fibres began in the 1940s, in most cases production being stimulated by the demands of the Second World War. Most of the materials produced at that time are still being manufactured today.

Nearly 20 million tonnes of ethene are produced each year by the chemical industry in Western Europe. The second most important organic chemical is propene, with about 12 million tonnes produced each year. Both alkenes are very important starting materials in the petrochemical industry especially for the manufacture of polymers, i.e. they are both monomers.

Ethene undergoes addition polymerisation to form poly(ethene) commonly known as polythene:

$$nCH_2{=}CH_2 \rightarrow {-}[CH_2{-}CH_2]_n{-}$$
$$\text{ethene} \qquad\qquad \text{poly(ethene)}$$

Low density poly(ethene) or LDPE ($0.92\,g\,cm^{-3}$) is made at very high pressures and at temperatures around 150–300°C in the presence of an initiator. Extensive branching of the chains occurs during manufacture, forming a more flexible material suitable for use in film packaging and electrical insulation.

High density poly(ethene) or HDPE ($0.96\,g\,cm^{-3}$) is manufactured at much lower pressures and temperatures in the presence of catalysts containing either titanium or chromium compounds. The chains have very few branches so that the resulting polymer is stronger and more rigid than LDPE. It is used to make pipes, gutters, containers for household chemicals and in industrial packaging.

Figure 11.1 Some retailers are now using recycled poly(ethene) for packaging

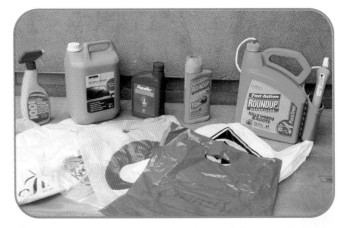

Figure 11.2 Poly(ethene): HDPE is used for containers; LDPE is used to make 'plastic' shopping bags

Another form known as linear low density poly(ethene) or LLDPE is made by a similar process to HDPE but involves co-polymerisation of ethene with monomers such as butene to introduce side chains into the polymer structure. About 100 000 tonnes of LLDPE are manufactured in Scotland per year. It has the same uses as LDPE.

Ethene is used to make other monomers for the plastics industry. An important example is its reaction with benzene during the production of styrene (or phenylethene) which is then polymerised to form polystyrene (or poly(phenylethene)).

$$n\ CH = CH_2 \longrightarrow \left[CH - CH_2 \right]_n$$

phenylethene
(styrene)

poly(phenylethene)
(polystyrene)

Ethene (ethylene) is manufactured by a process called steam cracking from two different feedstocks, one being a gas, the other a liquid. The feedstocks are cracked at high temperatures (about 800°C) in the presence of steam. As well as ethene, fuel gases (namely, methane and hydrogen), propene and gasoline are produced.

The gaseous feedstock is usually ethane which has been separated from the gas fraction obtained during the distillation of crude oil. The liquid feedstock is usually naphtha – another fraction obtained in the same way. It has a boiling point range of 70–180°C and contains hydrocarbons with 6–10 carbon atoms per molecule.

Both types of feedstock are cracked at Grangemouth and ethane is cracked at Mossmorran in Fife. Table 11.1 gives a comparison of the approximate percentages of the different products obtained on cracking these two feedstocks. While ethane provides a much greater percentage of ethene, naphtha produces higher proportions of valuable by-products.

| Product | Approximate percentage obtained from cracking | |
	Ethane	Naphtha
Ethene	80	30
Fuel gases	10	25
Propene	5	15
Gasoline	5	30

Table 11.1

The propene produced from these cracking processes is also a very important starting material for making polymers. It is polymerised by addition to produce poly(propene) as shown:

$$n\ \underset{\underset{H}{|}}{\overset{\overset{CH_3}{|}}{C}} = \underset{\underset{H}{|}}{\overset{\overset{H}{|}}{C}} \longrightarrow \left[\underset{\underset{H}{|}}{\overset{\overset{CH_3}{|}}{C}} - \underset{\underset{H}{|}}{\overset{\overset{H}{|}}{C}} \right]_n$$

propene

poly(propene)

Poly(propene) is used amongst other things to make marine ropes and hawsers, agricultural baler twine and carpet fibre. Propene can also be processed to form other monomers, e.g. $CH_2 = CH(CN)$, a compound called propenonitrile or acrylonitrile which is polymerised to give poly(acrylonitrile), an important synthetic fibre used in the textiles industry. It is usually known as acrylic fibre.

Figure 11.3 Mossmorran in Fife where ethane is cracked

Figure 11.4 Poly(propene) is used to make containers, flower pots and many other household goods

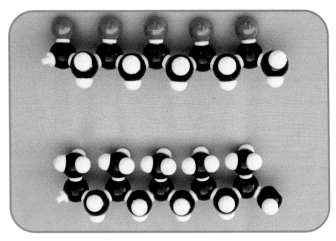

Figure 11.6 Poly(chloroethene) and poly(propene)

The production and major uses of ethene and propene are summarised in the flow diagram in Figure 11.5.

Question

1 Draw the structure of poly(acrylonitrile) showing three linked monomer molecules.

   Formula of acrylonitrile: $CH_2{=}CHCN$

Condensation polymers

Condensation polymers are made from monomers containing two functional groups per molecule. The monomers join in chains or networks with the elimination of a molecule of water at each new linkage formed.

Polyester

Polyester is a good example of a condensation polymer. It is synthesised by condensation of two monomers, one having two hydroxyl groups is called a **diol** and the other, with two carboxyl groups, is a **diacid**. Their structures are shown in Figure 11.7 and the structure of polyester, obtained by condensing two diol molecules and two diacid molecules, is shown in Figure 11.8. The alcohol groups lose an 'H' and the carboxyl groups lose an 'OH' in forming the ester linkages. The presence of two functional groups per molecule allows the chain to be continued indefinitely in each direction.

The dotted lines indicate the **repeating unit**. It is useful to start this unit between the 'O' and the 'C' of the — C—O— ester linkage. The repeating unit can be used to show the structures of the monomers forming

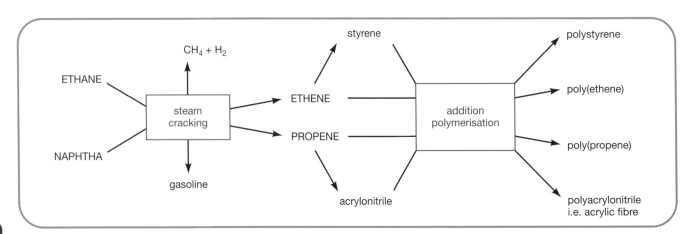

Figure 11.5 Production and uses of ethene and propene

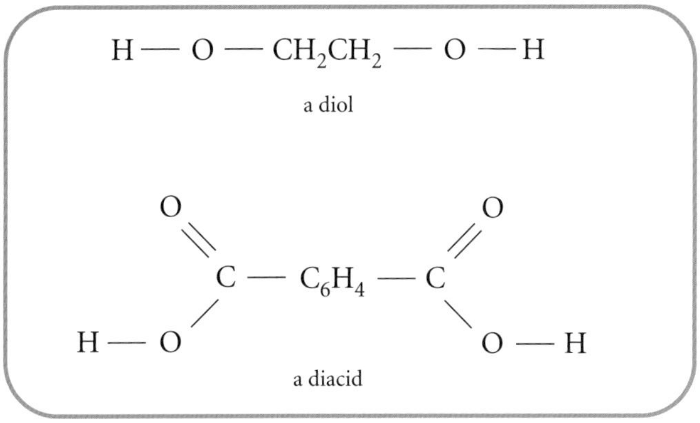

H — O — CH₂CH₂ — O —H

a diol

$$O \quad\quad\quad O$$
$$\parallel \quad\quad\quad \parallel$$
$$C — C_6H_4 — C$$

H — O $\qquad$ O — H

a diacid

Figure 11.7 Monomers for a polyester

Figure 11.9 Polyester resin is used to make 'plastic' bottles

the polymer chain. Start by adding an 'H' to the 'O' of the linkage, and an 'OH' to the 'C'. Repeat for the ester linkage at the other end of the repeating unit. Note that the repeating unit contains vestiges of both monomer molecules. The —C—O— ester linkage within it must be broken in the same way.

If the monomer molecules link alternately as above, a linear polymer suitable for making textile fibres used in 'non-iron' clothing, ropes, cords and sailcloth is formed. If quantities of other monomers, with additional —OH or —COOH groups, are added, some **crosslinking** occurs forming a three dimensional structure. This gives a thermoplastic resin which can be 'blown' into bottles for soft drinks. This can be recycled, when it can be used to make 'pile' or 'fleece' jackets. The recycled material is also being used to make fibre for crease-resistant trousers. Fourteen bottles make one pair of trousers!

The resin is called poly(ethyleneterephthalate), PET or PETE, whilst one of the commonest brands of the fibre is 'Terylene' from the trivial names of the monomers, *tere*phthalic acid and eth*ylene* glycol.

Figure 11.10 Bedding, fleeces, shirts, trousers and underwear all contain at least 50% polyester

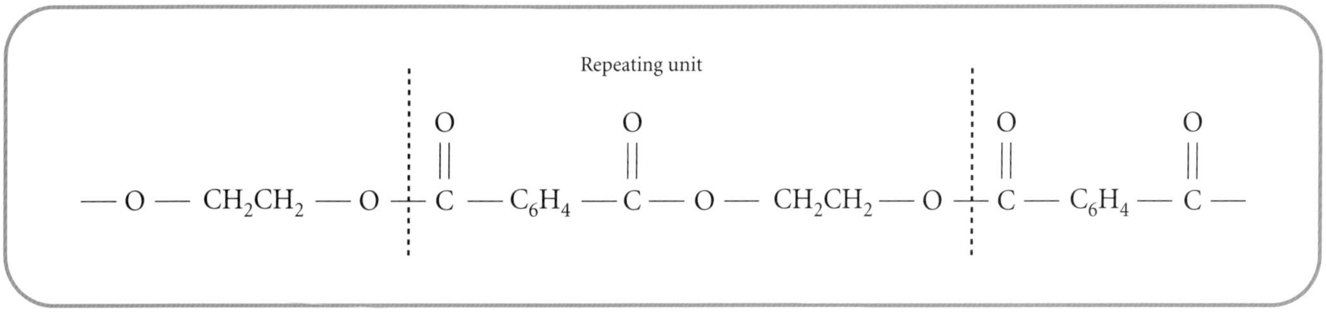

Repeating unit

$$O \qquad\qquad O \qquad\qquad\qquad\qquad O \qquad\qquad O$$
$$\parallel \qquad\qquad \parallel \qquad\qquad\qquad\qquad \parallel \qquad\qquad \parallel$$
— O — CH₂CH₂ — O — C — C₆H₄ — C — O — CH₂CH₂ — O — C — C₆H₄ — C —

Figure 11.8 A polyester structure

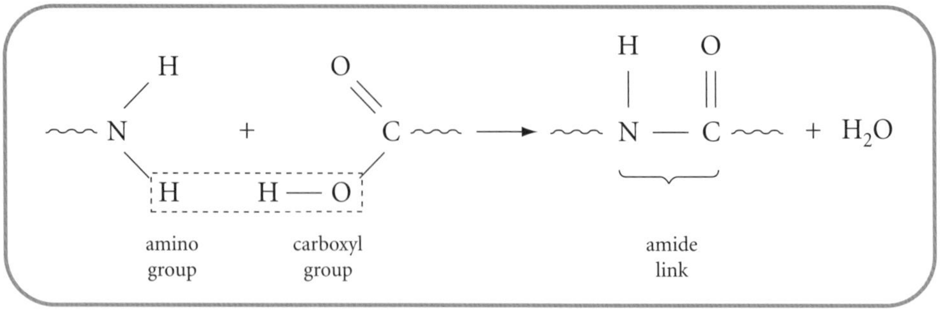

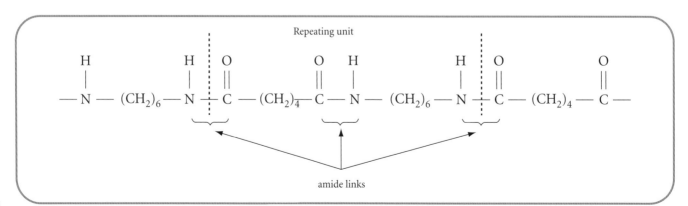

Figure 11.11 Formation of an amide link

Polyamides

Polyamides are also condensation polymers. Their monomers are **diamines**, containing two —NH$_2$, **amine** or **amino**, functional groups, and **diacids** with two carboxyl groups. These groups can react together (Figure 11.11) to form an amide link.

Figure 11.12 shows the structures of typical monomers for making a polyamide, and Figure 11.13 shows the resulting polyamide made from two molecules of each

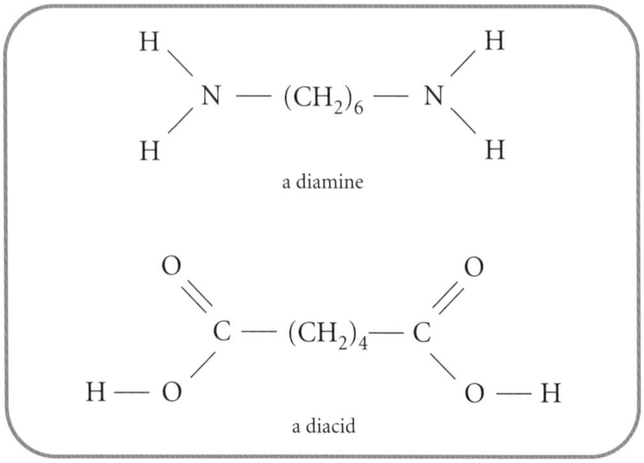

Figure 11.12 Monomers for a polyamide

monomer. Once again the dotted lines show the repeating unit, starting from the middle of the —NH—CO— amide linkage. As before, the structures of the monomers can be recreated by adding an 'OH' to the 'C', but this time the 'H' is added to the 'NH'. The amide link inside the repeating unit must be broken in the same way.

Different types of polyamide can be made by altering the number of carbon atoms in the original monomers. Polyamides are thermoplastic (unless they contain aromatic groups – see below).

Polyamides, usually called 'nylon' in Britain, are used to make textile fibres for clothing, climbing ropes and parachutes. Polyamides are also used in pulleys, machine gear wheels, bearings and bushes. When mixed with fillers and reinforcing materials they can be used in many engineering applications requiring specific properties.

Many of the uses of nylon stem from its strength resulting from the fact that NH and CO groups are able to form hydrogen bonds between the polymer chains.

116

Figure 11.13 A polyamide structure

Figure 11.14 Polyester and polyamide models

Figure 11.15 Polyamide is used to make cord and a variety of clothing, including waterproof clothing, sportswear and underwear

Aramids

Aramids are a type of aromatic polyamide discovered in the 1960s. The structure of the polymer, with the tradename Kevlar, is shown in Figure 11.16.

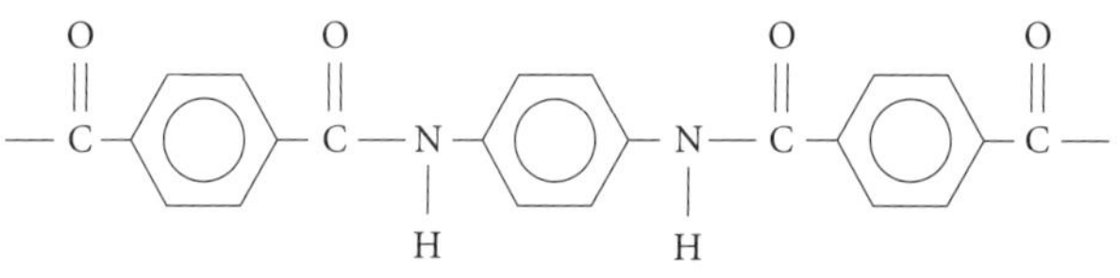

Figure 11.16 Kevlar structure

The chains line up parallel to each other and are held together by hydrogen bonds in sheets. The sheets then stack in a very ordered structure which weight for weight is five times as strong as steel. Kevlar is fire resistant and flexible. It does not melt so is not thermoplastic. It is made into fibre from solution in concentrated sulphuric acid by precipitation into water in which it is insoluble. Uses include tyre cords (instead of steel), body armour, safety clothing and space suits.

Question

2   Draw three repeating units of the structure of the polyamide made from the single monomer caprolactam. (*Hint:* a carbon to nitrogen bond must be broken before monomer molecules join.)

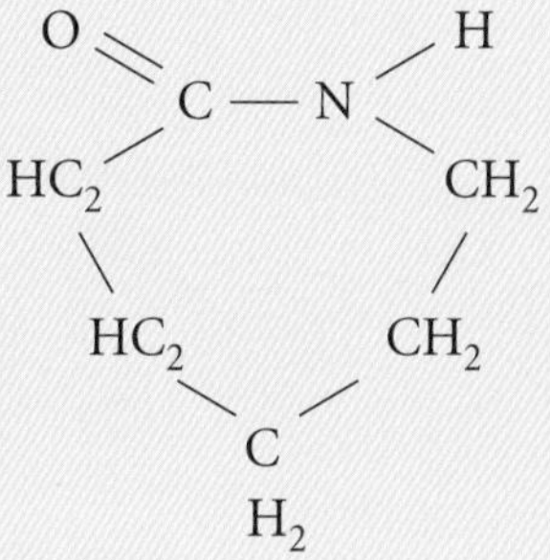

caprolactam

Figure 11.17 Kevlar is used in the construction of police body armour

---

Question

3   Draw the structural formulae of the two monomers for making Kevlar.

Methanal-based polymers

Methanal is an important feedstock in the manufacture of a number of thermosetting polymers. It, in turn, has as its feedstock methanol made industrially from synthesis gas, a mixture of carbon monoxide and hydrogen. In a process called steam reforming, methane and steam are passed over a nickel catalyst at around 800°C and 30 atmospheres pressure:

$$CH_4(g) + H_2O(g) \rightarrow CO(g) + 3H_2(g)$$

synthesis gas

The hydrogen content of synthesis gas is then adjusted and the mixture of carbon monoxide and hydrogen is passed over a second catalyst, usually zinc(II) oxide and chromium(III) oxide at a temperature of 300°C and a pressure of 300 atmospheres. The following reaction takes place:

$$CO(g) + 2H_2(g) \rightarrow CH_3OH(g)$$

methanol

The methanol is then oxidised to methanal by passing it with air over a silver catalyst at 500°C:

$$CH_3OH(g) + \tfrac{1}{2}O_2(g) \rightarrow HCHO(g) + H_2O(g)$$

methanal

Thermosetting polymers made with methanal (formaldehyde)	Uses
Urea-formaldehyde	Electrical fittings (white) Adhesives in chipboard Cavity-wall insulation
Bakelite (phenol-formaldehyde)	Electrical fittings (brown) Oven and pan handles 'Black' telephones

Table 11.2

Bakelite's formation is shown in Figure 11.18. The reaction does not stop at this stage. Further condensation results in crosslinking of the chains to give a three-dimensional network. A very rigid structure is formed which does not soften on heating. Urea-formaldehyde is also highly crosslinked, a feature usually resulting in thermosetting properties.

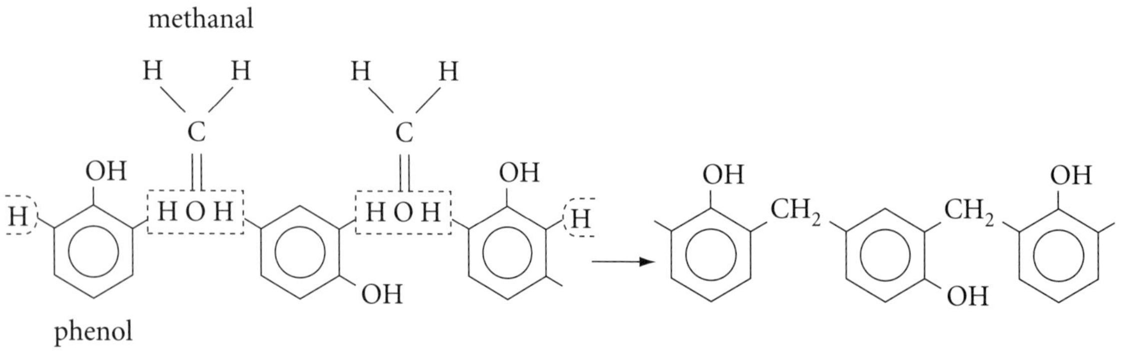

Figure 11.18 Formation of bakelite

Polymers with special properties

Solubility in water

Poly(ethenol) is a plastic which dissolves readily in water. It is made from another polymer poly(ethenyl ethenoate) by reacting it with methanol in a process called **ester exchange** shown in Figure 11.19.

The temperature of the process or the time allowed controls the extent of ester exchange, an important factor governing the solubility of the product which increases with the number of ester groups replaced by —OH groups. The —OH groups attached to the chain can hydrogen bond with water molecules, taking it into solution.

Uses of poly(ethenol) include the manufacture of hospital laundry bags which can be loaded straight into washing machines without laundry workers handling dirty linen. The bags dissolve during washing. Some surgical sutures are made of soluble polymers and are eventually absorbed into the body. New cars are sometimes coated with soluble polymers for protection whilst stored in the open. The substance can be hosed off before the car goes into the showroom. 'Slime', available from novelty shops, is made from partially cross-linked poly(ethenol).

Figure 11.20 The hospital laundry bag in this photo is made from poly(ethenol)

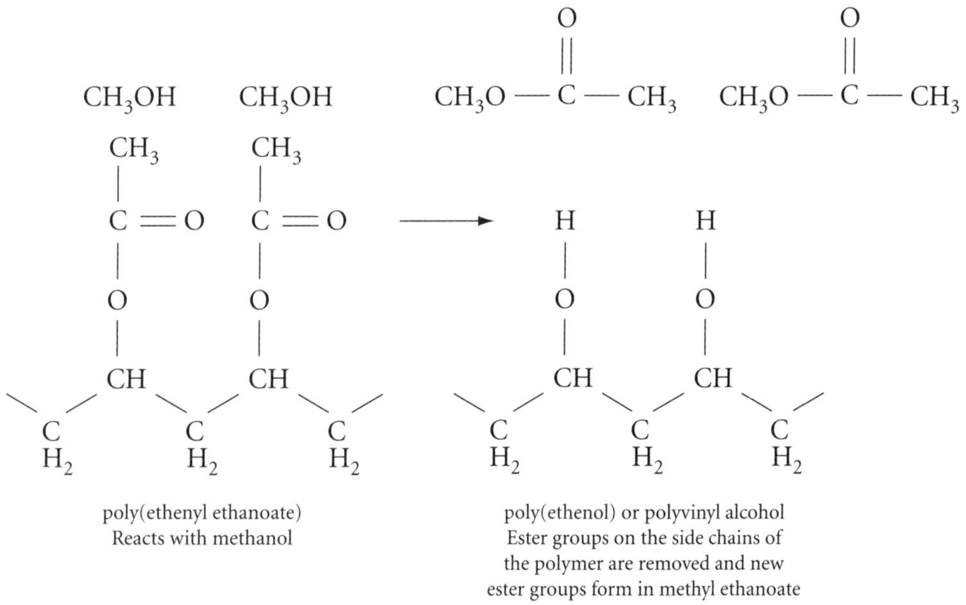

poly(ethenyl ethenoate)
Reacts with methanol

poly(ethenol) or polyvinyl alcohol
Ester groups on the side chains of the polymer are removed and new ester groups form in methyl ethanoate

Figure 11.19 Formation of poly(ethenol)

Electrical conductivity

Poly(ethyne) can be treated to make a polymer which conducts electricity.

$$HC \equiv CH + HC \equiv CH + HC \equiv CH$$

$$\downarrow$$

Showing the structure of poly(ethyne) with alternative single and double bonds is merely a convenience for drawing it. In fact the electrons forming the carbon–carbon double bonds are delocalised along the chain. This makes the polymer electrically conducting.

Questions

4 Write the formula of the hypothetical monomer ethenol. (In fact this monomer is unstable.)

5 Benzene also has delocalised electrons. Why is it *not* electrically conducting?

Photoconductivity

Poly(vinyl carbazole) is a polymer which becomes electrically conducting when exposed to light. It can be used in photocopiers which, when first developed, depended on the element selenium which has similar electrical properties but is poisonous. The monomer is shown below:

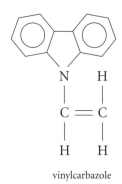

vinylcarbazole

Question

6 Draw the structure of poly(vinyl carbazole) showing three linked monomer molecules.

Degradability

The widespread use and the durability of most polymers is causing huge environmental problems. The practice of throwing ships' rubbish overboard has resulted in it being possible to detect grains of plastic amongst the sand grains on almost any beach on Earth. Prompted by this, research has now produced degradable plastics.

Figure 11.21 Some retailers are using degradable polymers for packaging

'Biopol' is a biopolymer, i.e. it is produced naturally (by certain bacteria). It is also biodegradable, i.e. able to be broken down naturally by micro-organisms. It is chiefly poly(hydroxybutanoate) or PHB. The structure is shown in Figure 11.22.

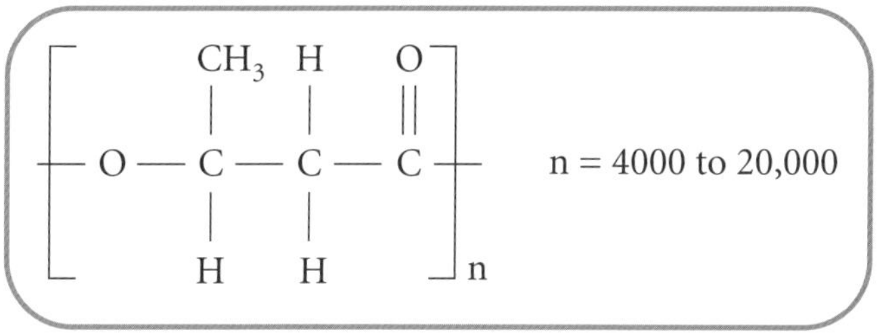

Figure 11.22 The structure of Biopol

Industrially Biopol is made by a fermentation process using the same bacteria as produce it naturally. It can also be produced by genetically-modified plants. Because it is thermoplastic, Biopol can be processed in normal ways into, for example, bottles for cosmetics and biodegradable oils. Since polyesters like Biopol are normally present in soil ecosystems, being released when bacteria die, other micro-organisms will metabolise them into simpler compounds. The compounds formed are CO_2 and water in aerobic conditions and methane in anaerobic conditions, for example a landfill site.

Work on Biopol is, however, being stopped for the following reasons:

♦ The cost of Biopol is substantially higher than conventional thermoplastics.

♦ Market acceptance of renewable materials has been disappointing in major trading areas.

♦ Environmentalists have been pressing for recyclability rather than biodegradability.

♦ Biodegradable material in plastic wastes makes them unsuitable for recycling.

Photodegradable low density polythene is LDPE (see page 112) which has been modified by including in its chains an absorber of ultra-violet light, usually carbonyl groups. In daylight UV is absorbed, breaking neighbouring bonds. The stable life of the polymer can be carefully controlled. The polymer is used in packaging such as carrier bags and 'six-pack' rings.

△1	PET	Polyethylene terephthalate	Used for fizzy drinks bottles and oven-ready meal trays
△2	HPDE	High-density polyethylene [High-density poly(ethene)]	Makes bottles for milk and washing-up liquids
△3	PVC	Polyvinyl chloride [Poly(chloroethene)]	Used for food trays, cling film, bottles for squash, mineral water and shampoo
△4	LDPE	Low-density polyethylene [Low-density poly(ethene)]	Makes carrier bags and bin liners
△5	PP	Polypropylene [Poly(propene)]	Used for margarine tubs, microwavable meal trays
△6	PS	Polystyrene [Poly(phenylethene)]	Makes yogurt pots, foam meat or fish trays, hamburger boxes and egg cartons, vending cups, plastic cutlery, protective packaging on electronic goods and toys
△7	Other	There are a variety of other plastics, for example Melamine	Melamine is used to make plastic plates and cups

Table 11.3 The symbols and initials stamped on recyclable polymers and common uses of these polymers

Study Questions

In questions 1–4 choose the correct word from the following list to complete the sentence.

> addition   bakelite   biodegradable   condensation
> diamines   diols   kevlar   water-soluble

1 Nylon is an example of a polymer made by _____ polymerisation.

2 _____ is the name of a polymer which can be used to make bullet-proof clothing.

3 Dicarboxylic acids and _____ are the types of monomer used in the manufacture of polyesters.

4 Hospital laundry bags can be made of poly(ethenol) because it is _____ .

5 Bakelite is a

A   thermosetting condensation polymer

B   thermoplastic addition polymer

C   thermosetting addition polymer

D   thermoplastic condensation polymer.

6 A certain polymer has the following structure:

$$—CH{=}CH—CH{=}CH—CH{=}CH—$$

Which statement is true about this polymer?

A   Its monomer is ethene.

B   It is soluble in water.

C   It is called poly(ethyne).

D   It is a condensation polymer.

7 Which polymer is used in photocopying as it is photoconductive?

A   Nylon

B   Kevlar

C   Poly(phenylethene)

D   Poly(vinyl carbazole)

8 Which polymer has a linear structure?

A   Kevlar

B   Bakelite

C   Polyester resin

D   Urea-methanal polymer

9* Sucrose can be represented by the structural formula:

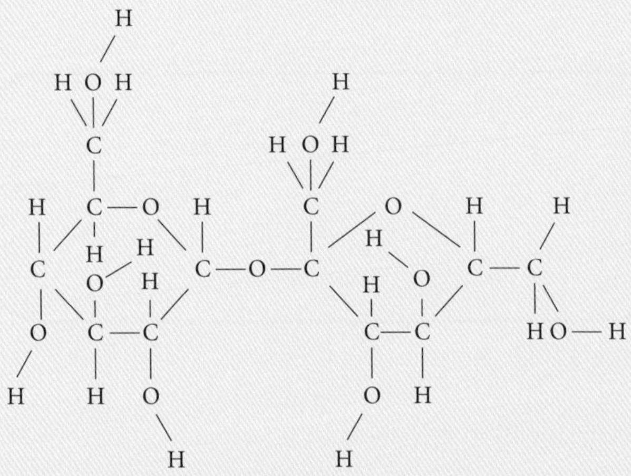

a) Copy the diagram and draw a circle around a primary alcohol group.

b) A fatty acid molecule can be represented thus:

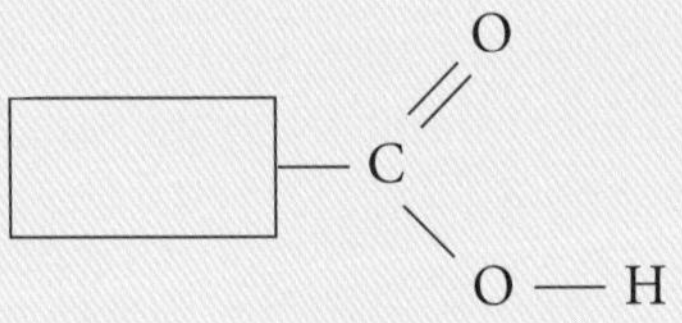

Molecules like this can form ester links with molecules like sucrose. Change the above structure to show *one* such ester link formed with *one* fatty acid molecule.

c) If you had continued to complete all the possible sites in sucrose with this ester linkage, you would have drawn the structural formula of 'sucrose polyester'. 'Sucrose polyester' is an ideal ingredient in food for slimmers, since it is not digested in our bodies.

Suggest why 'sucrose polyester' is *not* hydrolysed by the enzymes in our digestive systems.

d) Over.

Study Questions (continued)

d)

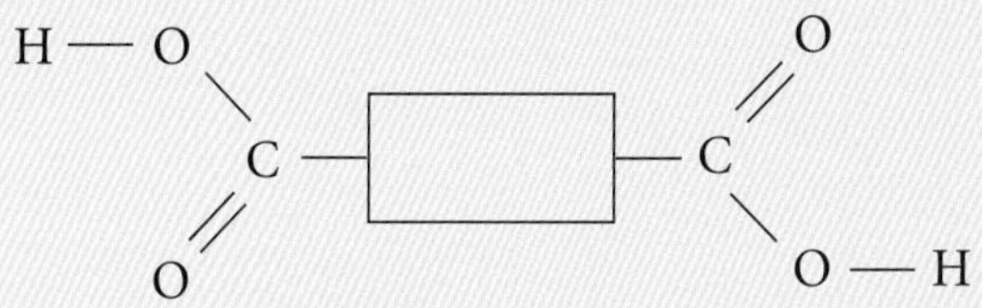

A

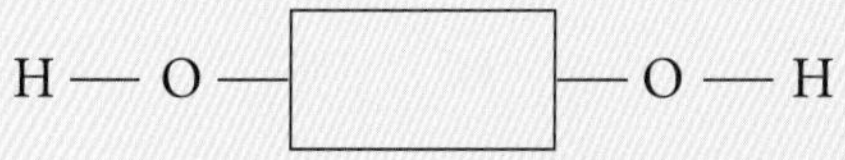

B

Molecules like A and B can be used to make a linear polyester.

i) Explain why molecules like A and B *cannot* be used to make a polyester resin.

ii) Explain why sucrose polyester is neither a linear polyester *nor* a polyester resin.

10 Kevlar is a polymer containing chains with the following structure:

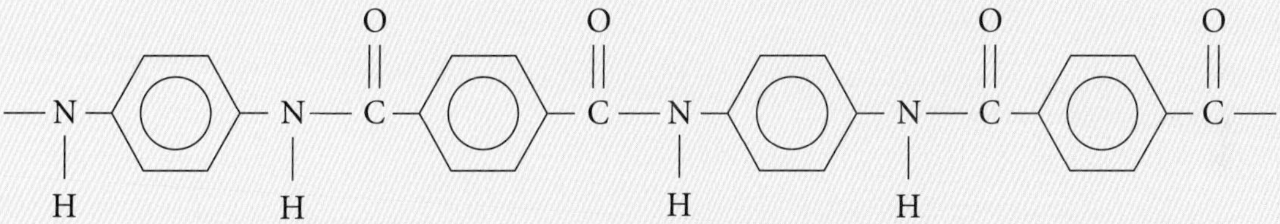

a) What type of polymerisation occurs when Kevlar is formed?

b) Which of the following terms can be applied to Kevlar?

Polyester; polyalkene; polyamide; polysaccharide

c) Kevlar's strength arises from alignment between the polymer chains so that interactions can occur as shown by dotted lines in the diagram below. What are these interactions called?

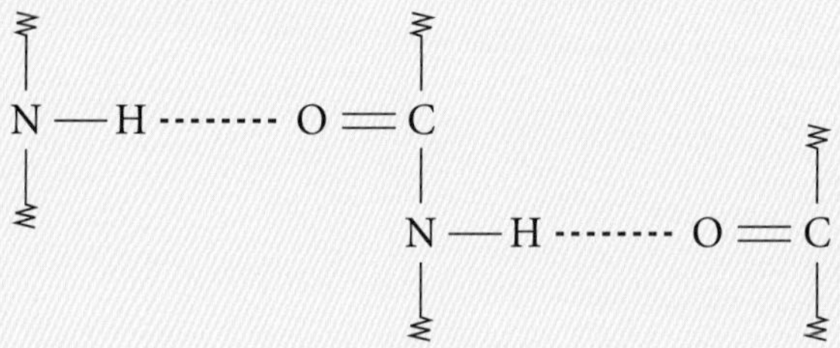

11* Bucrylate is an ester which is used in surgery for repairing torn tissue.

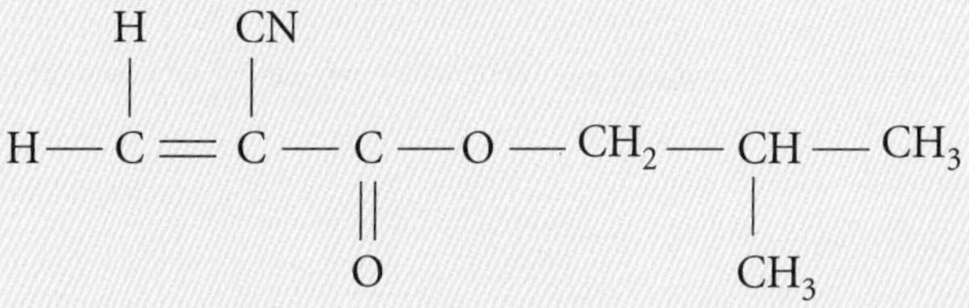

It instantaneously polymerises when it comes in contact with ionic solutions.

a) What type of polymerisation will bucrylate undergo?

b) Draw the structure of the *repeating unit* in polybucrylate.

Study Questions (continued)

12 Poly(buta-1,3-diene), a form of synthetic rubber, has the following structure

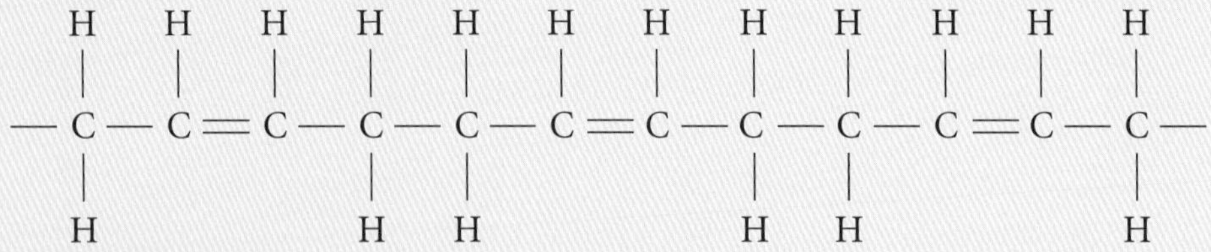

a) Draw the repeating unit of this polymer.

b) Draw the structural formula of the monomer. The systematic name of the polymer should help you.

13 Terephthalic acid is a dicarboxylic acid. Its structural formula is:

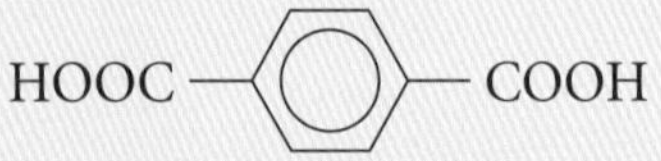

a) What is its molecular formula?

b) *Terephthalic acid is used in the production of terylene. Terylene is a linear textile fibre commonly known as a polyester. Part of a terylene molecule is shown below.

    i) What type of polymerisation has taken place in the formation of terylene from terephthalic acid?

    ii) Draw the full structural formula for the other monomer used in terylene manufacture.

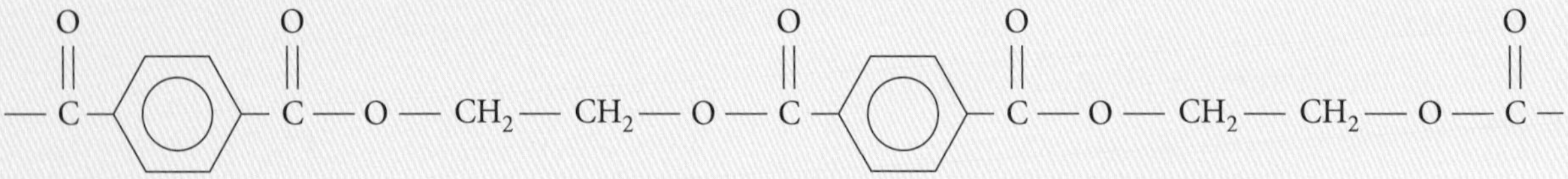

    iii) How would the structure of a cured polyester *resin* differ from the structure of a linear polyester?

12 Natural Products

From previous work you should know and understand the following:

★ The meaning of unsaturation in carbon compounds
★ The result of the addition of hydrogen to alkenes
★ The meaning of condensation polymerisation
★ Carbohydrates as a source of energy for the body
★ Condensation and hydrolysis of carbohydrates
★ The function of enzymes

★ Alcohols
★ Carboxylic acids
★ Esters
★ Hydrogen bonding
★ The nitrogen cycle.

Fats and oils

Fats and oils can be of animal, vegetable or marine origin. Some of them are listed below.

Vegetable	Animal	Marine
Soyabean oil	Pork fat (lard)	Whale oil
Palm oil	Mutton fat	Cod liver oil
Olive oil	Beef fat	
Rapeseed oil		
Linseed oil		
Castor oil		

Table 12.1

Vegetable oils are usually pressed out of the seed of the appropriate plant – rapeseed oil is commonly produced in the UK. Whale oil is now only produced commercially by two countries since international agreements to protect the whale were introduced. Animal fats are extracted by 'rendering down' fatty parts of the appropriate animal. Suet is such a material.

Oils and fats differ physically only in that oils are above their melting points and hence are liquid whereas fats are, of course, solid. Under cold conditions some vegetable oils may become cloudy, or even solidify.

The lower melting points of oils compared with fats of similar molecular mass are caused by more of their molecules being unsaturated, i.e. containing carbon–carbon double bonds. Their '**degree of unsaturation**' is higher than that of fats. Shaking of oils with bromine water results in the bromine water being readily decolourised. The bromine's decolourisation is the standard test for unsaturation. If solutions of fats in a saturated organic solvent are treated this way, decolourisation is not so rapid. The test can be modified to estimate quantitatively the degree of unsaturation of various oils and fats.

Fat and oil molecules are roughly 'tuning-fork' shaped, with the three limbs consisting of hydrocarbon chains (Figures 12.4 and 12.5). If the chains are saturated, the molecules can pack neatly together, even at quite high temperatures.

If the chains contain one or more double bonds, the zig-zag chains become more distorted and close packing of molecules is less easy. This results in weaker van der Waals' attractions. Unless the substance is cooled to remove more thermal energy it will not solidify (i.e. oils have lower melting points).

Since oils differ chemically from fats by being unsaturated, hardening of oils can be carried out by addition of hydrogen across the double bonds. Ethene can be converted to ethane by addition of hydrogen:

Figure 12.1 A field of oil-seed rape. A typical scene in Scotland in May

Figure 12.2 Harvested rape seed

Figure 12.3 Extra virgin rapeseed oil

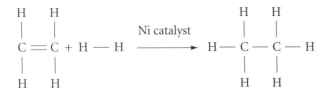

Similarly oils can be converted to fats by the use of hydrogen with nickel catalysts (see Figure 12.6).

The degree of unsaturation is now less than that of the original oil.

Margarine is made by hydrogenation, i.e. partially hardening vegetable oils using hydrogen and a nickel

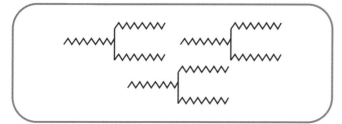

Figure 12.4 Diagrammatic representation of the structure of fat molecules

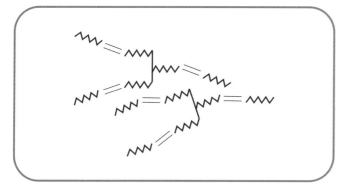

Figure 12.5 An exaggerated picture of oil molecules

catalyst in the warm oil. The catalyst is then removed from the product which hardens on cooling.

There is evidence that saturated fats are more liable to cause heart disease than unsaturated fats. Recently suggestions have been made that fish oils may be especially beneficial in reducing heart disease. General opinion seems to be that too much fat is to be avoided and it is wiser to err on the side of unsaturated rather than saturated fat. Fats and oils do, however, provide a source of fat soluble vitamins A and D. These are naturally present in dairy products, and are now a legally required additive in margarines to prevent vitamin deficiency problems such as 'rickets'.

The major dietary function of fats and oils is to provide energy in the body. They provide more energy than carbohydrates, but it is released much more slowly (Table 12.2).

Foodstuff	Energy yield kJ/100 g
Sucrose ('sugar' – carbohydrate)	1672
Sunflower oil	3700
Cooking margarine	3006
'Low fat' spread	1569
Butter	3140
White bread (carbohydrate)	961

Table 12.2

In terms of controlling 'calories' it is obviously beneficial to cut down on fatty food rather than on carbohydrates.

Various vegetable oils are now being used in the production of biodegradable detergents. A major use of oils is in paints. Their unsaturated nature allows them to harden by the uptake of oxygen. Linseed oil, from flax, is especially important in this context.

In recent years, rapeseed oil has been increasingly used as a source of 'bio-diesel'. This can be used in diesel engines without the need to modify them and, since it is produced from an annual crop, it is a source of renewable energy which does not contribute to global warming. Land used to produce bio-diesel is decreasing the area available for food production, however.

Hydrolysis of fats and oils

When most fats or oils are treated with superheated steam they break up, or **hydrolyse**, into a substance

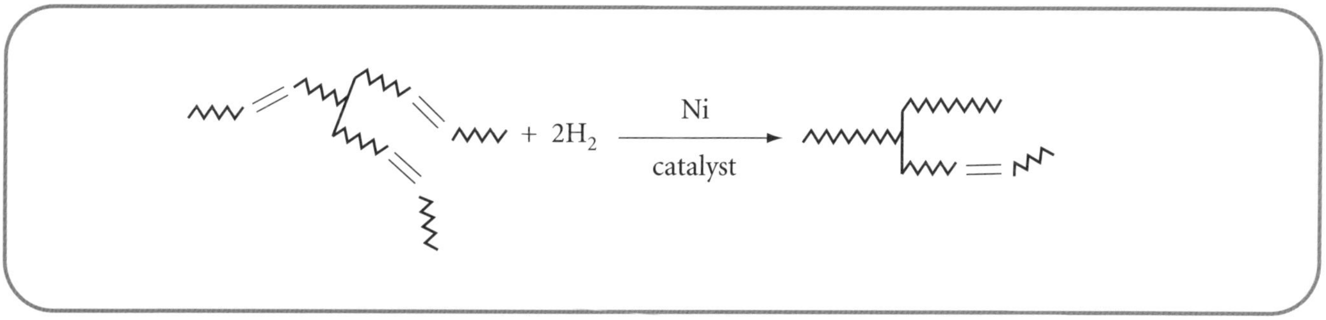

Figure 12.6 Partial hydrogenation of an oil molecule

known as glycerol and various 'fatty acids'. Glycerol is an alcohol with three —OH groups per molecule – a **trihydric** alcohol.

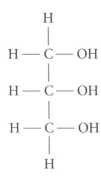

propane -1,2,3-triol or glycerol

The fatty acids are straight chain carboxylic acids which may be saturated (alkanoic acids) or unsaturated (alkenoic acids) and which contain even numbers of carbon atoms ranging most often from C_{14} to C_{24} but primarily C_{16} and C_{18}. Some contain —OH substituent groups.

Examples of fatty acids include:

♦ Palmitic acid $CH_3(CH_2)_{14}COOH$

♦ Stearic acid $CH_3(CH_2)_{16}COOH$

♦ Oleic acid $CH_3(CH_2)_7CH=CH(CH_2)_7COOH$

♦ Linoleic acid
$CH_3(CH_2)_3(CH_2CH=CH)_2(CH_2)_7COOH$

♦ Ricinoleic acid
$CH_3(CH_2)_5CH(OH)CH_2CH=CH(CH_2)_7COOH$.

The above names are 'trivial' names. Stearic acid is, for example, systematically named octadecanoic acid.

No fat or oil is a pure 'triglyceride', as these compounds are called, but a mixture of many. Indeed in any triglyceride the three acid molecules combined with one molecule of glycerol may or may not be identical, e.g. one triglyceride is:

$$CH_3(CH_2)_{14}\overset{\overset{\displaystyle O}{\|}}{C}-O-CH_2$$
$$H-\overset{\overset{\displaystyle O}{\|}}{C}-O-C(CH_2)_7CH=CH(CH_2)_7CH_3$$
$$CH_3(CH_2)_{16}\overset{\overset{\displaystyle O}{\|}}{C}-O-CH_2$$

This is called 'Palmito-oleo-stearin' and the number of possible glycerides is very large.

Fats and oils contain glycerol combined with fatty acids in the ratio of one mole of glycerol to three moles of fatty acid. This structure appears very complicated, but in fact fats and oils are complex examples of **esters** dealt with in Chapter 10. The triglyceride shown above has three ester linkages in its structure.

$$-\overset{\overset{\displaystyle O}{\|}}{C}-O-$$

ester linkage

Question

1 Write the structural formula of the triglyceride made from propane-1,2,3-triol and ethanoic acid. Circle the ester linkages.

Soaps

It is important to realise that soaps are made by the alkaline hydrolysis of fats and oils by sodium or potassium hydroxide. The glycerol liberated is an important raw material and the fatty acids are produced in the form of their sodium or potassium salts. These salts are 'soap'. Since they are soluble they must be extracted from the hydrolysis mixture by adding a large excess of sodium chloride, after which the soap can be filtered off. *Palmolive* obviously takes its name from palm oil and olive oil. *SR toothpaste* was apparently named from sodium ricinoleate, a soap, made from castor oil, which was described as an ingredient in early TV advertising.

Soaps are useful for cleaning because of the structure of the ion liberated when they dissolve in water (Figure 12.7).

The covalent 'tail' bonds to the similar greasy material on fabric or skin, the ionic head is attracted to the polar covalent water molecules. An emulsion of globules of grease in water is formed as shown in Figure 12.8.

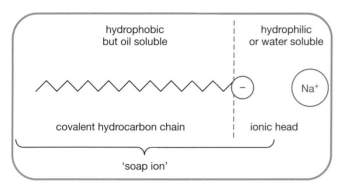

Figure 12.7 The structure of a soap

Proteins

Proteins make up a huge variety of animal and plant material of which the following are examples:

◆ peas and beans

◆ meat

◆ fish

◆ cheese

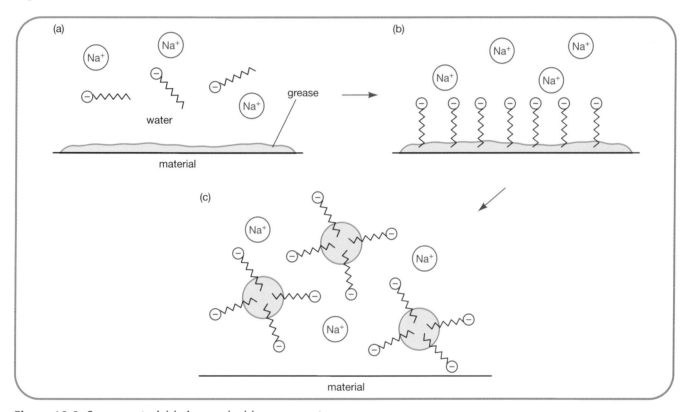

Figure 12.8 Greasy material being washed by soapy water

Figure 12.9 A fat (glyceryl trioctadecanoate) and a soap (sodium octadecanoate)

◆ eggs

◆ hides and skin

◆ wool and silk.

Although proteins can be grouped into categories, such as those above, there can be large structural variations within each division. For example, meat would include such proteins as beef, chicken, mutton and pork which are all different from each other. Variation also occurs within the individual animal itself, for example muscle, kidney and liver proteins.

All proteins yield acrid-smelling alkaline gases on heating with soda lime (Figure 12.10). (Soda lime is a very strongly alkaline mixture of sodium and calcium hydroxides.) Such gases are all amines or ammonia, i.e. nitrogen-containing substances. Hence we can deduce that all proteins contain nitrogen and we can begin to understand why plants require soluble nitrogen compounds as an essential nutrient. Without them plants cannot synthesise protein.

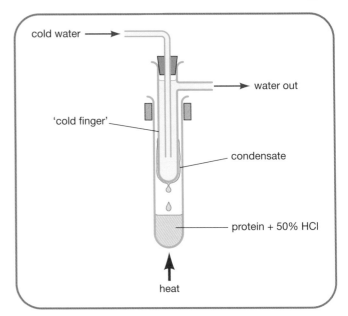

Figure 12.11 Protein being hydrolysed with 50% hydrochloric acid

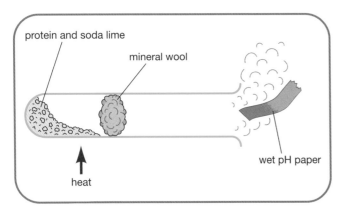

Figure 12.10 Protein mixed with soda lime and then heated

Animals cannot synthesise protein from simple nitrogen compounds but they can reconstruct vegetable or other animal protein from food which they have eaten. All protein derives, therefore, from the fixation of atmospheric nitrogen as nitrates in soil.

If protein is hydrolysed using 50% hydrochloric acid, refluxing the mixture for several hours, then the protein breaks down into its constituents called amino acids. Figure 12.11 illustrates suitable apparatus for a small scale experiment. Alternatively the reflux apparatus of Figure 10.7 on page 107 can be used.

These amino acids have a typical structure:

$$H_2N - \underset{\underset{H}{|}}{\overset{\overset{R}{|}}{C}} - C \overset{\displaystyle O}{\underset{\displaystyle O - H}{\Big\langle}}$$

R represents a variable organic group (or hydrogen)

amine group   carboxyl group

The simplest amino acids are glycine and alanine:

$$H_2N - \underset{\underset{H}{|}}{\overset{\overset{H}{|}}{C}} - C \overset{\displaystyle O}{\underset{\displaystyle O - H}{\Big\langle}}$$

aminoethanoic acid (glycine)

$$H_2N - \underset{\underset{H}{|}}{\overset{\overset{CH_3}{|}}{C}} - C \overset{\displaystyle O}{\underset{\displaystyle O - H}{\Big\langle}}$$

2-aminopropanoic acid (alanine)

Twenty amino acids occur frequently in proteins, another six occur occasionally. The amino acids produced by hydrolysis of proteins can be identified, with difficulty, by chromatography as in Figure 12.12.

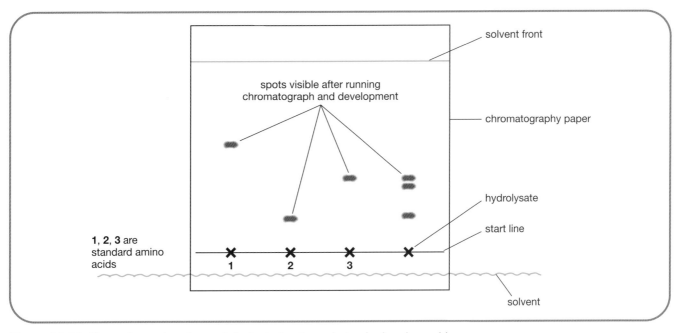

Figure 12.12 Chromatography of a protein hydrolysate and standard amino acids

The diagram shows the results of simple ascending chromatography. The protein hydrolysate is spotted onto filter paper, with solutions of known amino acids alongside. The paper is placed in a tank containing a solvent of propanol and water. When the solvent front has almost reached the top of the paper, the paper is removed and then the colourless amino acids are made visible by spraying with ninhydrin. The paper is then heated after which pink, blue or brown spots appear on the paper. In the example shown, the protein hydrolysate appears to contain

amino acids '2' and '3' and some other amino acid which is not '1'.

In practice the separation achieved by this method is not good so two-way chromatography is used with two different solvents. Figure 12.13 illustrates the result of such an experiment.

The same protein hydrolysate is used on its own and the paper placed for a fixed time in each solvent in turn. After removal from the second solvent, development with ninhydrin is carried out. To identify

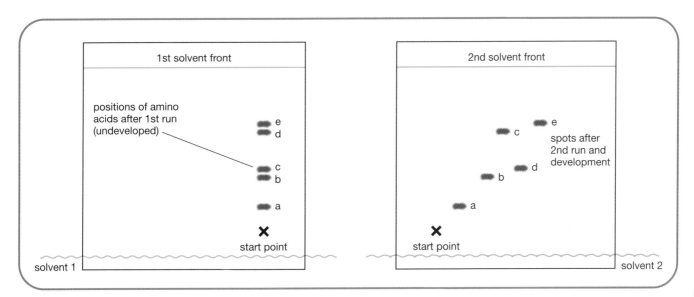

Figure 12.13 Chromatography of a protein hydrolysate by the two-way method

the resultant spots, the same process is repeated under identical conditions using each amino acid in turn on separate sheets of paper. Spots are identified by superimposing these separate chromatograms on the original hydrolysate chromatograph.

The formation and breakdown of proteins

The amino acids in a protein are linked similarly to polyamides (page 116) in chains.

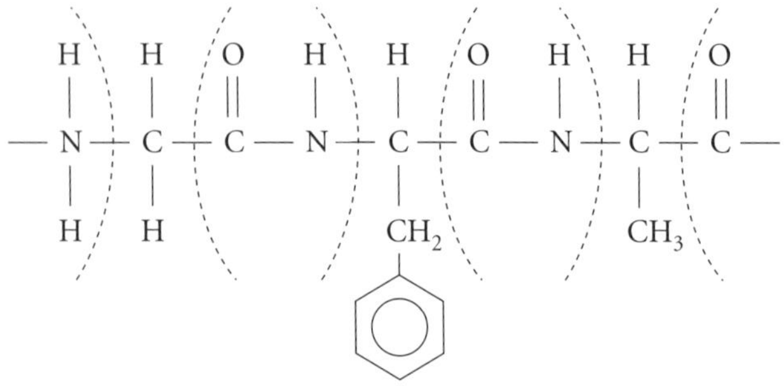

The groups in brackets are called **amide** or peptide links and when hydrolysis occurs, the links break with the addition of an —H and an —OH at these positions. The three amino acids formed from the above protein fragment are:

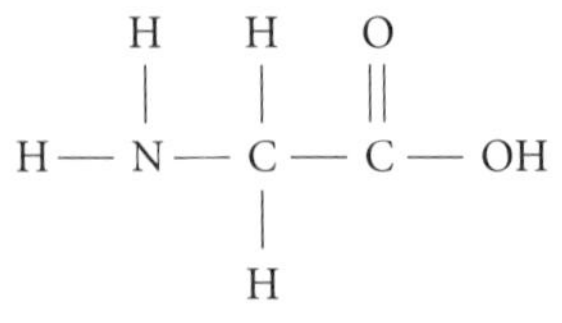

glycine

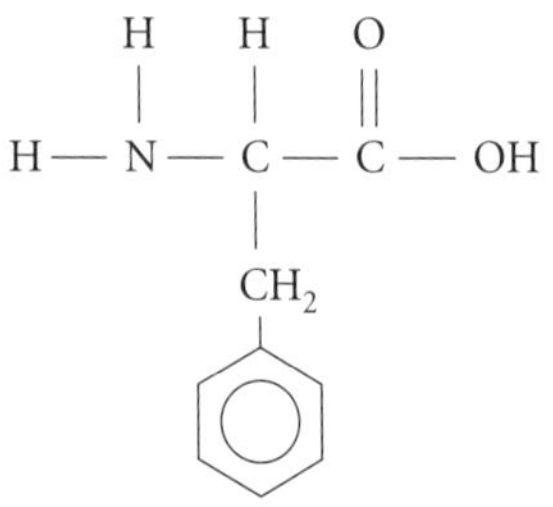

phenylalanine

+

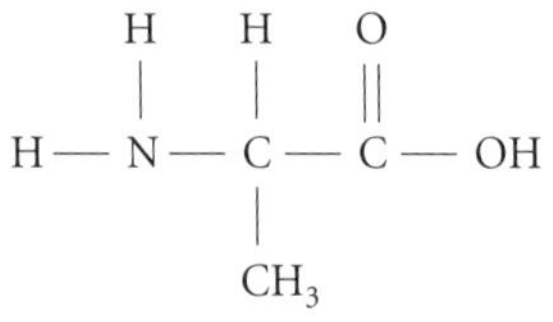

alanine

Conversely, when amino acids come together to form protein, a molecule of water is lost for each peptide link

formed. This is therefore an example of a **condensation polymerisation** reaction. (See Figure 12.14)

Question

2 Cysteine has the formula:

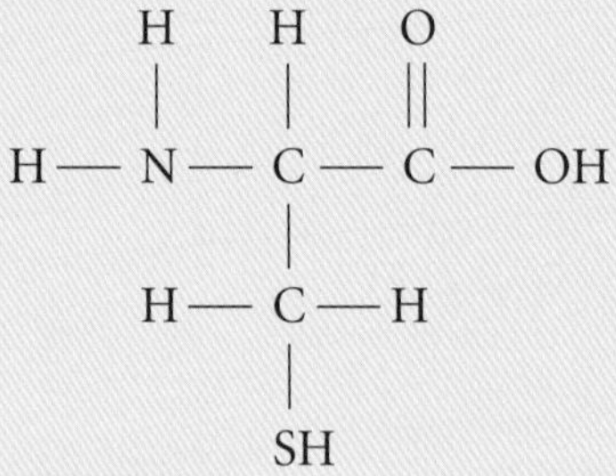

a) What type of compound is this?

b) Show part of a protein chain in which one molecule of cysteine is joined to one of glycine and one of alanine.

c) Your answer to **b)** is not the only possible structure. Why not?

d) The residue from the reaction of this protein with soda lime yields H_2S when treated with hydrochloric acid. What can be deduced about any protein which does so?

Protein chains can contain several thousand amino acids and hence protein molecules have huge molecular masses. The variety of proteins is caused by the possibility of arranging up to 26 amino acids in varying numbers and in varying orders into these long chains. In effect, an alphabet of amino acids is assembled into a dictionary full of proteins.

Proteins comprise a large part of an animal's diet. During digestion the animal and vegetable proteins in foodstuffs are hydrolysed into their component amino acids which are small enough to pass into the bloodstream. Proteins required for the body's specific needs are built up from amino acids in the body cells according to information supplied by DNA in the cell nuclei.

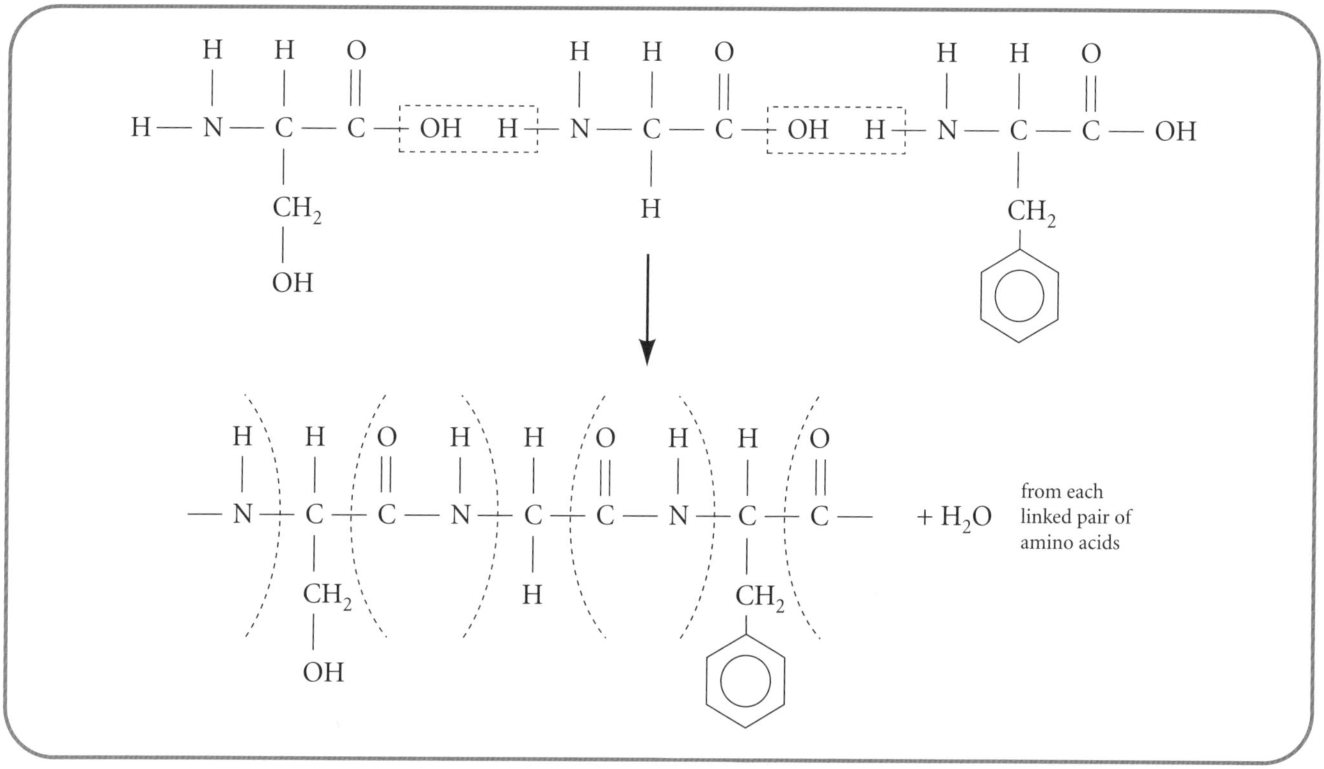

Figure 12.14 Condensation of amino acids to form amide links in a protein

Some of the amino acids required can be synthesised in the body, but others, called **essential amino acids,** have to be present in dietary protein which is one important reason for consuming a wide enough variety of foodstuffs.

The structure of proteins

The chains formed from amino acids have the following polar groups along their length.

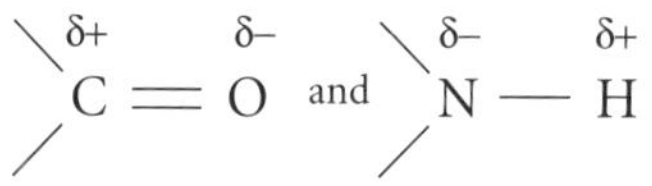

$$\overset{\delta+}{\underset{}{C}} = \overset{\delta-}{O} \quad \text{and} \quad \overset{\delta-}{\underset{}{N}} - \overset{\delta+}{H}$$

These groups can hydrogen bond with other groups in the same chain to produce a **helix,** or with groups in adjacent chains to produce 'pleated' sheets (Figures 12.15 and 12.16).

The helices or sheets then assemble into more complicated structures classified as either **fibrous** or **globular** proteins. Fibrous proteins are generally regarded as those where the length is at least five times their width.

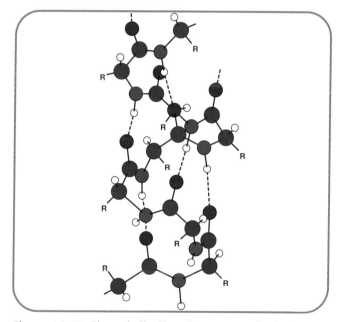

Figure 12.15 The α-helix (See Figure 12.16 for key)

Fibrous proteins are the major structural materials in animal tissue. Examples are **keratin** found in horn, hoof and hair, **collagen** found in tendons and, when mineralised, in bone, silk and muscle proteins.

Globular proteins are involved in the maintenance and

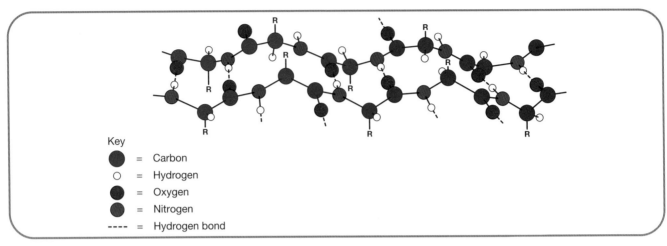

Key

● = Carbon
○ = Hydrogen
● = Oxygen
● = Nitrogen
---- = Hydrogen bond

Figure 12.16 The β-pleated sheet

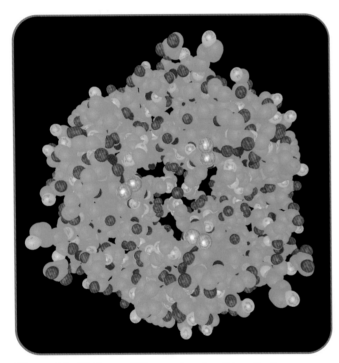

Figure 12.17 Space-filling molecule of insulin hexamer

regulation of life processes. The proteins are able – by virtue of their shape and their ability to form various types of bond, specifically hydrogen bonds – to bind to other substances. Examples of globular proteins are:

♦ hormones (regulators of metabolic reactions). An example is **insulin** which is secreted by specialised cells in the pancreas. It regulates carbohydrate metabolism and a deficiency of it results in diabetes.

♦ **immunoglobins** are secreted by the body to deal with antigens – unusual molecules present in, or on the surface of, foreign bodies like bacteria and viruses.

♦ **haemoglobin** which has iron-containing molecules (haem) bonded to protein molecules (globin). It allows oxygen molecules to bind to it relatively weakly so that they are released when required. Carbon monoxide, however, bonds to it strongly, preventing the oxygen joining, and therefore has a high toxicity.

♦ enzymes – these are dealt with below.

Figure 12.18 A single globin unit in haemoglobin. The black dots are the carbon atoms next to the C=Os of the amide links of the protein chain. The chain is hydrogen bonded into a helix, which is then contorted into the globular structure. The arrow indicates the position of the haem group

Enzymes

Enzymes are organic catalysts and all contain protein. Some are wholly protein while others require another organic or inorganic group – a co-enzyme – to activate them. Each enzyme catalyses a specific reaction which is demonstrated in a fermentation sequence from starch to ethanol:

$$\text{amylose} \xrightarrow{\textit{amylase}} \text{maltose} \xrightarrow{\textit{maltase}} \text{glucose} \xrightarrow{\textit{'zymase'}} \text{ethanol}$$
(starch)

'Zymase' is actually a complex mixture of enzymes, each of which assists a step in the reaction sequence from glucose to ethanol.

Generally enzyme names end in '-ase'. They are grouped as 'lipases' – fat digesting enzymes, and 'proteinases' – protein digesting enzymes.

It is believed that enzymes work by the 'lock and key' principle. The complex shape of the protein which is the enzyme only allows certain 'substrate' molecules to fit into it. Only these molecules have their reactions influenced by the enzyme (Figure 12.19).

One enzyme is lysozyme which is a globular protein with a long narrow cleft in its side. A long cellulose-type substrate molecule fits into this cleft. Whilst there, the chain of the substrate is broken and the parts are then released. Lysozyme is present in tears and other body secretions and is capable of breaking down the material of bacterial cell walls.

It is noteworthy that the activity of enzymes is lost at a little above 40°C and at low pH. This is not surprising since all proteins are denatured (i.e. have their structures changed) if temperatures rise or if they are acidified. Cooking an egg, for example, is intended to denature protein with a view to making it more palatable. A fried egg is a denatured egg.

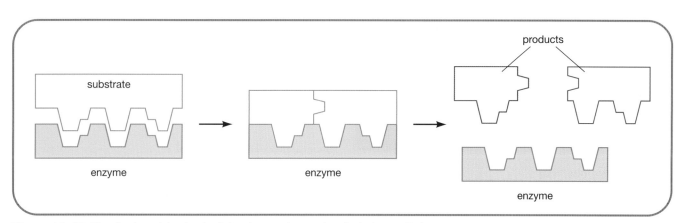

Figure 12.19 The 'lock and key' principle. A complex molecule being split by an enzyme

Prescribed Practical Activity

Enzymes do not function effectively under all conditions. It is possible to investigate the factors affecting the activity of **catalase**, an enzyme which catalyses the breakdown of hydrogen peroxide into water and oxygen, with the apparatus shown.

The basic procedure is to place three discs of potato (a source of catalase) in the side-arm tube with 5 cm³ of water. 1 cm³ of 30 volume hydrogen peroxide solution is added with a syringe and the tube is stoppered. The number of bubbles of oxygen emerging from the delivery tube in 3 minutes is counted.

To investigate the effect of temperature, the experiment is carried out at temperatures close to 20°C, 30°C, 40°C, 50°C and 60°C, by heating the water bath. Fresh potato, water and hydrogen peroxide are used each time. The potato and water are left for a few minutes to reach the chosen temperature before the peroxide is added.

To investigate the effect of pH, the water is replaced in turn by buffer solutions of pH 4, 7 and 10 and by 0.1 mol l⁻¹ hydrochloric acid (pH 1) and 0.1 mol l⁻¹ sodium hydroxide solution (pH 13). This time the water bath is kept at room temperature.

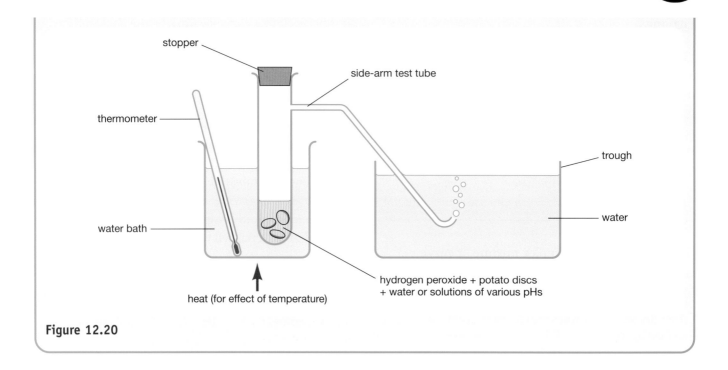

Figure 12.20

Study Questions

In questions 1–4 choose the correct word from the following list to complete the sentence.

> condensation   denatured   fibrous   digested
> globular   greater   hydrolysis   lesser

1 Vegetable oils have a _____ degree of unsaturation than fats.

2 _____ is the type of reaction involved when soap is made from fats and oils.

3 Enzymes are proteins which have a _____ structure.

4 Changing the temperature or pH can cause a protein to become _____.

5 Which of the following produces the most energy per gram?

   A Granulated sugar   B Cooking margarine

   C White bread   D Low-fat spread

6 Which type of reaction occurs when margarine is made from vegetable oils?

   A Condensation   B Hydration

   C Oxidation   D Hydrogenation

7 Which of the following contains the smallest amount of protein?

   A Butter   B Cheese

   C Eggs   D Dried milk

8 Proteins can be classified as

   A polyesters   B amino acids

   C polyamides   D carboxylic acids.

9* Both fats and oils are mixtures of triglycerides. Each triglyceride molecule contains three ester linkages and the majority of the molecules contain carbon–carbon double bonds.

   a) Why does the triglyceride molecule contain *three* ester linkages?

   b) Oils can be converted into fats by a process called hardening. What happens to the triglyceride molecules when oils are hardened?

   c) On exposure to air, oils turn rancid more rapidly than fats. In this process, the triglyceride molecules are broken up into smaller, 'foul-smelling' molecules.

   At what functional group within the triglyceride molecule is the breaking likely to occur?

Study Questions (continued)

10

Hydrolysis of the triglyceride shown above yields an alcohol and carboxylic acids.

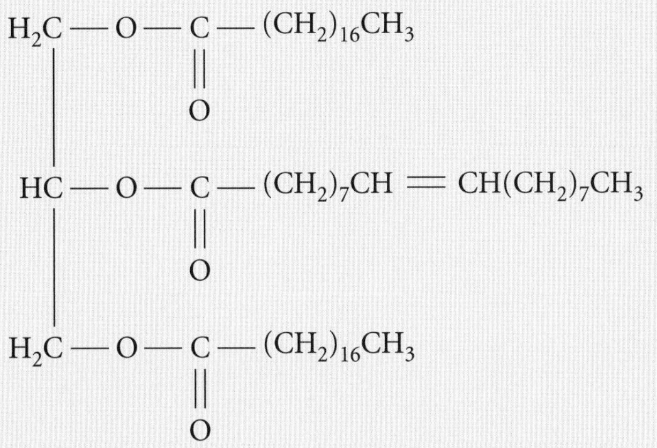

$$H_2C — O — C — (CH_2)_{16}CH_3$$
$$\parallel$$
$$O$$

$$HC — O — C — (CH_2)_7CH = CH(CH_2)_7CH_3$$
$$\parallel$$
$$O$$

$$H_2C — O — C — (CH_2)_{16}CH_3$$
$$\parallel$$
$$O$$

a) How many different acids will be produced?

b) Give the molecular formula of the saturated acid produced on hydrolysis.

c) Name the alcohol produced and give its structural formula.

11* Biodiesel is a mixture of esters which can be made by heating rapeseed oil with methanol in the presence of a catalyst.

a) Name compound X.

b) A typical diesel molecule obtained from crude oil has the molecular formula $C_{16}H_{34}$ (hexadecane). Other than the ester group, name a functional group present in biodiesel molecules which is **not** present in hexadecane.

c) Vegetable oils like rapeseed oil are converted into fats for use in the food industry. What name is given to this process?

12* This dipeptide forms the two amino acids – aspartic acid and phenylalanine – when it is hydrolysed.

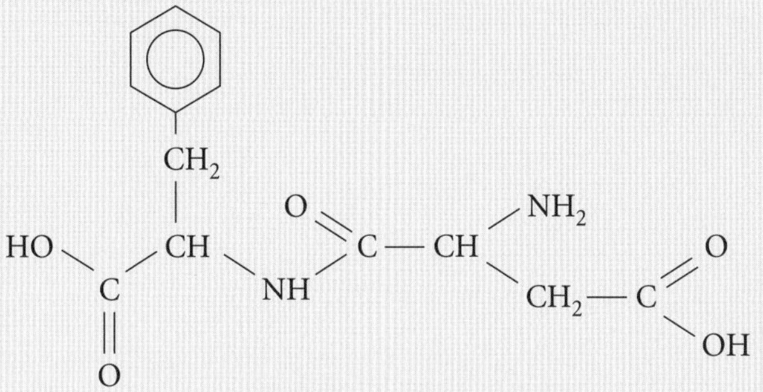

a) Identify the amide link in the above structure by placing brackets around it.

b) Identify the phenyl group in the above structure by circling it.

c) Draw the structure of the amino acid phenylalanine.

d) The artificial sweetener, aspartame, is a methyl ester of the dipeptide shown above. Its sweetness depends on the shape and structure of the molecule.

Suggest a reason why aspartame is *not* used in food that will be cooked, but is used in cold drinks, for example.

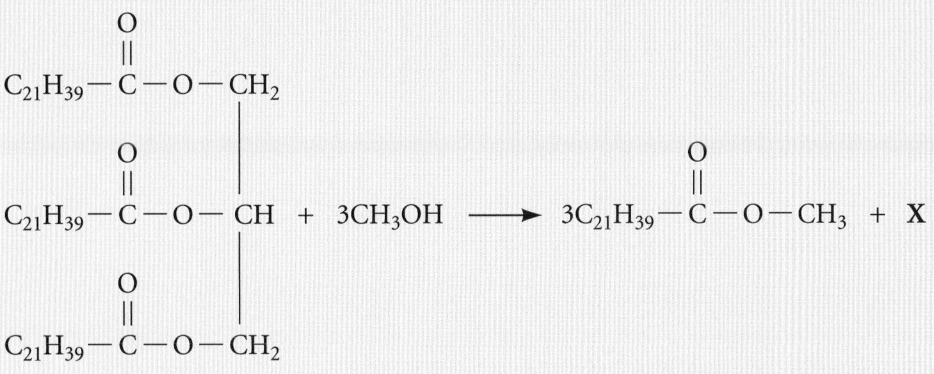

$$O$$
$$\parallel$$
$$C_{21}H_{39} — C — O — CH_2$$

$$O$$
$$\parallel$$
$$C_{21}H_{39} — C — O — CH \quad + \quad 3CH_3OH \quad \longrightarrow \quad 3C_{21}H_{39} — C — O — CH_3 \quad + \quad \textbf{X}$$

$$O$$
$$\parallel$$
$$C_{21}H_{39} — C — O — CH_2$$

a triglyceride in rapeseed oil   methanol   a component of biodiesel

Study Questions (continued)

13* The structure shown is part of a protein molecule.

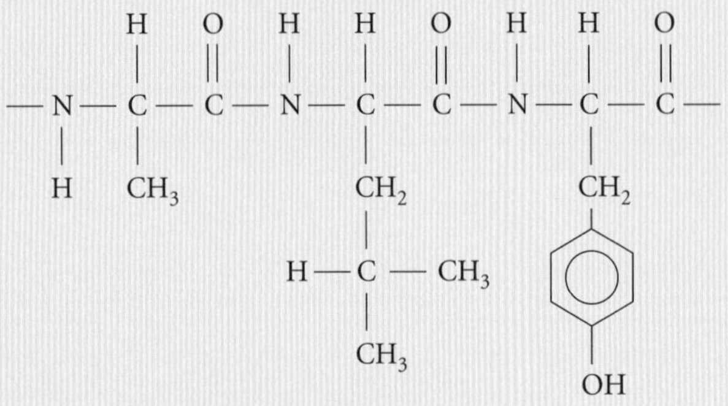

$$-N-C-C-N-C-C-N-C-C-$$

with substituents:

- N: H (below), H (above)
- first C: CH$_3$
- middle C: CH$_2$ — H—C—CH$_3$ with CH$_3$ below
- last C: CH$_2$ — benzene ring with OH

a) Circle a peptide (amide) link in the structure.

b) Draw the structural formula for *one* of the molecules which would be formed if the protein was hydrolysed.

c) Enzymes are proteins.

   i) Why does the enzyme maltase catalyse the hydrolysis of maltose, but not the hydrolysis of sucrose?

   ii) Maltase has optimum activity in alkaline conditions. Why does maltase lose its ability to act as a catalyst in acid conditions?

14* All enzymes are globular proteins.

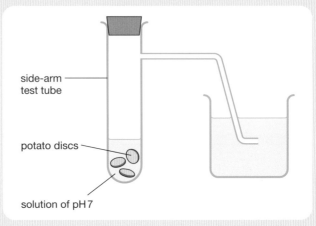

side-arm test tube

potato discs

solution of pH 7

a) What term is used to describe proteins which are **not** globular?

b) Catalase is an enzyme, contained in potatoes. A student was studying the effect of varying pH on the activity of catalase. The following apparatus was set up and left for 3 minutes.

   i) What must be added to the side-arm test tube to study the enzyme activity at this pH?

   ii) Describe how the enzyme activity at this pH can be measured.

Unit 3

Chemical Reactions

The first chapter introduces you to various issues arising when chemistry is applied on an industrial scale. This is followed by the application of enthalpy changes in Hess's Law. The study of reversible reactions and chemical equilibrium lead on to the meaning of pH and the importance of strong and weak acids and bases. Redox reactions and nuclear chemistry are the concluding topics in the Higher course.

13 The Chemical Industry

From previous work you should know and understand the following:

★ The Haber Process for the manufacture of ammonia

★ The fractional distillation of crude oil

★ The cracking of long-chain alkanes

★ The manufacture of addition polymers

★ The manufacture of condensation polymers

★ The manufacture of fuels.

Introduction

The UK chemical industry makes a major contribution to the quality of life in this country. Some idea of the influence of chemicals in everyday life can be gained by a close look at Table 13.1 which lists the main applications of products of the chemical industry.

Plastics	Cosmetics & toiletries
Paints	Disinfectants
Detergents	Domestic polishes & cleaners
Explosives	Veterinary health products
Adhesives	Inks, dyestuffs & pigments
Pharmaceuticals	Chemicals used in treating
Aerosols	water, metals, paper & fabrics
Fertilisers	Intermediates for making
Pesticides	synthetic fibres
Herbicides	

Table 13.1

Progress has been most noticeable since the mid-nineteenth century. Table 13.2 gives a selection of major discoveries and developments relevant to the chemical industry during this period.

In Scotland the chemical industry is indebted to the pioneering work of James 'Paraffin' Young in obtaining many useful materials from shale oil during the latter half of the nineteenth century. These included motor spirit, fuel oil, lamp oil and paraffin wax. Many of the techniques he developed are still used in the oil industry. Ammonia was also extracted and reacted with sulphuric acid to make ammonium sulphate for use as fertiliser.

The UK chemical industry also plays a vital role in our national economy. Within the UK it is one of the largest manufacturing industries, contributing about 2% of the Gross Domestic Product. The chemical industry's share of the total gross value added, i.e. the money raised in converting raw materials into end-products, achieved by all manufacturing industries in the UK, was 11% in 2003. This equalled the production of electrical and optical equipment. The only industries with a greater share that year were i) food, drink and tobacco (15%) ii) paper, printing and publishing (13%) and iii) transport equipment (12%).

As can be seen from Figure 13.1 the average growth rate of the chemical industry, 2.9% per year between 1994 and 2004, is nearly six times that of all manufacturing industries in the UK. Furthermore, the chemical industry has maintained a positive trade balance, i.e. exports exceeding imports, for many years. This is illustrated in Figure 13.2 which also gives the comparison with all other UK manufacturing industries for the period 1994–2004.

Year	Development	Industry
1850	Anaesthetic use of chloroform	
	Mauve – the first synthetic dyestuff	Dyestuffs (based on coal tar)
1860	Solvay Process – manufacture of sodium carbonate	
	Contact Process – manufacture of sulphuric acid	
1880	Manufacture of chlorine and sodium hydroxide by electrolysis	
	Aspirin	Pharmaceuticals
1900	Bakelite	
	Haber Process – manufacture of ammonia	Fertilisers
	Ostwald Process – manufacture of nitric acid	
1920	Polystyrene	Synthetic fibres
	Nylon	
	Polyethylene* (polythene)	Modern plastics
1940	PVC; synthetic rubber	
	Penicillin	
	DDT; selective weedkillers	Agrochemicals
1950	Pure silicon	Electronics
	Beta blockers	
1970	Biotechnology	
	Genetic engineering	
	Environmental legislation	
	Recycling	
2000		

The chemical industry still uses many traditional names for its feedstocks and products.

Table 13.2

In the world scene, the UK chemical industry was seventh after USA, Japan, China, Germany, France and Italy in 2004 in terms of total sales of chemicals. However, in Europe only France had a higher average

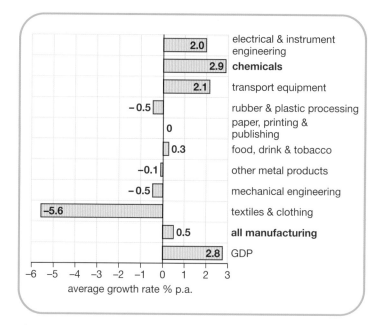

Figure 13.1 UK industrial growth rate comparisons for the period 1994–2004

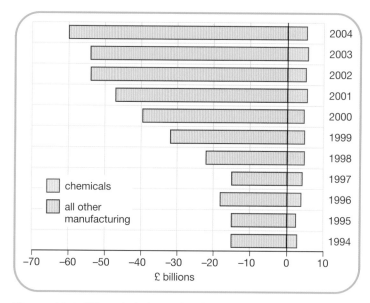

Figure 13.2 UK trade balance in chemicals and all other manufactures

growth rate between 1994 and 2004 for its chemical industry compared with manufacturing industry.

The chemical industry also makes a significant contribution to our 'invisible' exports. Income from licensing chemical processes developed in the UK exceeds £1 billion, several times payments made to overseas countries for their technology. For all other manufacturing industry there is an overall licensing *deficit*.

141

Manufacturing chemicals

Once the need for a new product has been recognised and its market potential assessed there are several stages which a chemical company will employ before production can begin. Some of these stages are shown in the form of a flow diagram in Figure 13.3, which includes a brief description of what each stage involves.

The manufacturing process

A chemical manufacturing process usually involves a sequence of steps. The following two examples illustrate this point.

Example 1

Although UK consumption of **sulphuric acid** is no longer as high as it once was, it is still a vital substance in the chemical industry. It is manufactured from sulphur by the Contact Process in a series of steps outlined below.

Step 1: Sulphur burning

$$S(l) + O_2(g) \rightarrow SO_2(g) \qquad \Delta H = \text{is negative}$$

Molten sulphur is sprayed into a furnace and burned in a blast of dry air to produce sulphur dioxide. The emerging gas mixture, containing about 10% each of sulphur dioxide and oxygen by volume, needs to be cooled from over 1000°C to near 400°C for the next step.

Figure 13.4 A sulphuric acid plant

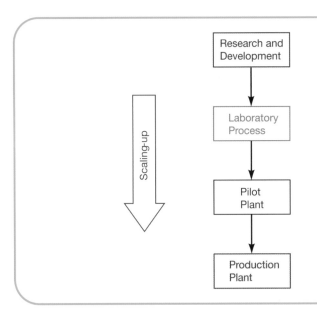

Research and Development	This includes selection of the most suitable route for producing the new chemical.
Laboratory Process	Small-scale investigation to review and modify the selected route.
Pilot Plant	Medium-scale investigation of product quality; health hazards and engineering costs assessed.
Production Plant	Plant design, construction, commissioning and start-up.

(Scaling-up)

Figure 13.3 Stages in the manufacture of a new product

Step 2: Sulphur dioxide conversion

$$SO_2(g) + \tfrac{1}{2}O_2(g) \rightleftharpoons SO_3(g) \quad \Delta H = -94\,kJ\,mol^{-1}$$

In the converter, sulphur dioxide combines with oxygen at about 450°C in the presence of a catalyst, vanadium(V) oxide, to produce sulphur trioxide. Modern converters contain about four layers of catalyst in 'fixed beds'. For economic and environmental reasons 99.5% conversion to sulphur trioxide is desirable, so heat is removed after each catalyst bed and the sulphur trioxide produced by the third bed is removed for the next step. This increases the percentage of conversion of the remaining reactants when passing through the fourth catalyst bed.

Step 3: Sulphur trioxide absorption

$$SO_3(g) + H_2O(l) \rightarrow H_2SO_4(l)$$

Sulphur trioxide is absorbed in 98% sulphuric acid at 100–150°C and the concentration of acid is maintained by the addition of water. Water alone is not used for absorption since it reacts so quickly with the sulphur trioxide that a stable mist of sulphuric acid is formed. This cannot be condensed and would cause extensive pollution.

Since this process involves exothermic reactions, cooling is required after step 1 and during step 2. This is achieved by converting water into steam which can then be used elsewhere in the industrial plant thus reducing the cost of producing sulphuric acid.

Example 2

The manufacture of **phenol** from benzene by the Cumene Process is also an example of a multi-stage chemical process.

Step 1

Benzene reacts with **propene** in the presence of a catalyst at 100°C to form an alkylbenzene called **cumene** (systematic name: 1-methylethylbenzene).

Step 2

Cumene is oxidised by oxygen in air to form its hydro-peroxide.

Step 3

This compound is then split in an acid-catalysed reaction to give two products, phenol and **propanone** (acetone). These are separated by distillation.

This example also illustrates the point that a process may yield more than one product. The economic viability of such a process may depend on there being a sufficient market for the co-product.

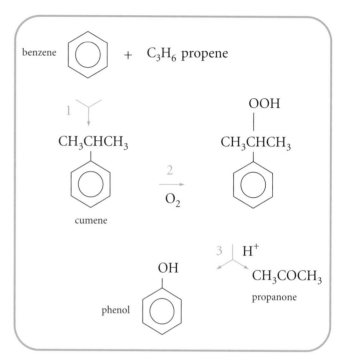

Figure 13.5 Manufacture of phenol from benzene

Raw materials and feedstocks

A feedstock is a reactant from which other chemicals can be extracted or synthesised. Feedstocks are themselves derived from **raw materials**. The major raw materials which are used in the chemical industry are:

♦ fossil fuels, i.e. coal, crude oil, natural gas

♦ metallic ores and minerals

♦ air and water

♦ organic materials of animal and vegetable origin such as vegetable oils, starch and gelatine.

Crude oil

Crude oil gives gives naphtha, a fraction obtained by distillation, which is used as an important feedstock for various chemical processes such as:

♦ steam cracking to produce ethene and propene for the manufacture of plastics

♦ reforming to produce aromatic hydrocarbons for the manufacture of dyes, drugs etc.

Ores and minerals

Two examples are given here:

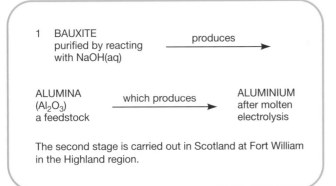

1 BAUXITE
   purified by reacting   produces ⟶
   with NaOH(aq)

   ALUMINA   which produces   ALUMINIUM
   (Al₂O₃)   ⟶   after molten
   a feedstock   electrolysis

   The second stage is carried out in Scotland at Fort William in the Highland region.

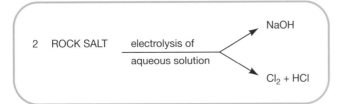

2 ROCK SALT   electrolysis of   ⟶ NaOH
                aqueous solution   ⟶ Cl₂ + HCl

Air

As a raw material, air provides:

♦ nitrogen for the production of ammonia

♦ oxygen for the oxidation of i) sulphur to sulphur oxides to manufacture sulphuric acid, and ii) ammonia to nitrogen oxides to manufacture nitric acid.

Water

Water is a raw material in the steam cracking of naphtha and ethane and in the hydration of ethene to produce ethanol.

Air and water have other functions in the chemical industry. Both can be used as coolants, while water is, of course, an important solvent.

Figure 13.6 Lochaber aluminium smelter

Question

1 Aluminium is extracted from its purified oxide by molten electrolysis. Suggest two advantages and two disadvantages of siting aluminium smelters in the Scottish Highlands.

Choosing a manufacturing route

Several factors may need to be taken into account when choosing a particular route to manufacture a product. These factors include the following:

♦ the cost, availability and suitability of the feedstock(s)

♦ the yield of product(s)

♦ whether or not unreacted materials can be recycled

♦ how marketable are the by-products

♦ the difficulty and cost of waste disposal

♦ energy consumption

♦ total emissions to the atmosphere.

There are several possible routes for the production of **ethanoic acid**. Two of them are summarised below.

Route 1: The direct oxidation of naphtha

naphtha + air → ethanoic acid + several by-products,
e.g. propanone, methanoic acid, propanoic acid.

Process conditions: 180°C; 50 atmospheres pressure.

The main advantage of this method is that it is a single-step process. The disadvantages are the low yield of ethanoic acid (<50%) and the large quantity of by-products. In the USA butane is the feedstock used. The yield of ethanoic acid is still low but the quantity of by-products obtained is much less.

Route 2: From methanol

$$CH_3OH(l) + CO(g) \rightarrow CH_3COOH(l)$$

Process conditions: 180°C; 30 atmospheres pressure; rhodium/iodine catalyst.

The main advantages of this method are the very high yield of products (>99%) and the flexibility of source of feedstock. Methanol can be made from coal, natural gas or oil via synthesis gas, see Chapter 11 page 118. The main disadvantage is that capital costs are high as special materials are needed for plant construction.

Owing to the low cost of feedstock and high yield of product, route 2 is the more likely choice for new production units.

Batch or continuous process

Manufacturing a chemical may be organised either as a batch process or a continuous process. In a batch process the reactants are added to the reactor. The reaction is started and its progress carefully monitored. At the end of the reaction the reactor is emptied and the product mixture passes on to the separation and purification stages. A batch reactor is usually a large cylindrical tank as shown in Figure 13.7.

In a continuous process reactants flow into the reactor at one end and the products flow out at the other end. The design of the reactor varies from one process to another. Catalytic cracking of long-chain hydrocarbons uses a cylindrical tank with another tank nearby where the catalyst is regenerated. This is shown diagrammatically in Figure 13.8. On the other hand, limestone is decomposed to produce quick lime for the

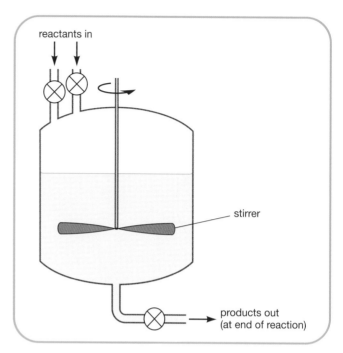

Figure 13.7 A batch reactor

manufacture of cement in a rotary kiln, which is a long metal tube several feet in diameter.

Batch and continuous processes each have their advantages and disadvantages. Some of these are outlined here.

♦ Where large quantities of a chemical, such as >1000 tonnes per year, are to be manufactured, a continuous process will usually be employed, while a batch process is more suited to small-scale production.

♦ Large-scale production should mean a cheaper product but this may depend on the process operating at full capacity.

♦ Filling and emptying a batch reactor takes time during which no product is being made.

♦ A small-scale chemical plant should be cheaper to build.

♦ A continuous process can be staffed by a smaller workforce.

♦ A batch reactor is more flexible in its use as other chemicals may be made in it. On the other hand if there is a regular demand for a high-tonnage chemical, lack of versatility of the continuous reactor is not a problem.

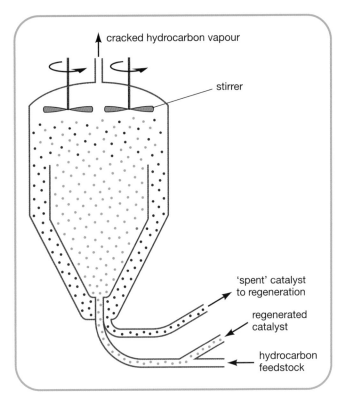

Figure 13.8 'Cat cracker' – a continuous flow reactor

◆ A continuous process is suited to fast, single-step reactions while a batch process is more appropriate for slow, multi-step reactions.

◆ A batch reactor can use reactants in any physical state. In a continuous process solids can present a problem as they are more difficult to pump from one part of a plant to another, unless the solid is finely powdered as is the case in catalytic cracking. The catalyst is fluidised by the reactant gases flowing through it and is moved from the reactor to the regenerator and back again.

◆ From a safety viewpoint a continuous process is at greatest risk when it is started up but this usually happens only once every few months or even years. The reaction can be kept under control by adjustment of flow-rates, temperature, etc. A batch process may be harder to control, especially if the reaction is highly exothermic.

Economic aspects

By now you should appreciate that each industrial chemical process has a characteristic set of conditions under which it operates. These conditions are chosen so

as to maximise economic efficiency. Making reactions go faster by raising the temperature, for example, is not always economic. Higher temperatures mean greater energy costs. High pressures mean stronger reactors and special equipment requiring regular maintenance. When you have studied the topic of equilibrium in Chapter 15 you will be in a better position to appreciate why, for example, the Haber Process operates at high pressure and moderate temperature.

The chemical industry is research-based. In 2003 £3.8 billion was spent on research and development by the UK chemical industry, equivalent to more than 10% of sales. In comparison, other UK manufacturing industry spent on average the equivalent of 2.4% of sales on research. The costs of research, pilot plant studies and the construction of production plants are all incurred before any product is sold. Sales must exceed these costs as well as production costs if they are to yield an overall profit for the company.

Manufacturing costs in the chemical industry can be divided into different categories as listed in Table 13.3.

Capital costs	Fixed costs	Variable costs
Research and development	Depreciation of plant	Raw materials
Plant construction	Labour	Energy
Buildings	Land purchase or rental	Overheads
Infrastructures	Sales expenses	Effluent treatment or disposal

Table 13.3

Variable costs relate to the chemical process involved. Raw material costs and energy costs are the most significant variable costs. These costs will not be incurred if production is halted but fixed costs will still have to be paid. A company will incur fixed costs whether it manufactures one tonne of product or thousands of tonnes. However, the effect of the fixed cost on the selling price of the product diminishes as the scale of operation increases.

Capital costs are recovered as depreciation included under fixed costs. Depreciation occurs as chemical plants frequently operate under severe and/or corrosive conditions or become obsolete through technological progress.

The UK chemical industry employs about a quarter of a million people from a wide variety of disciplines. Figure 13.9 shows how these are distributed in different parts of the UK. Chemists, chemical engineers and process operators directly involved in the manufacturing process need to be highly qualified and well trained.

The chemical industry can be described as capital intensive rather than labour intensive. In some processes, e.g. the manufacture of sulphuric acid, a large chemical plant may be operated by only a few people. The entire chemical industry employs less than 1.5% of the British workforce.

The chemical industry contributes to the wealth of the nation but any expansion does not in itself provide a means of reducing unemployment significantly, except during plant construction. Industries associated with the chemical industry, and the spending power of relatively well-paid, skilled chemical workers do provide employment, however.

The use of energy

As indicated previously, energy is a major variable cost. At times of international tension, the price of oil-derived energy has risen very rapidly. The chemical industry has responded by:

♦ switching where possible to processes which use less energy

♦ saving energy by using heat from exothermic processes elsewhere in the plant

♦ using 'waste' heat to generate electricity for the plant

♦ selling energy to supply district heating schemes for local housing.

Apart from its cost, wasted energy is causing needless pollution. When derived from fossil fuels it is contributing to global warming.

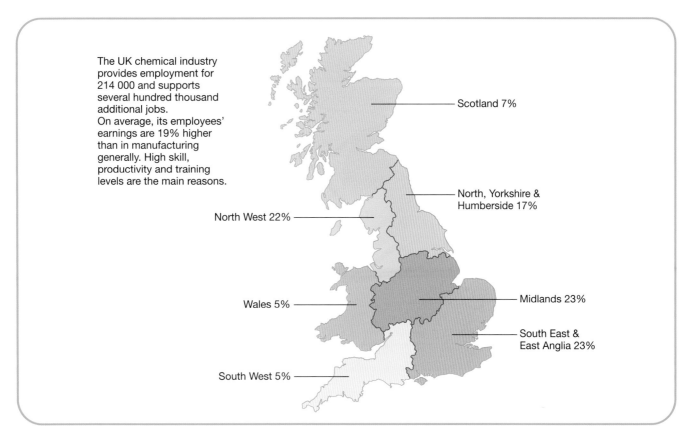

The UK chemical industry provides employment for 214 000 and supports several hundred thousand additional jobs.
On average, its employees' earnings are 19% higher than in manufacturing generally. High skill, productivity and training levels are the main reasons.

Scotland 7%

North, Yorkshire & Humberside 17%

North West 22%

Wales 5%

Midlands 23%

South East & East Anglia 23%

South West 5%

Figure 13.9 Chemical industry employment by region – percentage shares of GB total

The location of chemical industry

Major chemical manufacturing sites have been established as a result of historical and practical considerations. Two locations in the United Kingdom will be used to illustrate this.

Grangemouth chemical works in Scotland

Grangemouth Works started in 1919 as a dyeworks for a textile dyeing company operating in Carlisle and has since expanded to produce pharmaceuticals, agrochemicals, pigments and speciality chemicals. It was sited at Grangemouth for several important practical reasons:

♦ A large area of flat land was available.

♦ There was plenty of water for manufacturing processes.

Figure 13.10 An aerial view of Grangemouth

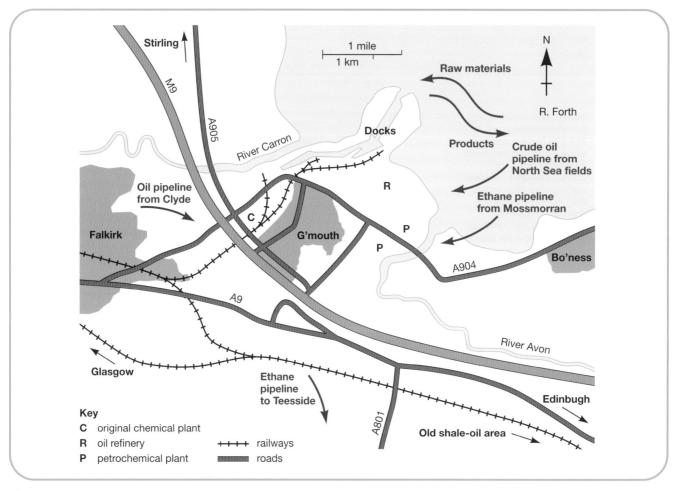

Figure 13.11 Location plan of Grangemouth. Some of the more recent developments in the area are shown

- There were good transport links by railway and by Grangemouth docks to import raw materials (from coal tar) and to export products.

- There was a pool of skilled labour with experience in the chemical industry because of the nearby shale-oil works.

- Effluent could be disposed of in the sea.

Even with these practical advantages, the incentive of a shortage of dyes, normally imported from Germany until the start of the First World War, was needed to cause a major factory to be set up in the UK. A crucial factor in its establishment at Grangemouth, only one of a number of suitable sites, was that the company's founder wanted to set up the works in his native Scotland. Many of the factors listed above were also responsible for the establishment of the neighbouring oil refinery, and, later, a petrochemicals plant. All three plants have expanded and diversified into additional products in recent years.

Teesside chemical works in Northern England

The works at Billingham began to make ammonia at the end of the First World War. The intention was to free Britain from dependence on imports of chemicals for fertiliser and explosives manufacture. The war had shown that the country was vulnerable to submarine blockade. Billingham expanded later, including the creation of the new Wilton works, to make sulphuric acid and petrochemicals used to make plastics, detergents and pharmaceuticals. The Billingham site was chosen because:

- Local coal provided a source of hydrogen for the Haber Process and was the original source of energy.

- A power station was being built nearby.

- The River Tees provided a port to import other raw materials and to export the products.

- There were good road and rail links.

- There was a good supply of water.

- The site was suitable for building a heavy plant without subsidence.

Figure 13.12 Billingham chemical plant

- Anhydrite, $CaSO_4$, was present underground and was a suitable raw material for sulphuric acid manufacture. Salt to manufacture alkalis and chlorine, and potash, a component of fertilisers, were also available nearby.

- Effluent could be disposed of to the sea, via the River Tees.

The two sites provide a contrast. The availability of raw materials was more important at Billingham where large scale manufacture was carried out. At Grangemouth, where smaller quantities of more specialised substances were produced, the available labour force was important. In both cases a war provided the spur for the establishment of the plant, and another war caused their first major expansion.

149

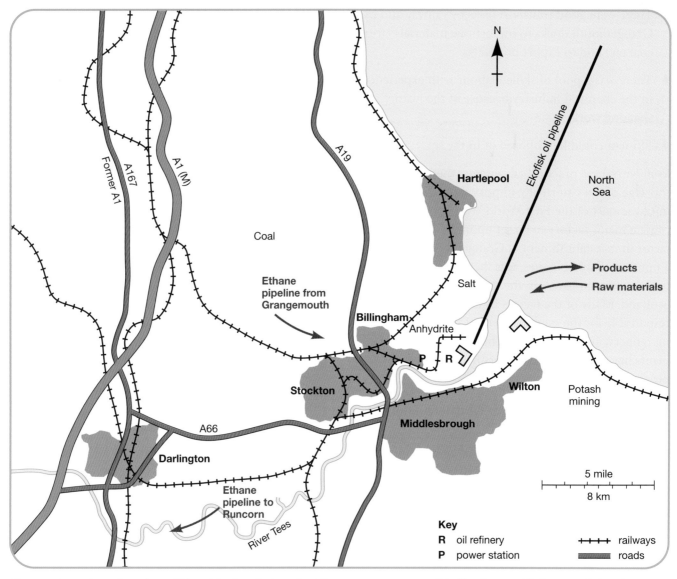

Figure 13.13 Location plan of Teesside chemical works. The resources responsible for the establishment of the site are shown, with more recent developments added

Environmental considerations

In the previous section, one reason for siting both plants was the ease of disposal of effluent, straight into the sea. This method of disposal is no longer acceptable and most chemical companies take pride in their concern for the environment. The later record of plants like Billingham shows how attitudes have changed.

♦ Over the period 1970–1989, discharges of phosphates, which cause algal growth and oxygen depletion in water, of cyanides and of heavy metals, which build up in the food chain and are toxic, declined at Billingham by about 90%.

♦ Waste sulphuric acid is recovered and reused rather than being pumped into the river, so reducing pollution and saving money.

♦ Reed beds are used to harbour microbes which break down waste organic chemicals in liquids which would otherwise deplete river water of oxygen, and the water reaching the river is clean. The reed beds provide the bonus of an excellent wildlife habitat.

Many chemical companies, in developed countries at least, are involved in similar schemes. There is, of course, a large degree of self-interest involved. Figure 13.15 shows the consequences that failing to develop a good environmental policy might have.

Figure 13.14 Chemical waste treatment using reed beds

Safety in the chemical industry

The chemistry industry, like all industries, has a duty to its employees and the public to operate without causing accidental injury and without causing risks to health.

There have, however, been some disastrous occurrences in the industry since 1970.

◆ **1976 Seveso, Italy.** A leak of gases containing carcinogenic dioxins caused large numbers of casualties with skin problems, but raised more fears of long-term effects, particularly for the children of women pregnant at the time of the accident.

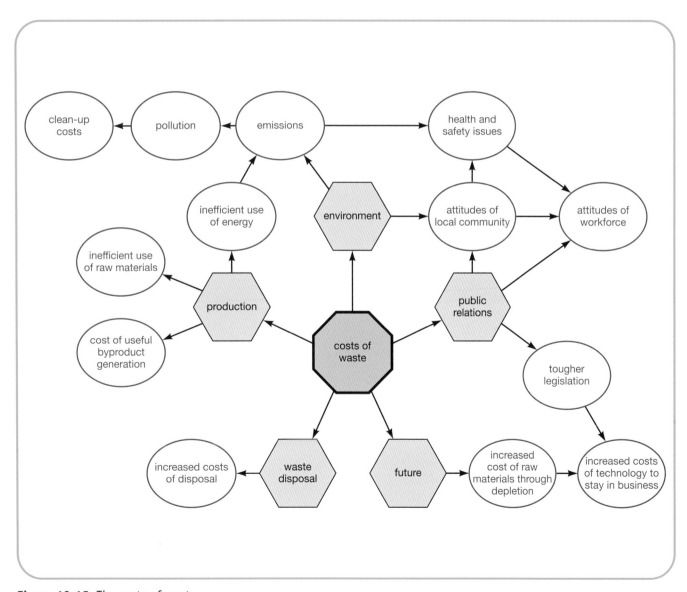

Figure 13.15 The costs of waste

♦ **1984 Bhopal, India.** A leak of toxic methyl isocyanate gas killed several thousand people in the worst chemical plant incident ever. The company concerned was forced to pay huge sums in compensation, its reputation throughout the world was damaged greatly and, as a consequence, the company had to be drastically reorganised to survive.

If only to prevent such financial damage, chemical companies work hard to prevent accidents, but chemists increasingly see themselves as members of an ethical profession with duties to care for other people and their environment.

Accidents have often resulted in major advances in techniques or plant design. The worst British chemical accident in the last 40 years was at Flixborough, Lincolnshire, in 1974. A plant making cyclohexane to be used in polyamide manufacture suffered an explosion and fire. Twenty-eight people were killed and 104 injured, some in neighbouring housing. The control room was destroyed along with its workers, and the plant could not be shut down quickly. Lessons learnt included the building of new plants away from housing and ensuring that control rooms were fire- and blast-proof. When the Fife Ethylene Plant was built at

Mossmorran a few years later these principles were put into effect.

The concern to prevent accidents is such that at the entrance to most plants, boards are displayed showing their (good) accident records. Between 1987 and 2004 in the UK there was a reduction in employee accidents by 59%. To put this matter in perspective, deaths on the roads in the UK exceed accidental deaths in the chemical industry by a factor of 1000.

Apart from dramatic accidents, the chemical industry has a history of causing long-term damage to the health of its workers. In the nineteenth century this was sometimes a consequence of lack of concern for the safety of workers in society as a whole. Later it may have been because of ignorance of harmful effects of particular chemicals. Now the rules governing exposure to harmful chemicals are very strict and rigorously enforced. The change in attitudes is illustrated in Table 13.4 by the data for vinyl chloride monomer. A long-term Maximum Exposure Level (MEL) for employees is derived from measurements made over an 8-hour working day and averaged. It must not be exceeded, and should be reduced if possible. The units are parts of substance per million parts of air.

Year	Observations	MEL/ppm
1930s	Narcotic effects noticed. Workers became drowsy.	No limit in force.
1962	To prevent narcotic effect, limit introduced.	500
1960s	Animal experimentation showed vinyl chloride affected liver, bones and kidneys.	
1971	New limit	200
1971	Three workers in US company died of liver cancer.	
1975	Health and Safety Executive (UK) suggest Reduced same year to	25 10
1978	Reduced to with a maximum averaged *annual* limit of	7 3

Table 13.4

In recent years, claims of many more deaths, and of many workers being treated for cancer, have increased pressure for further reductions in the MEL.

In the 1980s the chemical industry recognised that it needed to make a significant improvement in its safety, health and environmental performance. Through national chemical associations in the UK, the USA and Canada, a programme called Responsible Care has been adopted. All members committed themselves to these improvements, and to increase openness about their activities. The programme has been adopted in many other countries and has proved very successful.

Indications of the improvements that have been achieved in the UK are:

♦ Between 1990 and 2004, energy efficiency improved by 34%, with a similar reduction in the 'carbon footprint'.

♦ About 50% of hazardous waste was recycled in 2004, by energy recovery or reprocessing, compared with 39% in 2000.

♦ Transportation incidents were reduced from about 4 per million tonnes transported in 1990 to just over 1 per million tonnes in 2004.

Study Questions

In questions 1–3 choose the correct word(s) from the following list to complete the sentence.

batch   capital   continuous   feedstock   fixed
raw material

1 Bauxite, a metallic ore, is the _____ from which aluminium is obtained.

2 Large-scale production of a chemical is more likely to occur by a _____ process.

3 When setting up a chemical plant the cost of research and development is usually regarded as a _____ cost.

4 Which of the following is a feedstock in the chemical industry?

   A Methane

   B Ethene

   C Water

   D Air

5 Which of the following is regarded as a variable cost in the chemical industry?

   A Land purchase

   B Plant construction

   C Depreciation of plant

   D Treatment of effluent

6* Acetone, widely used as a solvent, is manufactured from cumene. Cumene is oxidised by air, and the cumene hydroperoxide product is then cleaved.

The mixture of acetone and phenol is separated by distillation.

a) Copy and complete the following flow chart to summarise the manufacture of acetone from cumene.

Study Questions (continued)

In your flow chart use ⬭ to represent chemicals

and ▭ to represent processes

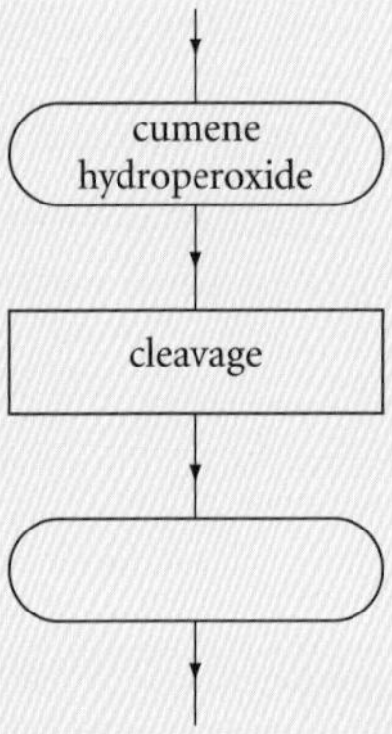

cumene hydroperoxide

↓

| cleavage |

↓

⬭

↓

b) For every 10 tonnes of acetone produced in this industrial process, calculate the mass of phenol (C_6H_5OH) produced.

c) Acetone can also be manufactured by oxidising propan-2-ol. In industry, several factors influence the decision as to which route might be used. Suggest *two* of these factors.

7 Natural gas, steam and air are the raw materials used to make ammonia in the UK. Preparation of the nitrogen–hydrogen mixture required to synthesise ammonia is a multi-stage process.

a) Using a suitable resource, find out about the two main stages involved, **i)** the manufacture of synthesis gas and **ii)** the 'shift reaction'. For each stage write an equation and describe the conditions used, i.e. temperature, pressure, catalyst, etc.

b) Oxygen is removed from the air to leave nitrogen. How is this achieved?

c) Before the first main stage, any sulphur, which may be present in trace amounts in compounds in natural gas, must be removed. Suggest why.

d) The carbon dioxide produced during the 'shift reaction' must be removed prior to synthesising ammonia from nitrogen and hydrogen. One way of achieving this is to react CO_2 with potassium carbonate solution. Water is also a reactant and the only product is potassium hydrogencarbonate solution. Write the balanced equation for this reaction.

8* The flow diagram below shows how vinyl chloride ($CH_2{=}CHCl$), an important feedstock, is made in industry.

a) i) What is the systematic name for vinyl chloride?

ii) Give a use for vinyl chloride.

b) Write the formulae for the *three* substances which are recycled.

c) Write the equation for the reaction taking place in the cracker.

d) Name the process taking place in the separator units.

e) Name the type of reaction taking place in the scrubber unit.

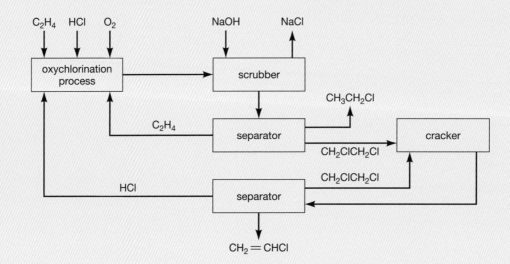

Study Questions (continued)

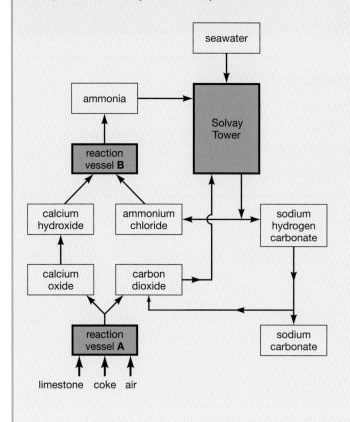

9* The flow diagram outlines the manufacture of sodium carbonate by the Solvay process.

a) Name the reactants in the reaction taking place in the Solvay Tower.

b) In reaction vessel **A**, carbon dioxide is produced by the following two reactions.

$$CaCO_3(s) \rightarrow CaO(s) + CO_2(g) \quad \Delta H = \underline{\hspace{2cm}}$$

$$C(s) + O_2(g) \rightarrow CO_2(g) \quad \quad \Delta H = \underline{\hspace{2cm}}$$

For each reaction, add a sign after the ΔH to show whether the reaction is endothermic or exothermic.

c) As well as ammonia, a salt and water are produced in reaction vessel **B**. Write a balanced equation for the production of ammonia in this reaction vessel.

d) The seawater used in the Solvay Process can contain contaminant magnesium ions. These can be removed by the addition of sodium carbonate solution. Why is sodium carbonate solution suitable for removing contaminant magnesium ions?

e) Using the information in the flow diagram, give *two different* features of the Solvay Process that make it economical.

14 Hess's Law

From previous work you should know and understand the following:

★ The enthalpy change in a chemical reaction is the difference in potential energy between the reactants and products involved

★ The symbol for enthalpy change is ΔH and it is measured in kJ mol^{-1}

★ If the reaction is exothermic, ΔH is negative; if endothermic, ΔH is positive.

see Chapter 2

In this chapter we will be using enthalpy changes when applying a very important rule in chemistry called Hess's Law. This law can be stated as follows:

'The enthalpy change of a chemical reaction depends only on the chemical nature and physical states of the reactants and products and is independent of any intermediate steps.'

In other words, Hess's Law states that the enthalpy change of a chemical reaction does not depend on the route taken during the reaction.

Verification of Hess's Law by experiment

Hess's Law can be tested by experiment as the following example shows. This involves converting solid potassium hydroxide into potassium chloride solution by two different routes, as illustrated in Figure 14.1.

The first route is a one-step process. Solid potassium hydroxide is added to dilute hydrochloric acid to produce potassium chloride solution. Let us call the enthalpy change for this reaction ΔH$_1$.

Route 1: KOH(s) + HCl(aq) → KCl(aq) + H$_2$O(l) ΔH$_1$

The second route involves two steps. Firstly the solid potassium hydroxide is added to water to form potassium hydroxide solution. This solution is

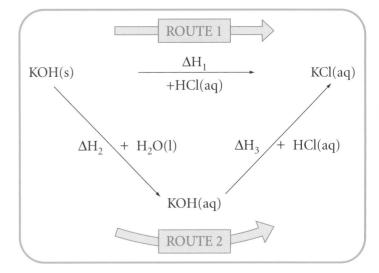

Figure 14.1

neutralised in the second step by addition to hydrochloric acid.

Route 2:

step 1: KOH(s) + (aq) → KOH(aq) ΔH$_2$

step 2: KOH(aq) + HCl(aq) → KCl(aq) + H$_2$O(l) ΔH$_3$

From work covered in Chapter 2 you should recognise that ΔH$_2$ is the enthalpy of solution of KOH(s) and that ΔH$_3$ is the enthalpy of neutralisation of hydrochloric acid by KOH(aq).

According to Hess's Law: ΔH$_1$ = ΔH$_2$ + ΔH$_3$

Prescribed Practical Activity

The aim is to confirm Hess's Law. The method can be summarised as follows.

Route 1: 25 cm³ of 1.0 mol l⁻¹ hydrochloric acid is measured out and its temperature noted. The acid is added to a known mass of solid potassium hydroxide (about 1.2 g) in a polystyrene cup and the mixture stirred. The highest temperature of the reacting mixture is noted.

Route 2 (step 1): 25 cm³ of water is measured out and its temperature noted. It is added to a known mass of solid potassium hydroxide (about 1.2 g) in a polystyrene cup. The mixture is stirred and its highest temperature is noted. This solution is allowed to cool before use in step 2.

Route 2 (step 2): 25 cm³ of 1.0 mol l⁻¹ hydrochloric acid is measured out and its temperature noted. The temperature of the potassium hydroxide solution prepared in step 1 is noted. These values are averaged to give the initial temperature. The solutions are mixed and the highest temperature recorded.

For each of these experiments, the energy released can be calculated using the relationship, $E_h = cm\Delta T$. The enthalpy change for each reaction is then obtained by dividing the energy released, E_h, by the number of moles of potassium hydroxide. If necessary you can refer to Worked Examples 2.2 and 2.3 on pages 23–25, respectively.

Essentially, Hess's Law is the application of the Law of Conservation of Energy to chemical reactions. This can be shown in a theoretical manner as follows. If, in the example given above, the first route was readily reversible and it was found that the sum of ΔH_2 and ΔH_3 was greater than ΔH_1, then it would be possible to create energy by carrying out steps 2 and 3 and then reversing the first route. This would violate the Law of Conservation of Energy, which states that energy cannot be created or destroyed but can only be converted from one form into another.

Calculations using Hess's Law

Chapter 2 describes a number of experiments to determine the enthalpy changes of certain reactions. Hess's Law is important because it often enables us to calculate enthalpy changes which are either very difficult or even impossible to obtain by experiment. The formation of carbon compounds from their constituent elements is just one example of such a reaction.

The following Worked Examples illustrate the application of Hess's Law in calculating enthalpy changes which cannot be found by experiment. In the first example, the enthalpy of formation of methane gas from carbon and hydrogen is calculated. The equation for this reaction is:

$$C(s) + 2H_2(g) \rightarrow CH_4(g)$$

Direct combination between carbon and hydrogen does not readily occur and, in any event, methane is not the only product. However the enthalpy change for this reaction can be calculated using Hess's Law. Each of the substances shown in the equation can be burned and their enthalpies of combustion can be found by experiment. Hence an alternative route can be devised as Figure 14.2 shows. The second Worked Example shows a similar calculation for ethanol.

Worked Example 14.1

$$C(s) + 2H_2(g) \rightarrow CH_4(g)$$

Calculate the enthalpy change of the above reaction using the enthalpies of combustion of carbon, hydrogen and methane.

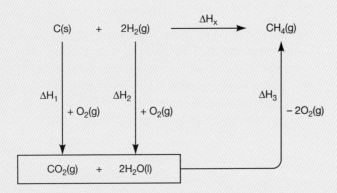

Figure 14.2

In the alternative route:

step 1 is the combustion of 1 mole of carbon,

$$C(s) + O_2(g) \rightarrow CO_2(g) \hspace{4cm} \Delta H_1 = -394 \, kJ$$

step 2 is the combustion of 2 moles of hydrogen,

$$2H_2(g) + O_2(g) \rightarrow 2H_2O(l) \hspace{3cm} \Delta H_2 = -(2 \times 286) \, kJ$$
$$= -572 \, kJ$$

step 3 is the *reverse* of combustion of 1 mole of methane (see Note below),

$$CO_2(g) + 2H_2O(l) \rightarrow CH_4(g) + 2O_2(g) \hspace{2cm} \Delta H_3 = +891 \, kJ$$

Adding these three equations gives the required equation, namely

$$C(s) + 2H_2(g) \rightarrow CH_4(g)$$

According to Hess's Law, the enthalpy change for this reaction is:

$$\Delta H_x = \Delta H_1 + \Delta H_2 + \Delta H_3$$
$$= -394 - 572 + 891$$
$$= -75 \, kJ \, mol^{-1}$$

Note: The Data Book gives the enthalpy of combustion of methane, namely

$$CH_4(g) + 2O_2(g) \rightarrow CO_2(g) + 2H_2O(l) \hspace{2cm} \Delta H = -891 \, kJ \, mol^{-1}$$

Step 3 is the reverse of this. Hence the numerical value of the enthalpy change for step 3 is the same, but the sign must be altered since the reaction has been reversed.

Worked Example 14.2

The following equation shows the formation of ethanol from carbon, hydrogen and oxygen.

$$2C(s) + 3H_2(g) + \tfrac{1}{2}O_2(g) \rightarrow C_2H_5OH(l)$$

Use the enthalpies of combustion of carbon, hydrogen and ethanol to calculate the enthalpy change of this reaction.

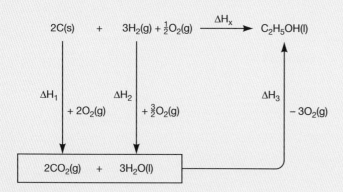

Figure 14.3

In the alternative route:

step 1 is the combustion of 2 moles of carbon,

$$2C(s) + 2O_2(g) \rightarrow 2CO_2(g) \qquad \Delta H_1 = -(2 \times 394)\ kJ$$
$$= -788\ kJ$$

step 2 is the combustion of 3 moles of hydrogen,

$$3H_2(g) + \tfrac{3}{2}O_2(g) \rightarrow 3H_2O(l) \qquad \Delta H_2 = -(3 \times 286)\ kJ$$
$$= -858\ kJ$$

step 3 is the *reverse* of combustion of ethanol,

$$2CO_2(g) + 3H_2O(l) \rightarrow C_2H_5OH(l) + 3O_2(g)$$
$$\Delta H_3 = +1367\ kJ$$

Adding these three equations gives the required equation, namely

$$2C(s) + 3H_2(g) + \tfrac{1}{2}O_2(g) \rightarrow C_2H_5OH(l)$$

According to Hess's Law, $\Delta H_x = \Delta H_1 + \Delta H_2 + \Delta H_3$
$$= -788 - 858 + 1367$$
$$= -279\ kJ\ mol^{-1}$$

Note: Oxygen is one of the elements present in ethanol but it is not involved in deriving the required enthalpy change. The calculation is based on enthalpies of combustion. Oxygen gas supports combustion; it does not have an enthalpy of combustion.

Questions

1   Use relevant data on enthalpies of combustion and the following equation to calculate the enthalpy of formation of ethyne, $C_2H_2(g)$.

$$2C(s) + H_2(g) \rightarrow C_2H_2(g)$$

2   Benzene reacts with hydrogen under certain conditions to yield cyclohexane according to the following equation:

$$C_6H_6(l) + 3H_2(g) \rightarrow C_6H_{12}(l)$$

Use relevant data on enthalpies of combustion to calculate the enthalpy change for this reaction.

(Enthalpy of combustion of cyclohexane, $\Delta H = -3924\ kJ\ mol^{-1}$)

The answer to question 1 is interesting as it shows that the formation of ethyne from its elements is highly endothermic. This is in contrast to the equivalent reaction for methane, which is exothermic, as shown in Worked Example 14.1. Figure 14.4 compares these two enthalpy changes in a potential energy diagram.

This suggests that ethyne is a much less stable compound than methane. Ethyne is a gas at room temperature and has a boiling point of −84°C. It is liable to explode if liquefied, but it can be stored safely in cylinders if it is dissolved in propanone under pressure.

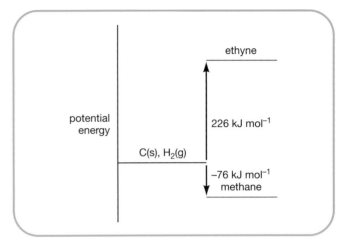

Figure 14.4

Study Questions

1 Benzene can be produced from ethyne. The equation for this reaction is:

$$3C_2H_2(g) \rightarrow C_6H_6(l)$$

The enthalpy change for this reaction in $kJ\,mol^{-1}$, using enthalpies of combustion from the data book, is

A +632   B −632   C +1968   D −1968.

2 Refer to the reaction paths shown below.

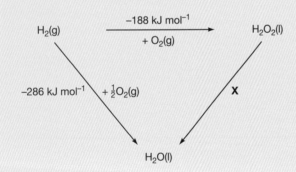

According to Hess's Law, the enthalpy change, in $kJ\,mol^{-1}$, for reaction **X** is

A −98   B +98   C −474   D +474

3* The following results are taken from the notebook of a pupil who was trying to confirm Hess's Law.

Experiment 1 – Addition of 1.6 g of sodium hydroxide solid to 50 cm³ of 1 mol l⁻¹ hydrochloric acid

$$NaOH_{(s)} + HCl_{(aq)} \rightarrow NaCl_{(aq)} + H_2O_{(l)}$$

Mass = 50 g
Initial temperature of HCl(aq) = 21.7°C
Highest temperature during experiment = 29.9°C
ΔT = 8.2°C

Experiment 2 – Addition of 25 cm³ of 2 mol l⁻¹ sodium hydroxide solution to 25 cm³ of 2 mol l⁻¹ hydrochloric acid

$$NaOH_{(aq)} + HCl_{(aq)} \rightarrow NaCl_{(aq)} + H_2O_{(l)}$$

Mass = 50 g
Initial temperature of HCl(aq) – T_1 = 21.7°C
Initial temperature of NaOH(aq) – T_2 = 22.1°C
Highest temperature during experiment = 28.6°C
ΔT =

a) i) In experiment 1, calculate which reactant is in excess. (Show your working clearly.)

ii) In experiment 1, calculate the enthalpy change during the reaction. You may wish to use the Data Book to help you.

Study Questions (continued)

b) Calculate ΔT for experiment 2.

c) Outline a third experiment which would have to be carried out in order to confirm Hess's Law.

d) Describe a precaution which should be taken to minimise heat loss during the experiments.

4 The equation for the formation of benzene from carbon and hydrogen is as follows:

$$6C(s) + 3H_2(g) \rightarrow C_6H_6(l)$$

a) Use enthalpies of combustion from your Data Book to calculate the enthalpy of this reaction.

b) Refer to Figure 14.4 on page 160 and indicate how the stability of benzene compares with that of methane and ethyne.

5 The equation for the complete hydrogenation of ethyne is shown below:

$$C_2H_2(g) + 2H_2(g) \rightarrow C_2H_6(g)$$

Calculate the enthalpy change of this reaction using enthalpies of combustion given in your Data Book.

6 Carbon disulphide, CS_2, is an inflammable liquid producing CO_2 and SO_2 when it burns.

The equation for the formation of carbon disulphide from carbon and sulphur is shown below along with the enthalpy change for the reaction.

$$C(s) + 2S(s) \rightarrow CS_2(l) \qquad \Delta H = +88 \text{ kJ mol}^{-1}$$

Use this information and the enthalpies of combustion of carbon and sulphur to calculate the enthalpy of combustion of carbon disulphide.

7 The enthalpy of combustion of butan-1-ol, C_4H_9OH, is $-2673 \text{ kJ mol}^{-1}$.

Use this information and data from your Data Book to calculate the enthalpy change for the following reaction:

$$4C(s) + 5H_2(g) + \tfrac{1}{2}O_2(g) \rightarrow C_4H_9OH(l)$$

8* Hess's Law can be used to calculate the enthalpy change for the formation of ethanoic acid from its elements.

$$2C(s) + 2H_2(g) + O_2(g) \rightarrow CH_3COOH(l)$$

Calculate the enthalpy change for the above reaction, in kJ mol^{-1}, using information from the Data Book and the following data:

$$CH_3COOH(l) + 2O_2(g) \rightarrow 2CO_2(g) + 2H_2O(l)$$
$$\Delta H = -876 \text{ kJ mol}^{-1}$$

15 Equilibrium

From previous work you should know and understand the following:

★ The Haber Process used to synthesise ammonia

★ Catalysts and other factors affecting reaction rates

★ Enthalpy changes, ΔH and its sign convention

★ Writing equations representing enthalpy changes.

Reversible reactions and dynamic equilibrium

Reversible reactions attain a state of equilibrium when the rate of the forward reaction is equal to the rate of the reverse reaction. Consider a saturated solution containing excess of the undissolved solute, in this case sodium chloride.

$$NaCl(s) \rightleftharpoons Na^+(aq) + Cl^-(aq)$$

In this situation the undissolved sodium chloride is in equilibrium with the dissociated ions. There is constant interchange of ions between the solid and the solution, but the amount of undissolved sodium chloride remains the same. The rate at which ions are dissolving is equal to the rate at which other ions are precipitated.

In bromine water, familiar from tests for unsaturation, there is the following equilibrium:

$$Br_2(l) + H_2O(l) \rightleftharpoons Br^-(aq) + BrO^-(aq) + 2H^+(aq)$$

In each of the above, when the solution is being made up, initially only the forward reactions occur.

$$NaCl(s) \rightarrow Na^+(aq) + Cl^-(aq)$$

$$Br_2(l) + H_2O(l) \rightarrow Br^-(aq) + BrO^-(aq) + 2H^+(aq)$$

The rates of these forward reactions will be related to the high initial concentrations of the solutes. As the reactions proceed, products will be formed. These products form the reactants for the reverse reactions. The reverse reactions can now start. These reactions are initially slow to proceed as *their* reactants are still only in low concentration, i.e.:

$$Na^+(aq) + Cl^-(aq) \rightarrow NaCl(s)$$

$$Br^-(aq) + BrO^-(aq) + 2H^+(aq) \rightarrow Br_2(l) + H_2O(l)$$

The rates of the forward reactions will decrease as their reactants are consumed, and the rates of the reverse reactions will increase as their reactants increase in concentration.

Eventually the rates of the forward and reverse reactions will be the same, and equilibrium is attained.

It is important to realise, however, that the reaction does *not* stop when equilibrium is attained. When a saturated solution of a salt such as NaCl is formed, an equilibrium is set up in which as many ions are passing into solution as are being redeposited on the solid crystals, i.e. the rate of solution equals the rate of precipitation. At equilibrium these processes do not cease. For this reason, chemical equilibrium is described as being *dynamic*.

It is also important to note that *when equilibrium is reached this does not imply that the equilibrium mixture consists of 50% reactants and 50% products*. This will only very rarely be the case. The actual position of equilibrium can be influenced by a number of factors as we shall see later in this chapter.

Under similar conditions, the same equilibrium can be arrived at from two different starting points. This can be shown (Figure 15.1) using the fact that iodine is soluble in trichloroethane, $C_2H_3Cl_3$, and also in aqueous potassium iodide solution. Tubes X and Z represent the two starting positions. In X iodine is dissolved in $C_2H_3Cl_3$ only and in Z it is dissolved in KI

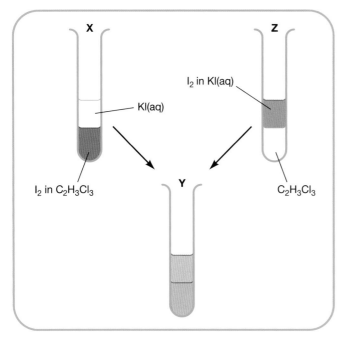

X

KI(aq)

I_2 in $C_2H_3Cl_3$

Z

I_2 in KI(aq)

$C_2H_3Cl_3$

Y

Figure 15.1 Equilibrium can be reached from either direction

solution only. Tube Y represents the equilibrium mixture which is obtained. Equilibrium can be attained quickly by shaking the tubes or slowly by allowing them to stand.

Many reactions appear to be irreversible, e.g. neutralisation of a strong acid and strong base or precipitation such as:

$$Ag^+(aq) + Cl^-(aq) \rightarrow AgCl(s)$$

However, all reactions are, in fact, reversible to at least some extent. We can say that, in such reactions, the equilibrium lies so far in one direction that for practical purposes they can be considered as having gone to completion.

Changing the position of equilibrium

We shall now consider the factors which can alter the position of equilibrium in a reversible reaction. Equilibrium is reached when the opposing reactions occur at an equal rate. Hence, we should expect that any condition which changes the rate of one reaction more than the other should change the position of equilibrium, i.e. the relative proportions of reactants and products in the mixture.

This section deals with the influence of changing the concentration, the pressure and the temperature on the equilibrium position. The effect of these changes can be summarised by Le Chatelier's Principle, which states that:

> **If a system at equilibrium is subjected to any change, the system readjusts itself to counteract the applied change.**

Note that this statement *only* refers to reversible reactions which have reached equilibrium.

Changing the concentration

Let us consider the following reaction at equilibrium:

$$A + B \rightleftharpoons C + D$$

An increase in the concentration of A (or B) will speed up the forward reaction, thus increasing the concentration of C and D until a new equilibrium is obtained. A similar effect can be achieved by reducing the concentration of C (or D) in some way. These results agree with Le Chatelier's Principle, since the equilibrium has moved to the right to counteract the applied change. The following reactions can be used to illustrate these points.

Bromine water

Bromine dissolves in water forming a red-brown solution which contains a mixture of Br_2 molecules (responsible for the colour), H_2O molecules, H^+, Br^- and BrO^- ions, as shown in the equation:

$$Br_2(l) + H_2O(l) \rightleftharpoons Br^-(aq) + BrO^-(aq) + 2H^+(aq)$$

The equilibrium can be adjusted as shown in Figure 15.2. The addition of OH^- ions removes H^+ ions to form water, and the equilibrium shifts to the right. Adding H^+ ions moves the equilibrium back to the left.

Iron(III) ions + thiocyanate ions (CNS⁻)

When separate solutions containing iron (III) ions and thiocyanate ions, respectively, are mixed, a deep blood-red solution is formed due to the presence of complex ions such as $[Fe(CNS)]^{2+}$. This reaction is reversible:

$$Fe^{3+}(aq) + CNS^-(aq) \rightarrow [Fe(CNS)]^{2+}(aq)$$

pale yellow   colourless   red

163

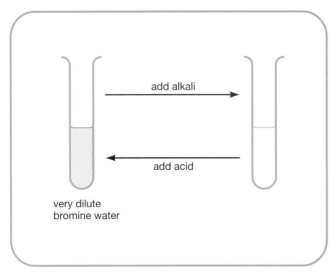

Figure 15.2 Changing the position of an equilibrium using bromine water

Question

1   When chlorine is dissolved in water the following equilibrium is set up:

$$Cl_2 + H_2O \rightleftharpoons 2H^+ + ClO^- + Cl^-$$

The hypochlorite ion, ClO^-, is responsible for the bleaching action of the solution. What effect would each of the following have on the bleaching action?

a) Adding dilute nitric acid.

b) Adding sodium chloride crystals.

c) Adding potassium hydroxide solution.

Tip: Consider the effect of the *ions* in each solution.

The intensity of the colour may be taken as an indication of the position of equilibrium. Some of the blood-red solution is diluted until an orange colour is obtained, and this solution is poured into four test tubes until each is half-full. Tube A is kept for reference; crystals of iron(III) chloride, potassium thiocyanate and sodium chloride are added to tubes B, C and D, respectively. The results are shown in Figure 15.3.

An increase in the concentration of Fe^{3+} ions or CNS^- ions shifts the equilibrium to the right and results in the formation of more of the complex ions. The addition of NaCl removes Fe^{3+} ions due to complex formation with Cl^- ions and the equilibrium shifts to the left to compensate.

Changing the temperature

In a reversible reaction, if the forward reaction is exothermic, then the reverse reaction must, of course, be endothermic. If a system at equilibrium is subjected to a change in temperature, the equilibrium position will adjust itself to counteract the applied change, according to Le Chatelier's Principle. Thus, a rise in temperature will favour the reaction which absorbs heat, i.e. the endothermic process, and a fall in temperature will favour the exothermic reaction. This can be seen in the following example.

Nitrogen dioxide

Brown fumes of nitrogen dioxide are formed when most metal nitrates are decomposed thermally or when copper is added to concentrated nitric acid. The gas

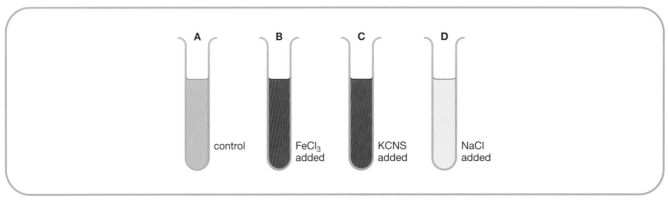

Figure 15.3 Changing the position of an equilibrium using iron(III) and thiocyanate ions

produced is, in fact, an equilibrium mixture of nitrogen dioxide, NO_2 (a dark brown gas) and dinitrogen tetroxide, N_2O_4 (a colourless gas). This is represented in the following equation. The forward reaction is endothermic.

$$N_2O_4(g) \rightleftharpoons 2NO_2(g)$$
colourless   dark brown

Figure 15.4 illustrates the results obtained on subjecting samples of this gas mixture to different temperature conditions. An increase in temperature favours the endothermic reaction and so the proportion of NO_2 increases and the gas mixture becomes darker in colour. A drop in temperature favours the exothermic reaction and, hence, the gas mixture lightens in colour.

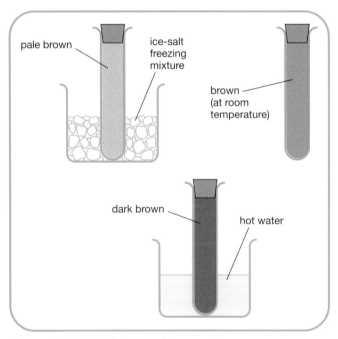

Figure 15.4 Equilibrium and temperature

Question

2 The Contact Process to make sulphuric acid involves the reaction:

$$2SO_2(g) + O_2(g) \rightleftharpoons 2SO_3(g) \qquad \Delta H = -197kJ$$

What would be the effect of raising the reaction temperature on

a) rate of reaction

b) yield of SO_3?

Changing the pressure

The pressure exerted by a gas is caused by the freely moving molecules bombarding the walls of the containing vessel. An increase in the number of molecules will be accompanied by an increase in pressure, the size of the container being kept constant. The effect of changes in pressure on a system involving gases is equivalent to the effect of changes in concentration on a system in solution.

The N_2O_4–NO_2 system is a suitable example to study in this connection.

$$N_2O_4(g) \rightleftharpoons 2NO_2(g)$$
1 mole   2 moles
1 volume   2 volumes (at the same T & P)

An increase in pressure will cause the system to readjust to counteract this effect, i.e. it will reduce the pressure within the system. Thus, the equilibrium will adjust to the left, forming more N_2O_4 molecules and reducing the number of molecules per unit volume. A suitable apparatus for the study of this effect is shown in Figure 15.5. The results are shown in Table 15.1.

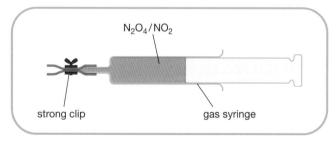

Figure 15.5 Equilibrium and pressure

Applied pressure change	Initial colour change	Final colour change
Increase (plunger in)	Darkens due to compression	Lightens as equilibrium shifts to the left
Decrease (plunger out)	Lightens due to expansion	Darkens as equilibrium shifts to the right

Authors' note: Teachers may wish to consult *School Science Review* 1978, 211, **60**: 309 for an alternative explanation of the changes observed in the NO_2–N_2O_4 system.

Table 15.1

Generally in a reversible reaction involving a gas or gases at equilibrium, an increase in pressure will cause the equilibrium to shift in the direction which results in a decrease in the number of gaseous molecules. In a system in which there is no overall change in the total number of gaseous molecules, changes in pressure will have no effect on the equilibrium position, e.g.

$$CO(g) \ + \ H_2O(g) \ \rightleftharpoons \ CO_2(g) \ + \ H_2(g)$$

1 mole   1 mole   1 mole   1 mole

Question

3 Reaction 1: $H_2(g) + I_2(g) \rightleftharpoons 2HI(g)$

Reaction 2: $2CO(g) + O_2(g) \rightleftharpoons 2CO_2(g)$

Reaction 3: $CH_3OH(g) \rightleftharpoons CO(g) + 2H_2(g)$

In which of the above reactions will an increase of pressure

a) shift the position of equilibrium to the right

b) have no effect on the equilibrium position?

Catalysts and equilibrium

A catalyst speeds up a reaction by lowering the activation energy. However, in a reversible reaction it reduces the activation energy for both the forward and reverse reactions by the same amount, as shown in Figure 15.6.

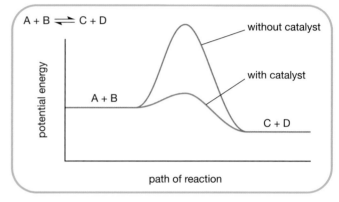

Figure 15.6 Potential energy: catalysed and uncatalysed reactions

Thus, a catalyst speeds up both the reactions to the same extent and does not alter the position of equilibrium. The use of a catalyst does not result in an increased yield of product. The advantage of using a catalyst in a reversible reaction is that it enables equilibrium to be reached more rapidly.

Summary

Change applied	Effect on equilibrium position
Concentration Addition of reactant or removal of product Addition of product or removal of reactant	Equilibrium shifts to the right Equilibrium shifts to the left
Temperature Increase Decrease	Shifts in direction of endothermic reaction Shifts in direction of exothermic reaction
Pressure Increase Decrease	Shifts in direction which reduces the number of molecules in gas phase Shifts in direction which increases the number of molecules in gas phase
Catalyst	No effect on equilibrium position; equilibrium more rapidly attained

Table 15.2

Questions

4 Propyl ethanoate is made by the reaction of ethanoic acid with propanol.

$$CH_3COOH + C_3H_7OH \rightleftharpoons CH_3COOC_3H_7 + H_2O$$

The reaction is catalysed by H^+ ions. If the very soluble gas hydrogen chloride, HCl, is passed into the reaction mixture what would be the effect on

a) the speed of reaching equilibrium

b) the yield of ester?

5 Look back at esterification, Chapter 10 on page 104, and suggest why the laboratory preparation of esters makes use of concentrated sulphuric acid.

Equilibrium and the Haber Process

In the Haber Process to synthesise ammonia, if a closed reaction vessel is used, an equilibrium in set up:

$$N_2(g) + 3H_2(g) \rightleftharpoons 2NH_3(g)$$

When the rates of the forward and reverse reactions are equal, equilibrium is reached. The same equilibrium is eventually reached whether starting from the nitrogen and hydrogen ('reactant') side or from the ammonia ('product') side. This is a general rule with any reversible reaction.

At equilibrium, the concentrations of reactants and products remain constant, but not necessarily equal. The relative concentrations of reactants and products can be affected by altering the conditions under which the reaction is taking place. This is illustrated in Table 15.3 using the reaction:

$$N_2(g) + 3H_2(g) \rightleftharpoons 2NH_3(g) \qquad \Delta H = -92\,kJ$$

In most industrial situations a continuous process is used rather than a small-scale 'batch' process. This means that a true equilibrium, which requires a closed system, is not achieved.

In practice, as described in Chapter 13, the operating conditions for most industrial processes involve compromises amongst yield, construction and maintenance costs, catalyst life and catalyst activity. The following illustrates some of these considerations for the Haber Process:

$$N_2(g) + 3H_2(g) \rightleftharpoons 2NH_3(g) \qquad \text{forward reaction}$$
$$\text{4 vol} \qquad\qquad \text{2 vol} \qquad \text{is exothermic}$$

The conditions for maximum yield are low temperature and high pressure. Figure 15.7 shows the percentage of NH_3 in equilibrium when reacting nitrogen and hydrogen in a $1:3$ mixture by volume at different temperature and pressures.

New conditions	Equilibrium change	Explanation
Increase pressure	To right, $[NH_3]$ increases	Forward direction involves a decrease in number of moles of gas (4 moles → 2 moles) and hence a decrease in volume. Decrease in volume is assisted by increase in pressure
Decrease pressure	To left, $[NH_3]$ decreases	Reverse direction involves an increase in number of moles of gas and hence in volume, assisted by reduced pressure
Increase temperature	To left, $[NH_3]$ decreases	Increasing temperature involves increasing energy of system. The reverse reaction is endothermic and will be assisted by providing energy
Decrease temperature	To right, $[NH_3]$ increases	Decreasing temperature removes energy from system, making reverse endothermic reaction less favourable, less NH_3 split up
Catalyst	No change	Both forward and reverse reactions are accelerated. The equilibrium is reached more rapidly

Table 15.3   [ ] is a recognised abbreviation for 'concentration of' e.g. $[NH_3]$ means concentration of ammonia.

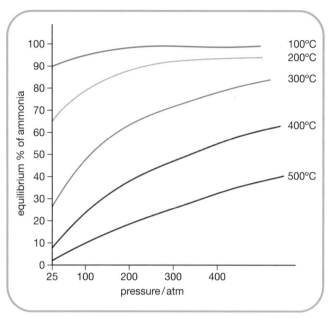

Figure 15.7 Haber Process: variation of yield with pressure and temperature

The lower the temperature is, the higher the percentage of NH_3, but the slower the reaction. The higher the pressure is, the higher the percentage of ammonia, but the greater the cost of equipment both in outlay and maintenance.

Commonly used operating conditions are: 380–450°C; high pressure, up to 200 atmospheres; iron catalyst with potassium hydroxide promoter. About 15% ammonia is produced in each pass over the catalyst. The ammonia is condensed on cooling and unreacted nitrogen and hydrogen are topped up and recycled, resulting in an overall maximum yield of 98%.

Study Questions

In questions 1–4 choose the correct word from the following list to complete the sentence.

> constant   decreases   endothermic   equal
> exothermic   left   increases   right

1 When a reversible reaction has reached equilibrium, the concentrations of reactants and products are _____.

2 Increasing the concentration of a product causes the equilibrium to move to the _____.

3 In the Haber process high pressure favours the production of ammonia since the number of moles of gas _____ going from reactants to product.

4 Raising the temperature shifts the equilibrium position in the direction of the _____ reaction.

5 $Cl_2(g) + H_2O(l) \rightleftharpoons Cl^-(aq) + ClO^-(aq) + 2H^+(aq)$

Which of the following substances is **least** likely to alter the position of equilibrium when added to an aqueous solution of chlorine?

A NaOH(aq)   B NaBr(s)   C HCl(aq)   D KCl(s)

6     $HCOOH + CH_3OH \rightleftharpoons HCOOCH_3 + H_2O$

The use of a catalyst in the reaction shown above will

A increase the equilibrium concentration of the ester

B decrease the time taken to reach equilibrium

C increase the activation energy

D decrease the enthalpy change.

7 The following equation shows one of the reactions involved in obtaining the reactants for the Haber process from natural gas and air.

$$CO(g) + H_2O(g) \rightleftharpoons CO_2(g) + H_2(g)$$

The forward reaction is exothermic.

The equilibrium can be moved to the right by

A increasing the pressure

B decreasing the pressure

C increasing the temperature

D decreasing the temperature.

Study Questions (continued)

8
$$SO_2(g) + \tfrac{1}{2} O_2(g) \rightleftharpoons SO_3(g)$$

$$\Delta H_{forward} = -94 \, kJ \, mol^{-1}$$

The ideal conditions for producing a high yield of sulphur trioxide are

A high pressure and low temperature

B low pressure and high temperature

C high pressure and high temperature

D low pressure and low temperature.

9* Ammonia is manufactured in industry by the reaction of nitrogen with hydrogen.

$$N_2(g) + 3H_2(g) \rightleftharpoons 2NH_3(g) \quad \Delta H = -92 \, kJ \, mol^{-1}$$

Typical conditions for an ammonia plant are:

Pressure	250 atmospheres
Temperature	380–450°C
Catalyst	Iron containing promoters to stop catalyst poisoning
Conversion	15% by volume of ammonia

a) Name the industrial process used to manufacture ammonia.

b) What is meant by catalyst poisoning?

c) Explain what would be expected to happen to the percentage conversion if the temperature of the ammonia plant was decreased.

d) High pressure favours conversion to ammonia. Suggest why pressures higher than 250 atmospheres are not used.

10* Synthesis gas, a mixture of hydrogen and carbon monoxide, is prepared as shown below. Nickel is known to catalyse the reaction.

$$CH_4(g) + H_2O(g) \rightleftharpoons 3H_2(g) + CO(g)$$

a) An increase in temperature increases the yield of synthesis gas. What information does this give about the enthalpy change in the forward reaction?

b) Using Le Chatelier's Principle explain how a change in pressure will affect the composition of the equilibrium mixture.

c) State how the rate of formation of synthesis gas will be affected by the use of the catalyst.

d) State how the composition of the equilibrium mixture will be affected by the use of the catalyst.

11 When bismuth(III) chloride is dissolved in water, an equilibrium is set up as shown in the following equation:

$$BiCl_3(aq) + H_2O(l) \rightleftharpoons BiOCl(s) + 2HCl(aq)$$

a) What effect, if any, will there be on the position of equilibrium on adding

i) potassium hydroxide,

ii) nitric acid,

iii) potassium nitrate?

b) Which of the reagents named in a), if added in sufficient quantity, would dissolve the precipitate?

12* Ammonia is now one of the world's most important chemicals, about two million tones being produced each year in the UK alone. It is manufactured by the direct combination of nitrogen and hydrogen by the Haber Process.

$$N_2(g) + 3H_2(g) \rightleftharpoons 2NH_3(g)$$

The graph shows how the percentage of ammonia in the gas mixture at equilibrium varies with pressure at different temperatures.

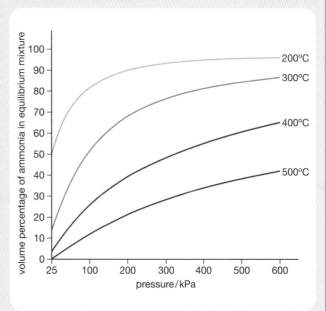

a) What does the term 'at equilibrium' mean?

Study Questions (continued)

b) Use the graph *and* the chemical equation to explain the conclusion that the reaction is exothermic.

c) i) Typical conditions for the Haber Process are approximately 400°C and 200 kPa.

Use the graph to estimate the percentage of ammonia which could be obtained if the mixture was left until equilibrium was reached at this temperature and pressure.

ii) In practice, the percentage of ammonia in the gas mixture never rises above 15%. Although the yield is low, the process is still profitable, Give *one* reason for this fact.

13* The graph shows the concentrations of reactant and product as equilibrium is established in a reaction.

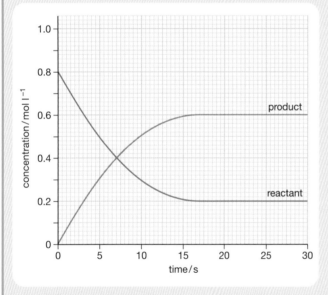

a) Calculate the average rate of reaction over the first 10s. (Show your working clearly.)

b) The equilibrium constant for a reaction is given the symbol ***K***. In this reaction ***K*** is given by:

$$K = \frac{\text{equilibrium concentration of product}}{\text{equilibrium concentration of reactant}}$$

Calculate the value of ***K*** for this reaction.

c) The reaction is repeated using a homogeneous catalyst.

i) What is meant by a *homogeneous* catalyst?

ii) What effect would the introduction of the catalyst have on the value of ***K***?

14* Ammonia is made by the Haber Process.

$$N_2(g) + 3H_2(g) \rightleftharpoons 2NH_3(g)$$

a) The Haber Process is normally carried out at 200 atmospheres pressure. Suggest *one* advantage and *one* disadvantage of increasing the pressure in the Haber Process beyond 200 atmospheres.

b) The activation energy (E_A) and enthalpy change (ΔH) for this reaction are 236 kJ mol^{-1} and -92 kJ mol^{-1}, respectively.

i) Use this information to draw the potential energy diagram for the reaction on axes like those given below.

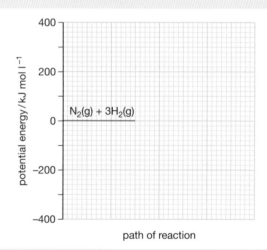

ii) Calculate the activation energy for the reverse reaction.

c) Over a period of time, 120 tonnes of hydrogen produced 88.4 tonnes of ammonia by the Haber Process. Calculate the percentage yield of ammonia. (1 tonne = 1000 kg)

Note to candidates: This question has been included here as an example of a past paper question which tests more than one topic – in this case, potential energy diagram from unit 1, percentage yield from unit 2 and equilibrium from unit 3. Several questions in Part 2 of the End of Course section beginning on page 211 test more than one topic at a time.

16 Acids and Bases

From previous work you should know and understand the following:

★ Acids and alkalis, the importance of H^+ and OH^- ions and the pH scale

★ Neutralisation and other reactions of acids

★ How to carry out calculations involving the mole and concentration.

The pH scale

The concentration of hydrogen ions in solution is measured in pH units. pH stands for the negative logarithm (to base 10) of the hydrogen ion concentration, i.e. $pH = -\log_{10}[H^+(aq)]$.

The symbol 'M' is often used as an abbreviation for moles l^{-1}, especially for prepared solutions. Square brackets [ ] denote concentrations of the relevant substance in mol l^{-1}.

Note: 'M' is not an abbreviation for mole or moles. The correct abbreviation for either of these is mol.

The pH scale is a continuous scale with values ranging from less than zero to more than 14. (A concentration of more than 1 mol l^{-1} of H^+ ions will give a positive $\log_{10}[H^+]$ and hence a negative pH value.) The scale is limited by the maximum concentration of H^+ and OH^- ions that can be achieved owing to the finite solubilities of acids and alkalis and the degree to which they dissociate into ions.

Figure 16.1 Using a lap-top as a pH meter

Integral values of pH can be related to hydrogen ion concentrations $[H^+]$ as in Table 16.1.

$[H^+]$ mol l^{-1}	$\log_{10}[H^+]$	$pH(-\log_{10}[H^+])$
1 (1 M)	0	0
1/10 (10^{-1} M)	-1	1
1/100 (10^{-2} M)	-2	2
1/1000 (10^{-3} M)	-3	3
1/10^7 (10^{-7} M)	-7	7
1/10^{14} (10^{-14} M)	-14	14

Table 16.1

For water and neutral solutions in water, the sole source of H^+ and OH^- ions is the very slight ionisation of some of the water molecules:

$$H_2O\ (l) \rightleftharpoons H^+(aq) + OH^-\ (aq)$$

An equilibrium is reached with the concentration of both H^+ and OH^- equal to 10^{-7} mol l^{-1} at 25°C.

i.e. $[H^+] = [OH^-] = 10^{-7}$ mol l^{-1}, pH = 7

then $[H^+][OH^-] = 10^{-7}$ mol $l^{-1} \times 10^{-7}$ mol l^{-1}

$= 10^{-14}$ mol^2 l^{-2}

This value is called the **ionic product of water**.

The equilibrium for the ionisation of water always exists in aqueous solutions. If the solution is acidic, i.e. there is an excess of H^+ ions, then the equilibrium adjusts to the left, with some of the excess H^+ ions combining with some of the OH^- ions from water. The ionic product remains 10^{-14} mol^2 l^{-2}. Similarly, excess OH^- ions in an alkaline solution combine with H^+

ions from water, maintaining the ionic product at 10^{-14} $mol^2 \, l^{-2}$.

It follows that we can calculate $[H^+]$, pH, $[OH^-]$ for acidic and alkaline solutions using the basic relationship:

$$[H^+] \, [OH^-] = 10^{-14} \, mol^2 \, l^{-2}$$

From which we derive:

$$[H^+] = \frac{10^{-14}}{[OH^-]} \, mol \, l^{-1}$$

and

$$[OH^-] = \frac{10^{-14}}{[H^+]} \, mol \, l^{-1}$$

Worked Example 16.1

What is the concentration of OH^- ions in a solution containing $0.01 \, mol \, l^{-1}$ of H^+ ions?

$$[H^+] = 10^{-2} \, mol \, l^{-1}$$

$$[OH^-] = \frac{10^{-14}}{[H^+]} = \frac{10^{-14}}{10^{-2}} \, mol \, l^{-1} = 10^{-12} \, mol \, l^{-1}$$

Worked Example 16.2

What is the concentration of H^+ ions in a solution containing $0.1 \, mol \, l^{-1}$ of OH^- ions?

$$[OH^-] = 10^{-1} \, moles \, l^{-1}$$

$$[H^+] = \frac{10^{-14}}{[OH^-]} = \frac{10^{-14}}{10^{-1}} = 10^{-13} \, mol \, l^{-1}$$

Worked Example 16.3

What is the pH of a solution containing $0.1 \, mol \, l^{-1}$ of OH^- ions?

As in Worked Example 16.2 above,

$$[H^+] = 10^{-13} \, mol \, l^{-1}$$

$$pH = -\log_{10}[H^+] = 13$$

Questions

1 What is the concentration of OH^- ions in a solution where $[H^+]$ is $10^{-3} \, mol \, l^{-1}$?

2 What is the pH of solutions which have **a)** $[H^+] = 10^{-4} \, mol \, l^{-1}$, **b)** $[OH^-] = 10^{-5} \, mol \, l^{-1}$?

Strong and weak acids

Strong acids are acids that dissociate completely into ions in solution, e.g.

$$HCl(aq) \rightarrow H^+(aq) + Cl^-(aq)$$

Strong acids include HCl, HNO_3 and H_2SO_4. For the first two the concentration of hydrogen ions is the same as the nominal concentration of the acid solution e.g. a $0.1 \, M$ solution of HCl has $[H^+] = 0.1 \, mol \, l^{-1}$. For H_2SO_4, a $0.1 \, M$ solution has $[H^+] = 0.2 \, mol \, l^{-1}$ since:

$$H_2SO_4(aq) \rightarrow 2H^+(aq) + SO_4^{2-}(aq)$$

Other acids dissociate only partially into ions in solution, i.e. an equilibrium exists between the ions and undissociated molecules. For example, for ethanoic acid:

$$CH_3COOH(aq) \rightleftharpoons CH_3COO^-(aq) + H^+(aq)$$

The equilibrium will move to the right with increasing dilution, i.e. dissociation is greater, or the degree of ionisation is greater with increasing dilution. Such incompletely dissociated acids are called **weak acids**.

(*Note to candidates:* 'Strong' and 'weak' refer to the inherent ability of acids to ionise. These words should not be used to refer to differences in concentration of solutions where 'concentrated' and 'dilute' are the only permissible terms.)

Solutions of strong and weak acids differ in pH, conductivity and reaction rates since their hydrogen ion concentrations are different. Solutions should be equimolar for a fair comparison.

For $0.1 \, M$ HCl and $0.1 \, M$ CH_3COOH, these differences can be summed up in Table 16.2.

It should be clear, however, since the undissociated molecules of weak acids are in equilibrium with their

	0.1 M HCl	0.1 M CH₃COOH
[H⁺]	0.1 mol l⁻¹	0.0013 mol l⁻¹
pH	1	2.88
Conductivity	high	low
Rate of reaction with Mg	fast	slow
Rate of reaction with CaCO₃	fast	slow

Table 16.2

ions, that as a reaction proceeds consuming H^+ ions, the equilibrium shifts in favour of more dissociation until eventually all the molecules are dissociated. Thus eventually the same amount of base is required to neutralise a certain volume of either 0.1 M HCl or 0.1 M CH_3COOH. Hence the **stoichiometry** (the mole ratio of reactants) of a neutralisation reaction is the same:

$$NaOH + HCl \rightarrow NaCl + H_2O$$
$$NaOH + CH_3COOH \rightarrow CH_3COONa + H_2O$$
$$1 \text{ mole acid} \equiv 1 \text{ mole alkali in each case.}$$

Questions

3 What is the pH of a 0.01 mol l⁻¹ solution of HCl? How does the pH of a solution of the same concentration of CH₃COOH compare?

4 How many moles of NaOH are needed to neutralise
a) 25 cm³ of 0.1 mol l⁻¹ HCl,
b) 20 cm³ of 0.1 mol l⁻¹ CH₃COOH,
c) 10 cm³ of 0.1 mol l⁻¹ H₂SO₄?

The influence of structure on acid strengths

The most common weak acids are the carboxylic acids containing the group:

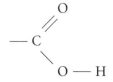

One example is ethanoic acid, with the following structure:

$$CH_3 - C \overset{\displaystyle O}{\underset{\displaystyle O - H}{<}}$$

These compounds ionise to a limited degree by the separation of a hydrogen ion from the carboxyl group:

$$-C \overset{\displaystyle O}{\underset{\displaystyle O-H}{<}} \rightleftharpoons -C \overset{\displaystyle O}{\underset{\displaystyle O^-(aq)}{<}} + H^+(aq)$$

The ionisation is assisted by the polarisation of the covalent bonds within the carboxyl groups as shown in Figure 16.2.

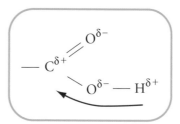

Figure 16.2

This results in electrons being pulled away from the hydrogen in the direction of the arrow, so weakening the O–H bond which then splits unevenly to leave H^+ ions.

Many similar weak acids are present in fruit juices and are responsible for their characteristic sour taste owing to the moderate concentration of hydrogen ions. Typical examples are shown in Table 16.3.

Weak inorganic acids which you may have come across earlier are derived from carbon dioxide and sulphur dioxide on solution in water.

Name of acid	Occurrence	Structural formula
Citric acid	In oranges and lemons	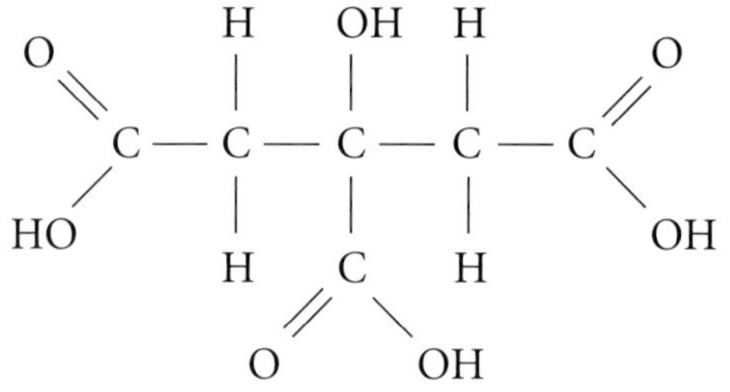
Malic acid	In apple juice	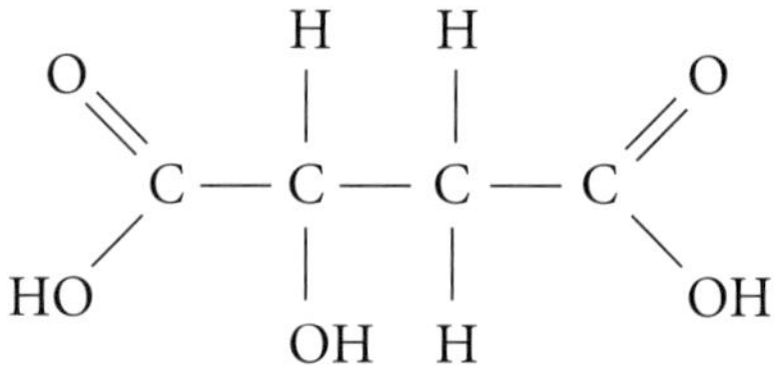
Oxalic acid	In rhubarb (mainly in leaves)	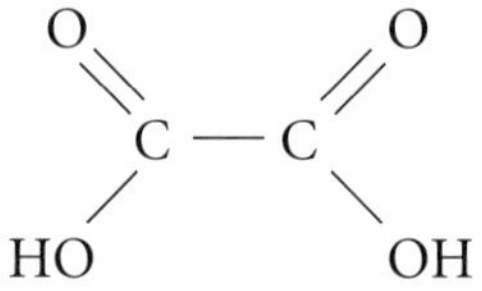

Table 16.3

$$CO_2(g) + H_2O(l) \rightleftharpoons H_2CO_3(aq) \rightleftharpoons H^+(aq) + HCO_3^-(aq) \rightleftharpoons 2H^+(aq) + CO_3^{2-}(aq)$$

carbonic acid   bicarbonate or   carbonate ion

hydrogencarbonate

ion

Carbonic acid is present in all rainfall and is responsible for the erosion of limestone on which it falls. This gives 'hard water', which is a solution of calcium hydrogencarbonate.

$$SO_2(g) + H_2O(l) \rightleftharpoons H_2SO_3(aq) \rightleftharpoons H^+(aq) + HSO_3^-(aq) \rightleftharpoons 2H^+(aq) + SO_3^{2-}(aq)$$

sulphurous acid   bisulphite or   sulphite ion

hydrogensulphite

ion

Sulphurous acid is one of the major compounds present in 'acid rain' and is rather stronger than carbonic acid, although fortunately weaker than sulphuric acid. It is strong enough to accelerate metal corrosion and to erode limestone much faster than carbonic acid.

Strong and weak bases

A base is any substance which will neutralise the H^+ ions of an acid to form water. Bases which are soluble in water, producing OH^- ions, are alkalis. Alkalis which are fully dissociated into ions are strong alkalis, those showing incomplete dissociation are weak alkalis. Strong alkalis include NaOH and KOH, weak alkalis include ammonia and amine solutions, e.g.

$$NaOH(aq) \rightarrow Na^+(aq) + OH^-(aq)$$
$$NH_3(aq) + H_2O(l) \rightleftharpoons NH_4^+(aq) + OH^-(aq)$$

An equilibrium, normally well to the left, is set up in the ammonia solution resulting in a low concentration of OH^- ions and the quite low pH of a weak alkali. A comparison of the properties of solutions of strong and weak alkalis, at the same molarity for fairness, is given in Table 16.4.

	0.1 M NaOH(aq)	0.1 M NH₃(aq)
[OH⁻]	0.1 mol l⁻¹	0.0013 mol l⁻¹
pH	13	11.12
Conductivity	high	low

Table 16.4

Note that for strong alkalis the nominal concentration of the solution matches $[OH^-]$ since there is complete ionisation.

The stoichiometry of neutralisations of strong and weak alkalis is the same, e.g.

$$NaOH + HCl \rightarrow NaCl + H_2O$$
$$NH_3 + HCl \rightarrow NH_4Cl$$

1 mole alkali $\equiv$ 1 mole acid in each case

The mechanism of the ionisation of ammonia is shown in Figure 16.3.

The weak alkali dissociates further as its OH^- ions react with acids. Eventually it will dissociate fully so that a certain volume of either 0.1 M NaOH or 0.1 M NH_3 solution will neutralise the same number of moles of acid.

> ## Question
>
> 5 How many moles of HCl will be needed to neutralise **a)** 20 cm³ of 0.1 mol l⁻¹ NaOH solution **b)** 10 cm³ of 5 mol l⁻¹ NH₃ solution?

pH of salt solutions

It is normally assumed that since salts can be made by neutralisation of an acid by an alkali, the pH will be 'neutral' i.e. 7. In fact, the measured pH of salts is often not 7.

In general terms, in solution:
Salts of strong acids and strong alkalis have pH 7
Salts of strong acids and weak alkalis have pH < 7
Salts of strong alkalis and weak acids have pH > 7

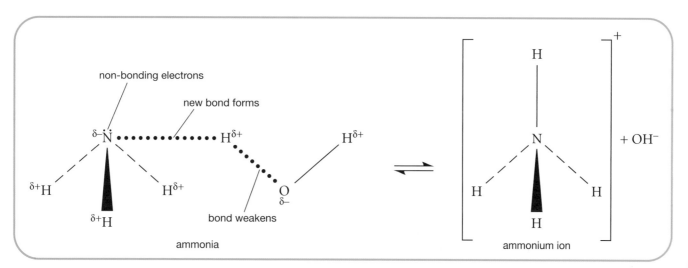

Figure 16.3 Ionisation of ammonia in water

For example:

♦ Sodium chloride and potassium nitrate are made from strong acids, HCl and HNO_3, and strong alkalis, NaOH and KOH. They have pH 7 in solution.

♦ Ammonium chloride is made from the strong acid HCl and the weak alkali NH_3 and has a pH of about 5 in solution.

♦ Sodium ethanoate is made from the strong alkali NaOH and the weak acid CH_3COOH and has a pH about 9 in solution.

The last two results can be explained as follows.

Ammonium chloride in water

Ions present initially:

$NH_4^+ + Cl^-$ from the salt (fully ionised)

$H^+ + OH^-$ from the water (a few only)

In the solution the NH_4^+ and OH^- ions' concentrations exceed those normally found in the weak alkali ammonia solution, and the ions form undissociated ammonia:

$$NH_4^+(aq) + OH^-(aq) \rightarrow NH_3(aq) + H_2O(l)$$

Removal of OH^- ions causes the water equilibrium to move to the right:

$$H_2O(l) \rightarrow H^+(aq) + OH^-(aq)$$

Excess H^+ ions are formed and the solution has pH less than 7.

Similar explanations apply to the chlorides, sulphates and nitrates of NH_4^+ and of metals not in groups I and II of the Periodic Table.

Sodium ethanoate in water

Ions present initially:

$Na^+ + CH_3COO^-$ from the salt (fully ionised)

$H^+ + OH^-$ from the water (a few only)

In the solution the CH_3COO^- and H^+ ions' concentrations exceed those normally found in the weak acid CH_3COOH, and the ions combine:

$$CH_3COO^-(aq) + H^+(aq) \rightarrow CH_3COOH(aq)$$

Removal of H^+ ions causes the water equilibrium to be disturbed:

$$H_2O \rightarrow H^+(aq) + OH^-(aq)$$

Excess OH^- ions are formed, and the solution has pH greater than 7.

Similar explanations apply to sodium and potassium salts of other carboxylic acids and of carbonic and sulphurous acids.

Soaps

The structure of soaps was described in Chapter 12 on page 129. Soaps are the salts of strong alkalis like NaOH and KOH, with weak long-chain carboxylic acids such as stearic and oleic acids. Soaps, therefore, are usually alkaline in solution.

Question

6 HCN, hydrocyanic acid, is a weak acid. Suggest a pH value for a dilute solution of sodium cyanide, NaCN. Explain your suggested value in terms of the equilibria existing in the solution.

Study Questions

In questions 1–4 choose the correct word from the following list to complete the sentence.

> faster   fully   higher   lower   partially
> slower   strong   weak

1   When comparing rates of reaction, dilute ethanoic acid reacts _____ than equimolar hydrochloric acid when added to calcium carbonate.

2   Ammonia solution is a weak base as it is _____ dissociated into ions.

3   Salts which form aqueous solutions in which $[H^+] = [OH^-]$, are made from a _____ acid and a strong base.

4   Dilute hydrochloric acid has a _____ pH than equimolar ethanoic acid.

5   What is the pH of a solution which has a $[H^+] = 0.0001$ mol l^{-1}?

    A 3   B 4   C 9   D 10

6   What is the concentration of OH^- ions, in mol l^{-1}, in a solution which has a pH of 6?

    A 1×10^{-4} B 1×10^{-6}
    C 1×10^{-8} D 1×10^{-10}

7   Excess magnesium ribbon was added to equal volumes of 0.5 mol l^{-1} hydrochloric acid and 0.5 mol l^{-1} ethanoic acid.

    Which statement is true about this experiment?

    A   Ethanoic acid is neutralised more quickly than hydrochloric acid.

    B   More magnesium reacts with hydrochloric acid than with ethanoic acid.

    C   The two reactions reach completion at the same time.

    D   The same mass of magnesium is used up in each reaction.

8   Which of the following compounds when dissolved in water gives a solution with pH = 7?

    A   Sodium ethanoate

    B   Potassium chloride

    C   Ammonium nitrate

    D   Potassium carbonate

9   Which solution has a $[H^+] > 1 \times 10^{-7}$ mol l^{-1}?

    A $NH_4Cl(aq)$ B $K_2SO_4(aq)$
    C $NaNO_3(aq)$ D $K_2SO_3(aq)$

10* You are asked to investigate the reaction between ethanoic acid (a weak acid) and sodium hydroxide solution. You are given the following indicator chart:

Indicator	Working pH range
	0 2 4 6 8 10 12 14
Methyl orange	
Methyl violet	
Bromocresol green	
Methyl red	
Phenolphthalein	

a) The indicator chart shows that the working range of methyl red extends to pH 6.

   i)   What is the hydrogen ion concentration in a solution of pH 6?

   ii) What is the hydroxide ion concentration in such a solution?

b) Explain why ethanoic acid is classified as a *weak* acid.

c) The working pH range of the indicator used in a titration should include the pH value of the salt solution formed in the reaction.

   i)   Name the salt formed in the reaction between ethanoic acid and sodium hydroxide.

   ii) Which of the indicators given in the chart would you use in the titration of ethanoic acid solution with sodium hydroxide solution? Explain your answer.

11*

Acid	pH of 2 M aqueous solution
A CCl_3COOH	0.50
B $CHCl_2COOH$	0.90

a) Which is the stronger acid? Explain your choice.

b) Acid A dissociates in water as follows:

$$CCl_3COOH(aq) \rightleftharpoons CCl_3COO^-(aq) + H^+(aq)$$

177

Study Questions (continued)

How would the equilibrium be affected by the addition of

i) solid NaOH

ii) solid NaCl

iii) solid CH_3COONa?

c) Explain your answer in the case of solid CH_3COONa.

12* The structural formulae for some acids containing oxygen are shown.

Acid	Strength	Structure
Carbonic	Weak	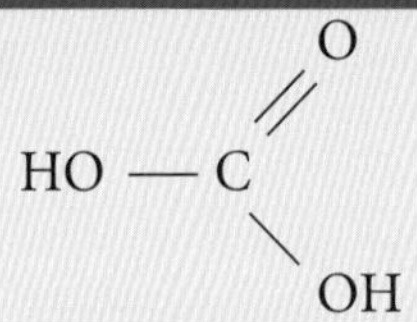 $HO - C$ with $=O$ and $-OH$
Ethanoic	Weak	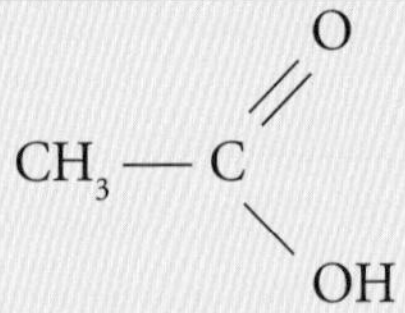 $CH_3 - C$ with $=O$ and $-OH$
Nitric	Strong	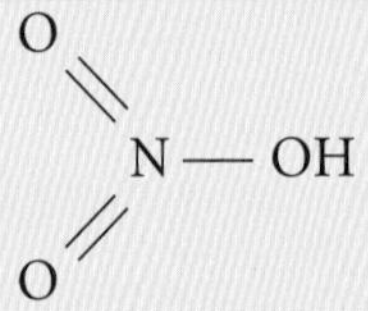 $O=N-OH$ with $=O$
Nitrous	Weak	$O=N-OH$
Sulphuric	Strong	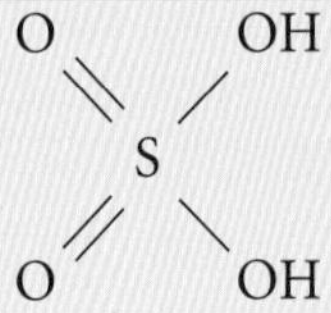 S with $=O$, $-OH$, $=O$, $-OH$
Sulphurous	Weak	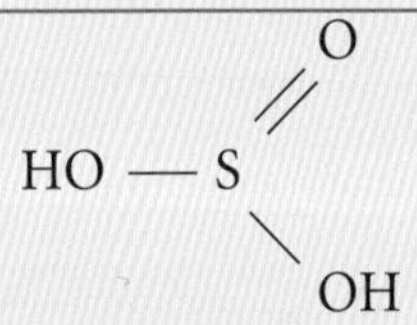 $HO-S$ with $=O$ and $-OH$

a) i) Describe *two* tests to distinguish between a weak acid and a strong acid, stating clearly the result of each test.

ii) State the *two* variables which must be controlled in *both* tests to make the tests fair.

b) What structural feature appears to determine the strength of these acids?

c) Chloric acid, $HClO_3$, is a strong acid. Draw its full structural formula.

d) Carbonic acid forms a salt, sodium carbonate. Explain why sodium carbonate solution is alkaline.

17 Redox Reactions

From previous work you should know and understand the following:

★ The processes of **oxidation** and **reduction** including reactions occurring at the electrodes during electrolysis

★ A **displacement reaction** involves oxidation and reduction and is therefore an example of a **redox reaction**

★ How to carry out volumetric titrations and calculations related to them. Question 5 on page 184 is a revision question on calculations involving acid–alkali titrations.

Oxidising agents and reducing agents

Reactions in which reduction and oxidation occur are called redox reactions. In a redox reaction electron transfer occurs between the reactants. One reactant is reduced while the other is oxidised. The box below summarises important definitions and gives examples of ion–electron equations. It also emphasises the mnemonic, OILRIG, to help you remember that oxidation involves electron loss and reduction involves electron gain.

> When a *reducing agent* reacts it loses electrons and, as a result, is itself *oxidised*, for example:
>
> $Mg \rightarrow Mg^{2+} + 2e^-$ and $Zn \rightarrow Zn^{2+} + 2e^-$
>
> **O**XIDATION **IS** **L**OSS of electrons
>
> When an *oxidising agent* reacts it gains electrons and, as a result, is itself *reduced*, for example:
>
> $Cu^{2+} + 2e^- \rightarrow Cu$ and $Ag^+ + e^- \rightarrow Ag$
>
> **R**EDUCTION **IS** **G**AIN of electrons
>
> '**O I L R I G**'

Displacement reactions

In earlier work you will have come across a type of reaction called displacement, in which one metal displaces another metal from a solution of its salt. This happens when the metal added has a greater tendency to lose electrons and form ions than the metal being displaced. In other words, the atoms of the metal added lose electrons to the ions of the metal being displaced

from solution. The relevant ion–electron equations can be combined to produce a balanced ionic equation for the overall redox reaction, as illustrated by the following two examples.

1 Magnesium displaces copper from a solution containing copper(II) ions

The following ion–electron equations show that each magnesium atom loses two electrons to a copper(II) ion. Since the number of electrons lost and gained is the same, the ion–electron equations can be added as follows to give a balanced ionic equation for the redox reaction.

Note that the redox equation does not contain electrons.

Oxidation:	$Mg(s)$	$\rightarrow Mg^{2+}(aq) + 2e^-$
Reduction:	$Cu^{2+}(aq) + 2e^-$	$\rightarrow Cu(s)$
Redox:	$Cu^{2+}(aq) + Mg(s)$	$\rightarrow Cu(s) + Mg^{2+}(aq)$

2 Zinc displaces silver from a solution containing silver(I) ions

In this example the ion–electron equations show that each zinc atom loses two electrons while each silver ion gains only one electron. Therefore, the second ion–electron equation must be doubled to balance the number of electrons lost and gained, so that the redox equation can be obtained in the same way as in the previous example.

Note that the total charge on each side of the redox equation is the same.

Oxidation:	$Zn(s)$	$\rightarrow Zn^{2+}(aq) + 2e^-$
Reduction:	$Ag^+(aq) + e^-$	$\rightarrow Ag(s) \qquad (\times 2)$
Redox:	$2Ag^+(aq) + Zn(s)$	$\rightarrow 2Ag(s) + Zn^{2+}(aq)$

In each of the above examples the negative ions present in the solutions have not been included or even referred to since they do not take part in the reaction. It is usual practice to omit such **spectator ions** from redox equations.

Question

1 For each of the displacement reactions described below write down the relevant ion–electron equations and use them to work out the redox equation. Do not include spectator ions.

a) Copper metal reacts with silver(I) nitrate solution to form copper(II) nitrate solution and silver.

b) Chromium metal reacts with nickel(II) sulphate solution to form chromium(III) sulphate solution and nickel.

Redox reactions involving oxyanions

So far we have dealt with relatively simple ion–electron equations which involve simple ions and atoms or molecules. Equations involving **oxyanions** are more complex. Oxyanions are negative ions which contain oxygen combined with another element. Examples of this type of ion include sulphite ions, SO_3^{2-}, and permanganate ions, MnO_4^-.

The following three examples of redox reactions are given to emphasise two main points, namely to show how to write:

♦ ion–electron equations which involve oxyanions

♦ balanced redox equations for more complex reactions.

State symbols have been omitted so as not to 'overload' the equations with information. **The first named is the oxidising agent in each case.**

1 **Bromine water** + **sodium sulphite solution**

Na^+ are spectator ions.

Bromine molecules are reduced to bromide ions.

$$Br_2 + 2e^- \rightarrow 2Br^-$$
$$\text{brown} \qquad\qquad \text{colourless}$$

Sulphite ions are oxidised to sulphate ions. Both ions are colourless.

$$SO_3^{2-} \rightarrow SO_4^{2-}$$

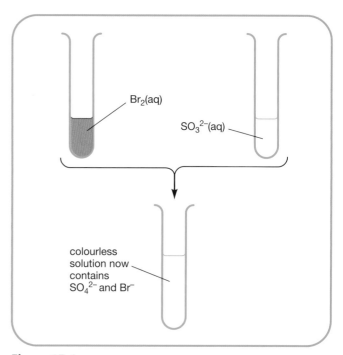

Figure 17.1

The formation of sulphate ions can be shown by testing the solution with barium chloride solution. The white precipitate of barium sulphate formed is insoluble in hydrochloric acid. Sulphite ions also form a white precipitate, $BaSO_3$, with barium chloride but it is soluble in $HCl(aq)$.

To complete the ion–electron equation for this change, firstly add H_2O to the left-hand side of the equation and $2H^+$ to the right-hand side to give

$$SO_3^{2-} + H_2O \rightarrow SO_4^{2-} + 2H^+$$

and secondly add two electrons to the right-hand side so that the charge is the same on each side of the equation giving

$$SO_3^{2-} + H_2O \rightarrow SO_4^{2-} + 2H^+ + 2e^-$$

The two ion–electron equations can now be combined to give the balanced redox equation as follows.

$$\begin{array}{l} Br_2 + 2e^- \quad\quad\quad\quad\quad \rightarrow 2Br^- \\ SO_3^{2-} + H_2O \quad\quad\quad \rightarrow SO_4^{2-} + 2H^+ + 2e^- \\ \hline Br_2 + SO_3^{2-} + H_2O \rightarrow 2Br^- + SO_4^{2-} + 2H^+ \end{array}$$

2 Acidified potassium permanganate solution + iron(II) sulphate solution

K^+ and SO_4^{2-} are spectator ions.

Iron(II) ions are oxidised to iron(III) ions.

$$Fe^{2+} \rightarrow Fe^{3+} + e^-$$

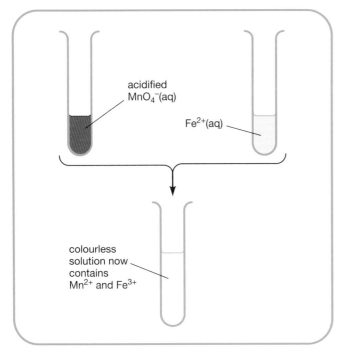

Figure 17.2

The formation of iron(III) ions can be shown by testing the solution with ammonium thiocyanate solution. This gives a dark blood-red solution in the presence of iron(III) ions.

Permanganate ions are reduced to manganese(II) ions.

$$\begin{array}{cc} MnO_4^- & \rightarrow Mn^{2+} \\ \text{purple} & \text{colourless} \end{array}$$

The ion–electron equation can be completed in a similar way to the previous example this time oxygen is being removed so that $4H_2O$ is added to the right-hand side and $8H^+$ to the left to give

$$MnO_4^- + 8H^+ \rightarrow Mn^{2+} + 4H_2O$$

Five electrons are then added to the left-hand side to balance the charge giving

$$MnO_4^- + 8H^+ + 5e^- \rightarrow Mn^{2+} + 4H_2O$$

This equation shows why the solution should be acidified.

The two ion–electron equations can now be combined to give the balanced redox equation. To balance the number of electrons lost and gained, the equation involving iron ions has to be multiplied by 5.

$$\begin{array}{l} 5Fe^{2+} \quad\quad\quad\quad\quad\quad\quad \rightarrow 5Fe^{3+} + 5e^- \\ MnO_4^- + 8H^+ + 5e^- \quad \rightarrow Mn^{2+} + 4H_2O \\ \hline MnO_4^- + 8H^+ + 5Fe^{2+} \rightarrow Mn^{2+} + 5Fe^{3+} + 4H_2O \end{array}$$

Charges:

$1-$	$8+$	$10+$	$2+$	$15+$	0
	$17+$			$17+$	

The total charge is the same on each side of the redox equation.

3 Dilute nitric acid + copper

Copper atoms are oxidised to copper(II) ions.

$$Cu \rightarrow Cu^{2+} + 2e^-$$

Nitrate ions are reduced to nitrogen monoxide, NO, a colourless gas which forms brown fumes of nitrogen dioxide in contact with air.

$$NO_3^- \rightarrow NO$$

Two oxygen atoms are being removed, so add $2H_2O$ to the right-hand side and $4H^+$ to the left.

$$NO_3^- + 4H^+ \rightarrow NO + 2H_2O$$

Finally three electrons are needed on the left-hand side to balance the charge.

$$NO_3^- + 4H^+ + 3e^- \rightarrow NO + 2H_2O$$

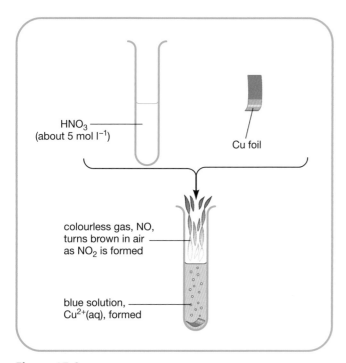

HNO₃
(about 5 mol l⁻¹)

Cu foil

colourless gas, NO, turns brown in air as NO_2 is formed

blue solution, Cu^{2+}(aq), formed

Figure 17.3

In order to write the balanced redox equation it is necessary to multiply the equation involving copper by 3 and the equation involving nitrate ions by 2.

$$3Cu \rightarrow 3Cu^{2+} + 6e^-$$
$$\underline{2NO_3^- + 8H^+ + 6e^- \rightarrow 2NO + 4H_2O}$$
$$3Cu + 2NO_3^- + 8H^+ \rightarrow 3Cu^{2+} + 2NO + 4H_2O$$

Note that the total charge $(6+)$ on each side of the redox equation is again the same.

It would appear that there are no spectator ions in this reaction but a closer look at the redox reaction shows a difference in the number of hydrogen ions and nitrate ions. Nitric acid contains equal numbers of these ions. The six extra nitrate ions not shown in the redox equation are spectator ions.

Summary

The writing of ion–electron equations involving oxyanions can be summarised as follows.

★ If a number of oxygen atoms have to be *added*, e.g.

$$SO_3^{2-} \rightarrow SO_4^{2-}$$

in example 1 (page 180), add the same number of water molecules to the left-hand side of the equation and twice that number of hydrogen ions to the right-hand side.

★ Conversely, if a number of oxygen atoms have to be *removed*, e.g.

$$MnO_4^- \rightarrow Mn^{2+}$$

in example 2 (page 181), add the same number of water molecules to the right-hand side of the equation and twice that number of hydrogen ions to the left-hand side.

★ Complete the equation by adding the number of electrons needed to balance the total charge. Add the electrons to the same side of the equation as the hydrogen ions.

Question

2 In each of the following examples complete the ion–electron equation and indicate whether it represents a reduction or an oxidation.

a) Manganese(IV) oxide, MnO_2(s), in water, changes to permanganate ions, MnO_4^-(aq).

b) Ferrate ions, FeO_4^{2-}(aq), in acidic solution are converted to iron(III) ions, Fe^{3+}(aq).

c) Vanadium(III) ions, V^{3+}(aq), are converted to vanadate ions, VO_3^-(aq).

Using the electrochemical series

Most data books provide a table known as the electrochemical series. In this table ion–electron equations are listed with electrons on the left-hand side, i.e. as reductions. In other words, each equation is written in the form:

Oxidising agent + Electron(s) → Reducing agent

When using equations from this series to write a redox equation it will always be necessary to reverse one of the equations as it appears on the table, since an oxidising agent can only react with a reducing agent and vice versa. To help you to decide which equation to reverse, remember that the starting materials, i.e. reactants, must appear on the left-hand side of the redox equation. These points are illustrated by the following examples.

1 Iron(III) chloride solution oxidises potassium iodide solution to form iron(II) ions and iodine. In this reaction Cl^- and K^+ are spectator ions. The relevant ion–electron equations obtained from the electrochemical series are shown below:

$$I_2 + 2e^- \rightarrow 2I^-$$
$$Fe^{3+} + e^- \rightarrow Fe^{2+}$$

Since the iodide ions are oxidised, the first equation must be reversed. The equation for the reduction of iron(III) ions must be doubled to balance the number of electrons transferred.

$$
\begin{array}{ll}
2I^- & \rightarrow I_2 + 2e^- \\
2Fe^{3+} + 2e^- & \rightarrow 2Fe^{2+} \\
\hline
2Fe^{3+} + 2I^- & \rightarrow 2Fe^{2+} + I_2
\end{array}
$$

2 Iron(II) sulphate solution reduces acidified potassium dichromate solution to form iron(III) ions and chromium(III) ions. In this reaction SO_4^{2-} and K^+ are spectator ions. The relevant ion–electron equations from the electrochemical series are:

$$Fe^{3+} + e^- \rightarrow Fe^{2+}$$
$$Cr_2O_7^{2-} + 14H^+ + 6e^- \rightarrow 2Cr^{3+} + 7H_2O$$

The first equation must be reversed and also multiplied by 6 to balance the number of electrons transferred.

$$
\begin{array}{ll}
6Fe^{2+} & \rightarrow 6Fe^{3+} + 6e^- \\
Cr_2O_7^{2-} + 14H^+ + 6e^- & \rightarrow 2Cr^{3+} + 7H_2O \\
\hline
Cr_2O_7^{2-} + 14H^+ + 6Fe^{2+} & \rightarrow 6Fe^{3+} + 2Cr^{3+} + 7H_2O
\end{array}
$$

Questions

3 Check the two redox equations above to see if the total charge on each side of the equations is balanced.

4 In each of the examples given in the table below use the information to

   i) identify the reducing agent, the oxidizing agent and the spectator ions

   ii) write the balanced redox equation for the overall reaction.

Reactants	Ion–electron equations
a) Iodine solution and sodium sulphite solution	$I_2 + 2e^- \rightarrow 2I^-$ $SO_3^{2-} + H_2O \rightarrow SO_4^{2-} + 2H^+ + 2e^-$
b) Iron(II) sulphate solution and hydrogen peroxide solution	$Fe^{2+} \rightarrow Fe^{3+} + e^-$ $H_2O_2 + 2H^+ + 2e^- \rightarrow 2H_2O$
c) Tin(II) chloride solution and potassium dichromate solution (acidified)	$Sn^{2+} \rightarrow Sn^{4+} + 2e^-$ $Cr_2O_7^{2-} + 14H^+ + 6e^- \rightarrow 2Cr^{3+} + 7H_2O$

Redox titrations

In previous work you will have carried out volumetric titrations involving acids and alkalis in which, for example, a fixed volume of alkali contained in a conical flask is titrated with an acid of known concentration contained in a burette. The volume of acid required to neutralize the alkali is found, a suitable indicator being used to determine the end-point of the titration. The balanced equation for the reaction is then used in calculating the concentration of the alkali.

> ## Question
>
> 5 In each of the following examples, write a balanced equation for the reaction and use it to calculate the concentration of the solution in italics.
>
>   **a)** 20 cm³ of 2 mol l⁻¹ KOH(aq) neutralises 16.0 cm³ of *HNO₃(aq)*.
>
>   **b)** 20.0 cm³ of *Na₂CO₃(aq)* neutralises 25.0 cm³ of 0.8 mol l⁻¹ HCl(aq).
>
>   **c)** 24.0 cm³ of 0.5 mol l⁻¹ sodium hydroxide solution neutralizes 20.0 cm³ of dilute *sulphuric acid*.

Volumetric analysis, as this procedure is also known, can be applied to redox reactions. For example, the concentration of a solution of a reducing agent can be determined using a solution of a suitable oxidising agent of known concentration provided that:

1 The balanced redox equation is known or can be derived from the relevant ion–electron equations.

2 The volumes of the reactants are accurately measured by pipette and burette.

3 Some method of indicating the end-point of the titration is available.

Two examples of redox titrations are given below. A brief outline of the experimental procedure is followed by a specimen calculation. In the first example a redox titration is used to find the mass of vitamin C in a tablet. In example 2 the concentration of a reducing agent, iron(II) sulphate solution, is found by titrating it against an oxidising agent, potassium permanganate solution, of known concentration. Worked Example 17.1 shows how the unknown concentration may be calculated from the results obtained.

Example 1

Prescribed Practical Activity

To determine the mass of vitamin C in a tablet by redox titration using an iodine solution of known concentration and starch solution as indicator.

A vitamin C tablet is dissolved in deionised water (about 50 cm³) in a beaker and transferred with washings to a 250 cm³ standard flask. The flask is made up to the mark, stoppered and inverted several times to ensure thorough mixing.

25 cm³ of this solution is transferred by pipette to a conical flask and a few drops of starch indicator are added. Iodine solution of known concentration is added from the burette until the first sign of a permanent blue-black colour is seen. The titration is repeated, with dropwise addition of the iodine solution as the end-point is approached, to obtain concordant titres.

A specimen calculation is given in Worked Example 17.1.

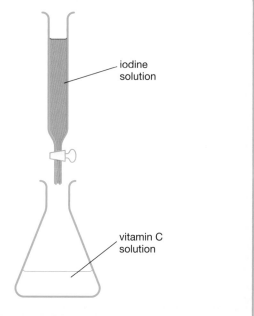

iodine solution

vitamin C solution

Figure 17.4

Worked Example 17.1

A solution of vitamin C was prepared as described in the Prescribed Practical Activity above. 25 cm³ of this solution was titrated against 0.031 mol l⁻¹ iodine solution using starch indicator. The average titre was 17.6 cm³. Calculate the mass of vitamin C (formula: $C_6H_8O_6$) in the original tablet.

Number of moles of iodine used in the titration,

$$n = C \times V = 0.031 \times \frac{17.6}{1000}$$

$$= 5.456 \times 10^{-4}$$

The redox equation is:

$$C_6H_8O_6 + I_2 \rightarrow C_6H_6O_6 + 2H^+ + 2I^-$$
$$\text{1 mole} \quad \text{1 mole}$$

Hence, number of moles of vitamin C in 25 cm³ (i.e. in the conical flask) $= 5.456 \times 10^{-4}$

Number of moles of vitamin C in 250 cm³ (i.e. in the standard flask) $= 5.456 \times 10^{-3}$

This is the number of moles of vitamin C in the original tablet.

Gram formula mass of
vitamin C, $C_6H_8O_6$ $= 176\,g$

Mass of vitamin C present in
the tablet $= 176 \times 5.456 \times 10^{-3}$

          $= 0.960\,g$

Example 2

To determine the concentration of an iron(II) sulphate solution by titration with a potassium permanganate solution of known concentration.

20 cm³ of iron(II) sulphate solution is transferred by pipette to a conical flask and excess dilute sulphuric acid is added. Potassium permanganate solution (0.02 mol l⁻¹) is added from the burette until the contents of the flask just turn from colourless to purple, initial and final burette readings being noted. The titration is repeated to obtain concordant titres.

Since the permanganate solution is so strongly coloured compared to the other solutions the reaction is self-indicating and the change at the end-point from colourless to purple is quite sharp.

A specimen calculation is given in Worked Example 17.2.

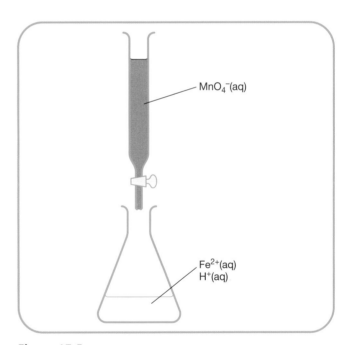

$MnO_4^-(aq)$

$Fe^{2+}(aq)$
$H^+(aq)$

Figure 17.5

Worked Example 17.2

Calculate the concentration of an iron(II) sulphate solution given that 20.0 cm³ of it reacted with 24.0 cm³ of 0.02 mol l⁻¹ potassium permanganate solution.

$$MnO_4^- + 8H^+ + 5Fe^{2+} \rightarrow 5Fe^{3+} + Mn^{2+} + 4H_2O$$

According to the equation, 1 mole of MnO_4^- oxidises 5 moles of Fe^{2+}.

Number of moles of MnO_4^- used, $n = C \times V = 0.02 \times \dfrac{24}{1000} = 4.8 \times 10^{-4}$

Worked Example 17.2 (continued)

Hence, number of moles of Fe^{2+} present $= 5 \times 4.8 \times 10^{-4} = 2.4 \times 10^{-3}$

This is contained in $20 \, cm^3$, i.e. 0.02 litres.

Hence, concentration of $Fe^{2+}(aq)$, $C = \dfrac{n}{V} = \dfrac{2.4 \times 10^{-3}}{0.02} = 0.12 \, mol \, l^{-1}$

Since 1 mole $FeSO_4(aq)$ contains 1 mole of $Fe^{2+}(aq)$, the concentration of $FeSO_4(aq) = 0.12 \, mol \, l^{-1}$

Alternatively, the concentration of iron(II) sulphate solution can be found by using the following relationship:

concentration $\times$ volume $\times$ number of electrons gained = Concentration $\times$ volume $\times$ number of electrons lost
per mole of oxidising agent   per mole of reducing agent

The two concentrations must be in the same units, i.e. $mol \, l^{-1}$, and the two volumes must also be expressed in the same units but do not have to be changed into litres, i.e. they can be in cubic centimetres or litres.

Let us now apply this relationship to the above example.

Oxidising agent (MnO_4^-):   Reducing agent (Fe^{2+}):

concentration $\times$ volume $\times$ number of electrons gained = concentration $\times$ volume $\times$ number of electrons lost

$\quad$ 0.02 $\quad \times \quad$ 24 $\quad \times \quad$ 5 $\qquad = \qquad$? $\quad \times \quad$ 20 $\quad \times \quad$ 1

Concentration of $Fe^{2+}(aq) = \dfrac{0.02 \times 24 \times 5}{20 \times 1} = 0.12 \, mol \, l^{-1}$

Hence concentration of $FeSO_4(aq) = 0.12 \, mol \, l^{-1}$

Using a relationship like the one above probably makes it easier to obtain the correct answer! However, ability to use it does not mean that you understand why it applies. Try to work out why it does apply by comparison with the previous method of calculation. You may find that it is better to use the first method if you are working from a balanced redox equation, while the second method may be more appropriate when you are given the separate ion–electron equations.

Questions

6 Calculate the concentrations of the solutions given in *italics* in the following examples. Either use the appropriate ion–electron equation provided in question 4 (page 183) or the redox equations you derived when answering that question.

 a) 12.5 cm^3 of an *iodine solution* reacts with 20.0 cm^3 of 0.1 $mol \, l^{-1}$ sodium sulphite solution.

 b) 25.0 cm^3 of an *iron(II) sulphate* solution reacts with 20.0 cm^3 of 0.5 $mol \, l^{-1}$ hydrogen peroxide solution.

 c) 15.0 cm^3 of a *tin(II) chloride* solution reacts with 25.0 cm^3 of 0.2 $mol \, l^{-1}$ potassium dichromate solution which has been acidified with dilute H_2SO_4.

Electrolysis

You should be familiar with the technique of electrolysis as a means of breaking down compounds to produce elements. Until now your study of the technique has focussed on **qualitative** electrolysis, i.e. identifying the kind of product without measuring its quantity. We return to this important topic at this stage to consider **quantitative** electrolysis.

During electrolysis, reduction occurs at the negative electrode (or cathode) since the positive ions are attracted and gain electrons. Conversely at the positive

Figure 17.6 Electrolysis cells at an aluminum smelting plant

Electrolyte	Cathode reaction	Anode reaction
1) Copper(II) chloride solution, $CuCl_2$(aq)	$Cu^{2+} + 2e^- \rightarrow Cu$	$2Cl^- \rightarrow Cl_2 + 2e^-$
2) Potassium iodide solution, KI(aq)	$2H^+ + 2e^- \rightarrow H_2$	$2I^- \rightarrow I_2 + 2e^-$
3) Molten sodium bromide, NaBr(l)	$Na^+ + e^- \rightarrow Na$	$2Br^- \rightarrow Br_2 + 2e^-$

Table 17.1

electrode (or anode) oxidation takes place as negative ions are attracted and lose electrons. Some example are given in Table 17.1.

In common with other redox reactions, the total number of electrons lost and gained during electrolysis must be the same. In example 3 in Table 17.1, the cathode reaction occurs twice as often as the anode reaction, i.e. twice as many sodium atoms as bromine molecules will be formed.

An alternative to this experiment is to use an electrolyte which will deposit a metal on the negative electrode. The negative electrode is cleaned and weighed before electrolysis. The time and current used during electrolysis are measured. At the end of the experiment the negative electrode is removed, rinsed, dried and

reweighed to obtain the mass of metal deposited. If the electrolyte is, for example, nickel(II) sulphate solution, the positive electrode should be made of nickel. The negative electrode on which nickel will be deposited during electrolysis can be made of copper. The ion–electron equation for the reaction at the negative electrode (cathode) is:

$$Ni^{2+} + 2e^- \rightarrow Ni$$

The correct value for the production of one mole of nickel is 1.93×10^5 C (or 193 000 C). This is also the correct value for the production of one mole of hydrogen gas by the electrolysis of dilute sulphuric acid.

$$2H^+ + 2e^- \rightarrow H_2$$

Table 17.2 summarises the results which would be obtained with other electrolytes.

Electrolyte	Ion–electron equation for cathode reaction	Quantity of electricity required to produce 1 mole of cathode product
$AgNO_3$(aq)	$Ag^+ + e^- \rightarrow Ag$	96 500 C
$CuCl_2$(aq)	$Cu^{2+} + 2e^- \rightarrow Cu$	193 000 C (or $2 \times 96\,500$ C)
NaCl(l)	$Na^+ + e^- \rightarrow Na$	96 500 C
Al_2O_3(l)	$Al^{3+} + 3e^- \rightarrow Al$	289 500 C (or $3 \times 96\,500$ C)

Table 17.2

Prescribed Practical Activity

The quantity of electricity required to produce a mole of electrode product can be found by experiment. Dilute sulphuric acid can be electrolysed using an apparatus such as that show in Figure 17.8. The following measurements need to be made:

1 the volume of hydrogen collected

2 the current used

3 the time during which the solution is electrolysed.

The quantity of electricity, **Q**, used during the experiment is calculated from the following relationship:
Q = I × t

> where **Q** is measured in coulombs (C)
> **I** is the current measured in amps (A)
> **t** is the time measured in seconds (s).

A specimen calculation is given in Worked Example 17.3.

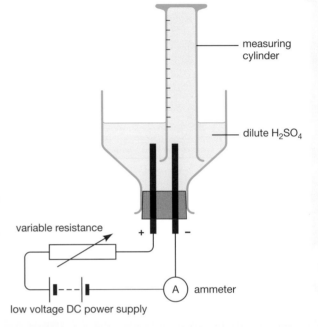

Figure 17.7

Worked Example 17.3

Dilute sulphuric acid was electrolysed using a current of 0.45 A for 6 minutes 50 seconds, i.e. t = 410s.

The volume of hydrogen collected at room temperature and pressure was 22.8 cm³.

Calculate the quantity of electricity required to produce 1 mole of hydrogen.

$$\text{Quantity of electricity used, } Q = I \times t$$
$$= 0.45 \times 410$$
$$= 184.5\,C$$

The molar volume of a gas at room temperature and pressure is approximately 24 litres mol⁻¹ (see Chapter 6 page 58).

22.8 cm³ (i.e. 0.0228 litres) of hydrogen was produced by 184.5 C

Hence, 1 mol (24 litres) of hydrogen would require

$$184.5 \times \frac{24}{0.0228} = 194\,210\,C$$
$$= 1.94 \times 10^5\,C$$

From the results in Table 17.2 it can be seen that the quantity of electricity needed to produce one mole of product at the cathode is **n** × 96 500 C, where n is the number of electrons in the appropriate ion–electron equation. The quantity of electricity, 96 500 C, is often referred to as the **Faraday** – named after the English scientist, Michael Faraday, who pioneered work on quantitative electrolysis. The Faraday is the quantity of electricity which is equivalent to one mole of electrons and has the value 96 500 C mol⁻¹ or 9.65 × 10⁴ C mol⁻¹.

Similar results would be obtained at the positive electrode (anode). The ion–electron equation for the reaction at the anode during the electrolysis of copper(II) chloride solution is:

$$2Cl^- \rightarrow Cl_2 + 2e^-$$

Hence, 193 000 C will be required to produce one mole of chlorine gas.

In Chapter 6 (page 55) it is stated that it is possible to determine the Avogadro Constant by experiment. One of the ways of doing this involves quantitative electrolysis. Let us consider the reaction occurring at the cathode during the electrolysis of silver(I) nitrate solution.

$$Ag^+ + e^- \rightarrow Ag$$

According to this equation, each silver ion gains one electron to form one silver atom. If we scale this up to one mole, then we can say that L silver ions require L electrons to form L silver atoms or one mole of silver. The quantity of electricity needed to achieve this is a Faraday. That is, 9.65×10^4 C is the total charge of one mole of electrons.

Consequently, we can calculate a value for the Avogadro Constant if the charge on one electron is known. This can be determined by what is often referred to as Millikan's oil drop experiment in which charged oil droplets are suspended between charged plates. Further details of this important experiment may be found in a suitable physics textbook.

Worked Example 17.4

Molten sodium chloride was electrolysed for 32 minutes 10 seconds using a current of 10 amps. Calculate the mass of product at each electrode.

Time, t, = 1930s.

Quantity of electricity used, $Q = I \times t$
$$= 10 \times 1930$$
$$= 19\,300\,C$$

a) At the negative electrode the product is sodium.

$$Na^+ + e^- \rightarrow Na$$
1 mol   1 mol
96 500 C   23 g

Hence, 19 300 C will produce $23 \times \dfrac{19\,300}{96\,500}$ g = 4.6 g of sodium

b) At the positive electrode the product is chlorine.

$$2Cl^- \rightarrow Cl_2 + 2e^-$$
1 mol   2 mol
71 g $2 \times 96\,500 = 193\,000\,C$

Hence, 19 300 C will produce $71 \times \dfrac{19\,300}{93\,000}$ g = 7.1 g of chlorine

---

The charge on one electron is 1.6×10^{-19} C.

Hence, the Avogadro Constant, L

$$= \frac{\text{Total charge on one mole of electrons}}{\text{Charge on one electron}}$$

$$= \frac{9.65 \times 10^4 \text{ C mol}^{-1}}{1.6 \times 10^{-19} \text{ C}} = 6.03 \times 10^{23} \text{ mol}^{-1}$$

This is close to the accepted value, namely 6.02×10^{23} mol^{-1}.

Calculations involving quantitative electrolysis

The mass of product obtained at an electrode during electrolysis can be calculated if:

1 the ion–electron equation for the reaction occurring at the electrode is known

2 the current and time during which electrolysis occurs are both given.

In other calculations the mass of product may be given so that the quantity of electricity used may be determined. Then, if the current is given, the time during which electrolysis has taken place can be calculated or vice versa. The following relationships should assist you.

$$\text{Hence, } Q = I \times t \text{ and } I = \frac{Q}{t} \text{ and } t = \frac{Q}{I}$$

Questions

7 Calculate the mass of silver deposited on passing a current of 0.5 A through silver(I) nitrate solution for 30 minutes.

8 When a constant current was passed through chromium(III) sulphate solution for 48 minutes 15 seconds, 2.08 g of chromium was produced. Calculate the current used.

(Ion–electron equation: $Cr^{3+} + 3e^- \rightarrow Cr$)

9 **a)** Calculate the quantity of electricity needed to produce 121.5 kg of magnesium by electrolysis of molten magnesium chloride.

b) If a current of 15 000 A is used, calculate the length of time (to the nearest hour) needed to produce this quantity of magnesium.

Study Questions

In questions 1–4 choose the correct word from the following list to complete the sentence.

> alkalis   current   less   metals   more
> oxidising   reducing   voltage

1 Redox reactions occur when acids react with
_____.

2 _____ agents are substances which accept electrons from other chemicals.

3 To calculate the quantity of electricity used during electrolysis it is necessary to measure the _____ used.

4 The electrolytic extraction of aluminium requires _____ than 250 000 coulombs per mole of metal.

5 Potassium permanganate solution, acidified with dilute sulphuric acid, is decolourised by iron(II) sulphate solution. Refer to the ion–electron equations in the Data Book.

Which ion is the reducing agent in this reaction?

A H^+

B Fe^{2+}

C MnO_4^-

D SO_4^{2-}

6
$$Br_2 + 2I^- \rightarrow I_2 + 2Br^-$$

Which substance involved in this redox reaction has been oxidised?

A Br_2

B I_2

C I^-

D Br^-

7
$$MnO_2 \rightarrow Mn^{2+}$$

Part of an ion–electron equation is shown above. When the equation is complete, the left hand side will include

A $4H^+ + 2e^-$

B $2H^+ + 4e^-$

C $4H^+ + 4e^-$

D $2H^+ + 2e^-$.

8 During chrome-plating, a solution containing chromium(III) ions is electrolysed.

For every 96 500 C used, the quantity of chromium metal plated on the negative electrode should be

A 26.0 g

B 0.67 mol

C 52.0 g

D 0.33 mol.

9* a) In acid solution, iodate ions, $IO_3^-(aq)$, are readily converted into iodine. Write an ion–electron equation for this half-reaction.

b) Use the equation to explain whether the iodate ion is an oxidising or reducing agent.

10 The 'clock reaction' referred to in Chapter 1 involves two redox reactions. From the equations given on page 5, work out the ion–electron equation for the reducing agent involved in each redox reaction.

11 The concentration of an aqueous solution of hydrogen peroxide can be determined in a redox titration with cerium(IV) sulphate solution, $Ce(SO_4)_2(aq)$.

The relevant ion–electron equations are

$$H_2O_2 \rightarrow O_2 + 2H^+ + 2e^-$$
$$Ce^{4+} + e^- \rightarrow Ce^{3+}$$

a) Write the balanced equation for the redox reaction.

b) Name the oxidising agent and the spectator ions present.

c) In an experiment 20.0 cm³ of hydrogen peroxide solution required 15.6 cm³ of 0.092 mol l⁻¹ cerium(IV) sulphate solution to reach the end-point. Calculate the concentration of the solution of hydrogen peroxide.

Study Questions (continued)

12* Seaweeds are a rich source of iodine in the form of iodide ions. The mass of iodine in a seaweed can be found using the procedure outlined below.

a) Step 1

The seaweed is dried in an oven and ground into a fine powder. Hydrogen peroxide solution is then added to oxidise the iodide ions to iodine molecules. The ion–electron equation for the reduction reaction is shown.

$$H_2O_2(aq) + 2H^+(aq) + 2e^- \rightarrow 2H_2O(l)$$

Write a balanced redox equation for the reaction of hydrogen peroxide with iodide ions.

b) Step 2

Using starch solution as an indicator, the iodine solution is then titrated with sodium thiosulphate solution to find the mass of iodine in the sample. The balanced equation for the reaction is shown.

$$2Na_2S_2O_3(aq) + I_2(aq) \rightarrow 2NaI(aq) + Na_2S_4O_6(aq)$$

In an analysis of seaweed, $14.9\,cm^3$ of 0.00500 mol l^{-1} sodium thiosulphate solution was required to reach the end-point. Calculate the mass of iodine present in the seaweed sample.

13* A copper compound was known to contain either copper(I) or copper(II) ions. The compound was dissolved in water and electrolysed. It was found that 0.32 g of copper was formed after the electrolysis cell had been operating for 16 minutes with a steady current of 1.0 A.

a) At which electrode would copper have been formed?

b) Using the above information, determine which copper ion was present. Working must be shown.

14 During electrolysis of a dilute acid, 0.25 g of hydrogen gas is collected. What mass of copper would be obtained if the same current is used for the same time in electrolysis of copper(II) chloride solution?

15 Aluminium is obtained by electrolysis of molten aluminium oxide. It is customary to use a very high current, often more than 100 000 amps.

a) Calculate the theoretical output in 24 hours of aluminium from an electrolytic cell which is operating with a current of 120 000 A.

b) In the electrolytic cell the positive electrode is made of carbon which is burned away by the oxygen released at this electrode.

Assuming complete combination of oxygen with carbon, calculate the mass of carbon needed per day in an electrolytic cell operating at 120 000 A. Use the equations:

$$2O^{2-} \rightarrow O_2 + 4e^-$$
$$C + O_2 \rightarrow CO_2$$

16* The most common method for the industrial purification of gold is the cyanide process.

a) The impure gold is first dissolved in sodium cyanide solution, NaCN, to give sodium gold cyanide, $NaAu(CN)_2$.

The partially balanced equation for the reaction is:

$$4Au + NaCN + O_2 + H_2O \rightarrow NaAu(CN)_2 + NaOH$$

Complete the balancing of this equation.

b) The gold is then obtained by a redox reaction using zinc.

$$2NaAu(CN)_2(aq) + Zn(s) \rightarrow Na_2Zn(CN)_4(aq) + 2Au(s)$$

Give another name for this type of reaction.

c) Gold may be purified by electrolysis. In an industrial process, a current of 10 000 A was passed through a gold solution for 25 minutes producing 10.21 kg of gold.

Calculate the charge on the gold ions in the solution.

17 Electrolysis of dilute sulphuric acid produces hydrogen and oxygen gases. The relevant ion–electron equations are

Negative electrode: $2H^+ + 2e^- \rightarrow H_2$

Positive electrode: $H_2O \rightarrow \frac{1}{2}O_2 + 2H^+ + 2e^-$

a) Calculate the length of time needed for a current of 0.5 A to be passed through dilute sulphuric acid to produce $40\,cm^3$ of hydrogen. $V_{mol} = 24$ litres mol^{-1}.

b) Use the equations to help you predict **i)** the volume of oxygen produced during the same time, and **ii)** what change, if any, there will be in the pH of the acid during electrolysis.

Study Questions (continued)

18* Chlorine and sodium hydroxide are important industrial chemicals. Both are produced by the electrolysis of sodium chloride solution.

a) The ion-electron equation for the formation of chlorine is:

$$2Cl^-(aq) \rightarrow Cl_2(g) + 2e^-$$

Calculate the volume of chlorine produced when a current of 350 000 A flows for 1 hour. (Take the molar volume to be 24 litres mol^{-1}.) Show your working clearly.

b) Bromine can be obtained by bubbling chlorine gas through a solution of bromide ions.

$$Cl_2(g) + 2Br^-(aq) \rightarrow Br_2(aq) + 2Cl^-(aq)$$

Name this type of reaction.

c) The hypochlorite ion, $ClO^-(aq)$, acts as a bleaching agent in solution.

i) Most household bleaches are made by reacting sodium hydroxide with chlorine. Sodium hypochlorite, sodium chloride and water are formed.

Write a balanced equation for the reaction.

ii) When chlorine is added to water, the following equilibrium is set up.

$$Cl_2(aq) + H_2O(l) \rightleftharpoons 2H^+(aq) + ClO^-(aq) + Cl^-(aq)$$

Why does the addition of sodium hydroxide increase the bleaching efficiency of the solution?

iii) When $ClO^-(aq)$ acts as bleach, it reacts to produce the $Cl^-(aq)$ ion.

$$ClO^-(aq) \rightarrow Cl^-(aq)$$

Complete the above to form the ion–electron equation.

Nuclear Chemistry

From previous work you should know and understand the following:

	Approximate mass	Charge	Location	Symbol
Proton	1	$+1$	Nucleus	$_1^1p$ or $_1^1H$
Neutron	1	0	Nucleus	$_0^1n$
Electron	1/2000	-1	Orbitals	$_{-1}^0e$

Table 18.1

★ Atomic structure – atoms consist of a nucleus and extranuclear electrons. The nucleus is made up of neutrons and protons. These particles are sometimes referred to as **nucleons**.

Atomic number = Number of protons (= Number of electrons in an uncharged atom)

Mass number = Number of protons + Number of neutrons

★ Isotopes:

1 are different types of atoms of the same element, i.e. the same atomic number but different mass number, owing to different numbers of neutrons

2 cannot be distinguished by chemical means, since chemical properties depend on electron arrangement

3 can be represented by the nuclide notation, for example, $_{92}^{238}U$ represents the uranium (atomic number 92) isotope of mass number 238. For most purposes, the word 'nuclide' is interchangeable with 'isotope'.

Radioactivity

In 1896, Becquerel found that compounds of uranium emit an invisible radiation which will penetrate opaque materials and fog photographic plates. The phenomenon became known as radioactivity. Compounds of thorium behave in a similar fashion, as do natural isotopes of some other elements and artificial isotopes of most elements. The radiation was found to be affected by a magnetic or electrical field as shown in Figure 18.1.

There are three types of radiation. Other experiments provide the information given in Table 18.2.

Gamma radiation is non-particulate. It is electromagnetic radiation (EMR), similar to X-radiation, but of a higher energy. It is emitted alone or

Name	Penetration	Nature	Symbol	Charge	Mass
α (alpha)	Few cm in air	He nucleus	$_2^4He$	2+	4
β (beta)	Thin metal foil	Electron	$_{-1}^0e$	1−	1/2000
γ (gamma)	Great thickness of concrete	EMR	None	None	None

Table 18.2

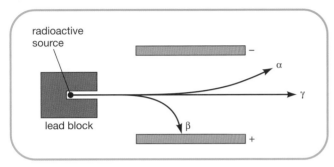

Figure 18.1 Radiation deflected by an electric field between charged plates

accompanying α or β radiation. **The radiation from an element is independent of the physical state of the element and is not affected when the element forms compounds.** This indicates that radiation is connected solely with the nucleus.

Radioactivity is the result of unstable nuclei rearranging to form stable nuclei with the emission of energy. It can be shown that the mass of any atom is less than the sum of the masses of protons, neutrons and electrons of which it is composed. This 'mass defect' is equivalent to the 'binding energy' of the nucleus, mass and energy being related by the well known equation:

$$E = mc^2$$

The binding energy for heavy atoms is generally greater than that for light atoms, since more nucleons are being held together. However, the binding energy per nucleon is not the same in all atoms. The most stable atoms have the greatest binding energy per nucleon. A graph of binding energy per nucleon of stable or near stable nuclides is shown in Figure 18.2.

The distance from the curve up to the x-axis is a measure of the energy per nucleon required to dismantle the nucleus. If we consider the starting and finishing isotopes of one of the natural radioactive series, $^{238}_{92}U$ and $^{206}_{82}Pb$, we can see that more energy is required to dismantle the nucleus for the Pb and it is therefore a more stable nucleus. In the change from $^{238}_{92}U$ to $^{206}_{82}Pb$ there is an emission of energy (possessed by the radiation and the kinetic energy of the various α and β particles emitted).

The stability of nuclei is related to the proton/neutron balance (Figure 18.3).

For small stable nuclei the numbers of neutrons and protons are approximately equal. Where a nucleus has

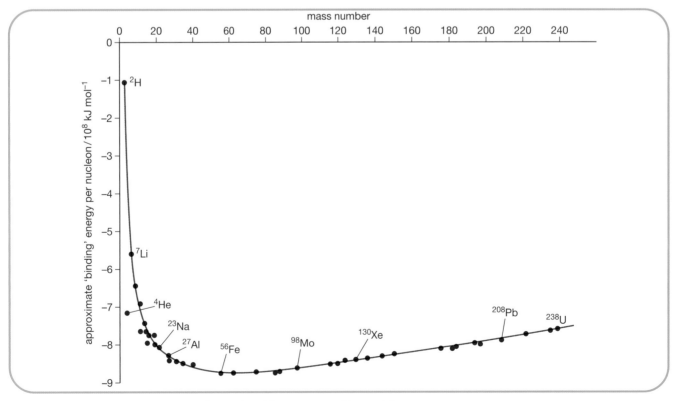

Figure 18.2 Binding energy per nucleon of stable or near stable nuclei

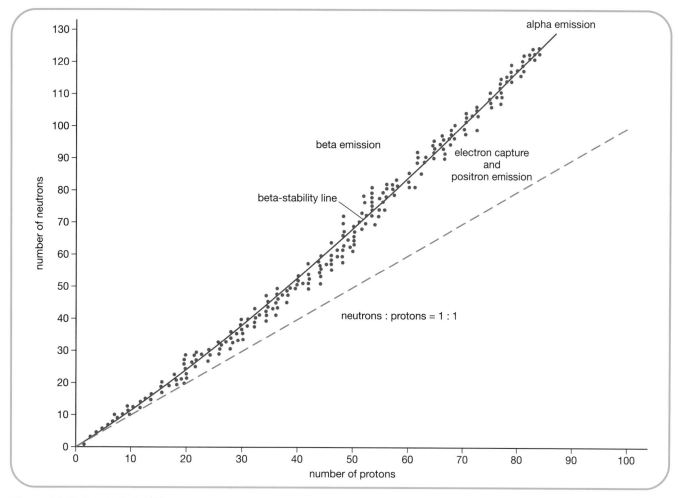

Figure 18.3 Band of stability

too many neutrons it can become stable if a neutron changes into a proton and an electron, in other words a β particle, is projected from the nucleus:

$$^1_0n \rightarrow \ ^1_1p + \ ^0_{-1}e$$

The proton/neutron ratio is reduced and a more stable nucleus may result. For example, carbon-14 emits β particles to form a stable isotope of nitrogen:

$$^{14}_6C \rightarrow \ ^{14}_7N + \ ^0_{-1}e$$

Where isotopes with low atomic number have too few neutrons, electron capture may take place. An electron is removed from the first electron layer to combine with a proton to form a neutron. The electrons rearrange to refill the first layer:

$$^1_1p + \ ^0_{-1}e \rightarrow \ ^1_0n$$

For example, argon-37 undergoes electron capture to form an isotope of chlorine:

$$^{37}_{18}Ar + \ ^0_{-1}e \rightarrow \ ^{37}_{17}Cl$$

Beyond atomic number 83, almost all isotopes are unstable. They gain stability if they decrease in mass; see Figure 18.2. Emission of α particles from the nucleus reduces the mass.

Chemical effect of radioactive disintegrations

The radiation originates in the nucleus, therefore the loss of α or β particles changes it considerably.

Loss of an α Particle $^4_2$He

This is equivalent to losing two protons (decreasing the atomic number by two) and two neutrons (overall decrease of mass number by four).

For example:

$$^{232}_{90}\text{Th} \rightarrow {}^{228}_{88}\text{Ra} + {}^{4}_{2}\text{He}$$

The total mass number (superscript) must be the same on each side of the equation, as must the total atomic number (subscript).

Note: Since the atomic number of the product is 88, it is now radium.

Similarly:

$$^{220}_{86}\text{Rn} \rightarrow {}^{216}_{84}\text{Po} + {}^{4}_{2}\text{He}$$

Question

1 Write equations for the α decay of $^{234}_{92}\text{U}$, $^{216}_{84}\text{Po}$ and $^{230}_{90}\text{Th}$.

Loss of a β particle $^{0}_{-1}\text{e}$

A β particle is an electron. Since the nucleus does not contain electrons, it is believed to be formed by:

$$^{1}_{0}\text{n} \rightarrow {}^{1}_{1}\text{p} + {}^{0}_{-1}\text{e}$$

Hence loss of the electron results in a gain of one unit of atomic number (one proton) without any change in mass number (since the mass of the proton is almost the same as the mass of the neutron). For example:

$$^{228}_{88}\text{Ra} \rightarrow {}^{228}_{89}\text{Ac} + {}^{0}_{-1}\text{e}$$

Again, total atomic number and mass number must be the same on each side of the equation. Since the product has atomic number 89 it is now actinium.

Similarly:

$$^{216}_{84}\text{Po} \rightarrow {}^{216}_{85}\text{At} + {}^{0}_{-1}\text{e}$$

In fact the various 'daughter nuclei' are usually radioactive themselves, so that a whole series of radioactive disintegrations occurs until a stable isotope is reached, frequently an isotope of lead.

Question

2 Write equations for the β decay of $^{214}_{82}\text{Pb}$, $^{228}_{89}\text{Ac}$ and $^{210}_{83}\text{Bi}$.

Artificial radioactivity

Many artificially-produced radioactive isotopes are known. These can be made by bombarding stable isotopes with neutrons in a nuclear reactor. Since neutrons have no charge, they are not repelled by the positive nucleus.

For example:

$$^{27}_{13}\text{Al} + {}^{1}_{0}\text{n} \rightarrow {}^{24}_{11}\text{Na} + {}^{4}_{2}\text{He}$$

The sodium isotope produced then decays by β emission:

$$^{24}_{11}\text{Na} \rightarrow {}^{24}_{12}\text{Mg} + {}^{0}_{-1}\text{e}$$

Similarly:

$$^{32}_{16}\text{S} + {}^{1}_{0}\text{n} \rightarrow {}^{32}_{15}\text{P} + {}^{1}_{1}\text{p} \text{ (a proton)}$$

The phosphorus isotope produced is a β emitter:

$$^{32}_{15}\text{P} \rightarrow {}^{32}_{16}\text{S} + {}^{0}_{-1}\text{e}$$

The 'target' material for the neutrons is usually a compound including the starting isotope. The other part of the compound may also be affected by the neutrons, so selection of the compound is a specialist task. With the development of high energy particle accelerators, many positively-charged particles can be used for bombardment, increasing the range of isotopes that can be made.

Question

3 Write equations for
   a) the neutron bombardment of $^{45}_{21}\text{Sc}$ which liberates an α particle.
   b) the α particle bombardment of $^{9}_{4}\text{Be}$ which liberates a neutron.
   c) the α particle bombardment of $^{14}_{7}\text{N}$ which liberates a proton.

Predictability of radioactive decay

The disintegration of any individual nucleus is a purely random event and, as already stated, is independent of chemical and physical factors. However, for a large enough population of unstable atoms, it is possible to calculate accurately how much of a radioisotope will be left after a given time.

Half-life

Radioactive decay of atoms of an isotope is such that the activity of the isotope decreases by half in a fixed time called the half-life of the isotope. Figure 18.4 shows the change of activity for an isotope with a half-life of 2 days. Half-lives vary from seconds to millions of years and are characteristic of particular isotopes.

The half-life of any isotope is independent of the mass of the sample being investigated. It is independent of temperature, pressure, concentration, presence of catalysts or chemical state of the isotope.

In the half-life, half of the atoms of the isotopes decay but it must be stressed that the process is completely random and it is not possible to predict the time of decay of any individual atom. Half-life is often abbreviated to $t_{\frac{1}{2}}$. After 'n' half-lives, the fraction of the original activity which remains is given by $\left(\frac{1}{2}\right)^n$.

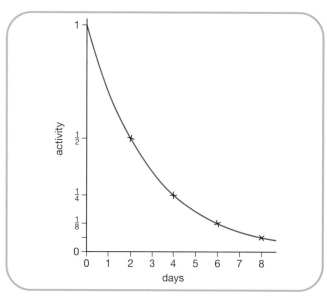

Figure 18.4 Change in activity with time for an isotope of half-life 2 days

Questions

4   Sodium-24 has a half-life of 15.0 hours. What fraction of a sample's activity will remain after 5 days?

5   A luminous watch dial containing $^{147}_{61}$Pm ($t_{\frac{1}{2}} = 2.5$ years) has only $\frac{1}{8}$ of its original 'glow'. How old is the watch?

Worked Example 18.1

What fraction of the activity of a sample of $^{106}_{45}$Rh will remain after three minutes? ($t_{\frac{1}{2}} = 30$ seconds)

$t_{\frac{1}{2}} = 30$ seconds, so 3 minutes = 6 $t_{\frac{1}{2}}$

Fraction of activity remaining $\left(\frac{1}{2}\right)^n = \left(\frac{1}{2}\right)^6 = \frac{1}{64}$

Worked Example 18.2

The activity of a sample of iodine, isotope $^{131}_{53}$I, is found to be only $\frac{1}{16}$ of the activity when it arrived at a hospital 32 days earlier. What is the half-life of $^{131}_{53}$I?

Fraction of activity remaining $= \frac{1}{16} = \left(\frac{1}{2}\right)^4$

i.e. the sample has disintegrated through four half-lives in 32 days, therefore its half-life is 8 days.

The uses of radioisotopes

This is an ever-expanding field, and so only certain important representative examples are given in the following sections.

Medical uses of radioisotopes

Radioactive isotopes have been used for a considerable time for the treatment of various types of cancer. For example the γ emitter $^{60}_{27}$Co is used for the treatment of deep-seated tumours, and the less penetrating β emitter $^{32}_{15}$P is used for treating skin cancer by direct application to the affected area.

Radioactive isotopes can be used to monitor processes occurring in the body. An example is the use of radioactive iodine to investigate possible disease of the thyroid gland. The gland is scanned and a plot of concentration made. Diseased areas can be identified. To reduce overall exposure, short half-life $^{132}_{53}$I and $^{123}_{53}$I are now used instead of $^{131}_{53}$I.

Industrial applications of radioisotopes

A source of very penetrating radiation, e.g. $^{60}_{27}$Co or $^{192}_{77}$Ir, is used to examine castings and welds for imperfections. A photographic film can be used as the detector, as in X-ray work.

Radioisotopes can also be used to make routine measurements. Continuous monitoring of thickness of sheet material, such as paper, plastic and thin metal sheet, can be carried out using β sources, or γ sources for thicker metal sheet. Two identical sources can be used. The radiation from one passes through reference material of the required thickness, the radiation from the other passes through the material under test. If its thickness is correct, the detectors register the same signal. If the thickness varies, the signals differ and this difference can be used automatically to correct the fault.

Domestic smoke alarms contain americium-241, an α emitter. Even low concentrations of smoke will change the normally steady level of radiation passing across an air gap sufficiently to trigger the alarm.

Flow patterns in estuaries, leaks in pipelines and ventilation flows can be investigated using small quantities of short half-life isotopes which have negligible residual activity.

Energy production

By far the most important method of obtaining energy from nuclear sources is **nuclear fission**. The basis of the method is the possibility of splitting some nuclei by bombarding them with slow-moving neutrons. The smaller nuclei produced are of elements in the centre of the Periodic Table. For example, one such change is

$$^{235}_{92}U + ^{1}_{0}n \rightarrow ^{236}_{92}U \rightarrow ^{140}_{54}Xe + ^{94}_{38}Sr + 2^{1}_{0}n$$

Figure 18.2 shows that the two new nuclei have a lower energy level than the original nucleus and the energy liberated is available for use in generating electricity. However, it is important to note that more neutrons are set free than are used, as shown in the above equation. This means that the reaction becomes a self-sustaining chain reaction, although in practice this only occurs if more than a certain 'critical mass' of $^{235}_{92}U$ is used. Figure 18.5 illustrates the fission of a $^{235}_{92}U$ nucleus.

A calculation based on the equation above shows that fission of one mole, 235 g, of $^{235}_{92}U$ yields about 19×10^6 MJ of energy. The combustion of carbon to produce the same amount of energy would need at least 60 tonnes of high quality coal and would produce some 220 tonnes of CO_2 in the atmosphere. There are advantages and disadvantages in the use of nuclear power in electricity generation. They are summed up in Table 18.3. Discussion of the peaceful uses of nuclear energy is often hampered by the association of nuclear energy with nuclear weapons.

In 2005, approximately 25% of the electricity produced in Scotland was from nuclear power stations, a larger proportion than in England and Wales. The UK

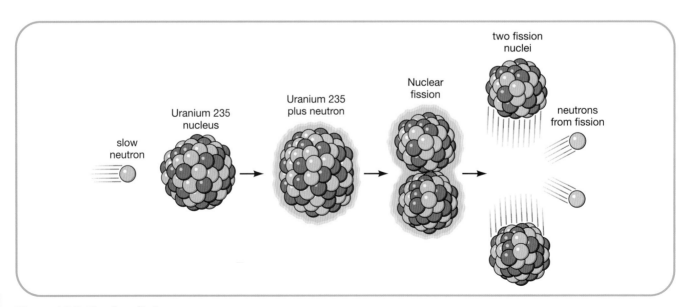

Figure 18.5 Uranium fission

Nuclear power generation

Advantages

No 'greenhouse' gases emitted

No SO_2 emitted to increase 'acid rain'

Fewer deaths and injuries in uranium mining than in coal mining

Fuel reserves will last longer than fossil fuel reserves

Power stations have less visual impact than coal- or oil-fired stations or wind farms

Alternative source of energy for countries with no fossil or renewable energy sources

Disadvantages

Finite (but low) probability of a disastrous accident

Contribution to 'background' radiation (see Figure 18.7)

Difficulty of disposing of spent fuel which remains radioactive for many years

Capital cost of plant is greater

Cost of decommissioning obsolete stations

Stations slow to respond to rapid changes in demand for power; conventional stations faster

Plutonium can be produced, possibly leading to proliferation of nuclear weapons

Table 18.3

Government is now reconsidering its policy of not building more nuclear power stations, but at present the Scottish Executive holds to the existing policy. The major factor which is influencing change is the security of supply of oil and gas which increasingly are having to be imported from countries which may not always be prepared to export to the UK. The ability of 'renewables', i.e. wind, wave and solar power, to fill an energy 'gap', although strongly advocated by some political parties and pressure groups, has yet to be convincingly demonstrated. The impact on the environment of large wind turbines and their attendant power lines, on pylons, is a major concern to those with an interest in the scenery of remote areas.

$^{235}_{92}U$ is only a small proportion of natural uranium (0.7%). The majority is the non-fissionable $^{238}_{92}U$. The remaining elderly British 'Magnox' reactors operate using natural uranium, the more modern AGR (Advanced Gas-cooled Reactor) uses uranium enriched with $^{235}_{92}U$ to 2.3% and the PWR (pressurised water reactor) uses enriched uranium to 3.2% $^{235}_{92}U$. The $^{238}_{92}U$ is not discarded – it 'captures' neutrons to produce $^{239}_{94}Pu$

$$^{238}_{92}U + ^{1}_{0}n \rightarrow ^{239}_{92}U \rightarrow ^{239}_{93}Np + ^{0}_{-1}e$$

$$^{239}_{93}Np \rightarrow ^{239}_{94}Pu + ^{0}_{-1}e \qquad t\frac{1}{2} = 2.33 \text{ days}$$

'Breeder' reactors were designed to surround the reactor core with a 'blanket' of $^{238}_{92}U$ to maximise this effect since plutonium, Pu, is itself fissionable and can be used in other reactors to produce energy.

The public mistrust of spent fuel reprocessing and the current ready availability of uranium have resulted in the closing down of the research breeder reactor at Dounreay, where commercial reprocessing is also being run down.

One British reactor type is shown in Figure 18.9. The Torness reactor is an Advanced Gas-cooled Reactor.

The possibility of exhausting supplies of $^{235}_{92}U$ but more importantly, the problem of disposing of fission products, has led to much research into energy production from **nuclear fusion**. Another look at Figure 18.2 will show that some of the very light nuclei can form heavier nuclei with the production of even more energy than fission generates. One such reaction under study is:

$$^{2}_{1}H + ^{3}_{1}H \rightarrow ^{4}_{2}He + ^{1}_{0}n$$

Figure 18.6 Torness nuclear power station. Location of UK nuclear stations on the coast allows cooling of turbine steam by sea water

Figure 18.7 Most coal-powered stations are located inland and require cooling towers. The visual impact is greater

Figure 18.8 Wind-powered turbines can have significant visual impact

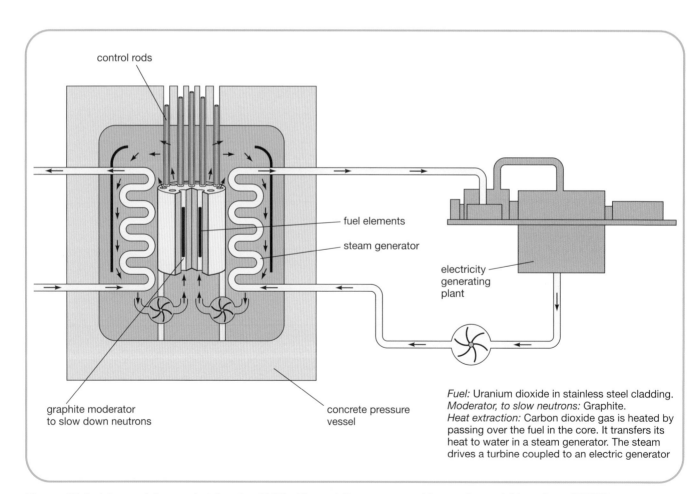

control rods

fuel elements

steam generator

electricity
generating
plant

graphite moderator
to slow down neutrons

concrete pressure
vessel

Fuel: Uranium dioxide in stainless steel cladding.
Moderator, to slow neutrons: Graphite.
Heat extraction: Carbon dioxide gas is heated by
passing over the fuel in the core. It transfers its
heat to water in a steam generator. The steam
drives a turbine coupled to an electric generator

Figure 18.9 Advanced Gas-cooled Reactor (AGR), Thermal Reactor – graphite moderated (data from UKAEA)

A 1000 MW power station would, it is claimed, require only one ton of fuel per year.

Another possible reaction, although this would require lithium rather than the more common hydrogen isotopes, is:

$$^{6}_{3}\text{Li} + ^{1}_{0}\text{n} \rightarrow ^{3}_{1}\text{H} + ^{4}_{2}\text{He}$$

This produces more energy than the previous reaction. Furthermore, the tritium produced could be used in bringing about the first reaction. It is in fact possible to devise several reactions which would form a 'chain', but the major difficulty is in bringing the nuclei together (against their electrostatic repulsion) for a long enough period for fusion to take place. Various very expensive devices are being tested, such as JET at Culham in England, for a European consortium. To date fusion has achieved the production of 16 MW at Culham, but this was only about 65% of the energy expended in producing the high-temperature plasma needed. The aim is to prolong the period of fusion and to use less energy in achieving fusion than is being produced by it. A new device, ITER, scheduled to operate from 2015, producing 500 MW, ten times the energy expended, is being constructed by a consortium including the EU, Japan, USA, South Korea, Russia, China and India at Cadarache in France.

Figure 18.10 The huge chamber of JET (Joint European Torus) at Culham in which fusion experiments are carried out

Agricultural uses

Tracer techniques are widely used in agriculture. One example is the use of $^{32}_{15}P$ in phosphate fertilisers to observe the use of phosphate by the growing plant. Another example is to use CO_2 labelled with $^{14}_{6}C$ to follow the way in which carbon is incorporated successfully into different plant constituents following photosynthesis.

More controversial is the use of γ irradiation of food crops to kill bacteria and moulds and increase storage life. The method is not popular in the UK, but in other European countries it is already widely used. It should be emphasised that there is no residual radiation from the source left on the crop.

Dating

Radioactive isotopes are constantly decaying at a known rate, so the age of materials containing them can be estimated by finding the present activity of the isotope. For example, $^{14}_{6}C$ is used to date archaeological specimens between about 600 and 10 000 years old. The method depends on the production of $^{14}_{6}C$ in the atmosphere by bombardment of nitrogen with neutrons formed from the effects of cosmic rays on other atoms.

$$^{14}_{7}N + ^{1}_{0}n \rightarrow ^{14}_{6}C + ^{1}_{1}p$$

This $^{14}_{6}C$, like $^{12}_{6}C$, is constantly taken up by growing plants and then by animals, so that their $^{14}_{6}C : ^{12}_{6}C$ ratio is known. When they die, however, the $^{14}_{6}C$ decays:

$$^{14}_{6}C \rightarrow ^{14}_{7}N + ^{0}_{-1}e \ (t_{\frac{1}{2}} = 5600 \text{ years})$$

By measuring the $^{14}_{6}C{:}^{12}_{6}C$ ratio, the time since the sample 'died' can be estimated. One of the most notable examples of carbon dating was the determination of the age of the linen of the Turin Shroud. The conclusion reached was that the linen was of early medieval date and could not, therefore, have been Christ's burial shroud, though this conclusion is disputed by some. The actual dating technique is not questioned, it is suggested the samples tested were from later repairs to the shroud.

There are several systems which use long half-life isotopes to date rocks, but for checking the age of vintage wines the short half-life tritium, $^{3}_{1}H$, has often been used ($t_{\frac{1}{2}} = 12.4$ years).

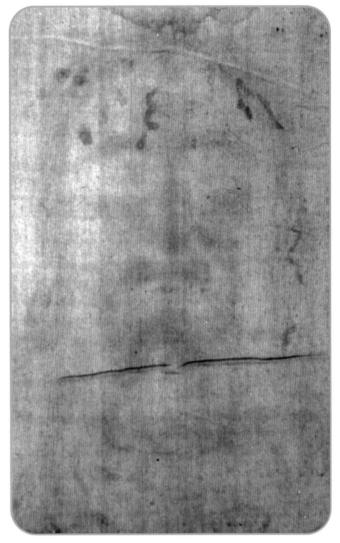

Figure 18.11 The image of the face on the Turin Shroud

Question

6   A sample of wine shows a β emission rate from tritium $^{3}_{1}H$ which is $\frac{1}{8}$ of that from wine bottled at the present day. If the half-life of tritium is 12.4 years, and its concentration in the atmosphere and in water exposed to the atmosphere is assumed to have remained constant, for how many years has the wine been bottled?

Background radiation

Everyone is exposed to radiation from various sources at all times. This radiation is background radiation. In addition, most of us are exposed to other sources for medical reasons and some of us because of our occupations. More controversial is the exposure of all of us to radiation from leakages from nuclear plants and disposal of radioactive waste. The numbers in Figure 18.12, which have not changed significantly in recent years, show the relative magnitudes of these radiation sources. The average individual dose to a person in Scotland is 2.4 millisievert per year, 85% being from natural sources. Any member of the public is only allowed to receive 1 mSv year after year from human activity (excluding medical exposure). 2 mSv per year gives a risk equivalent to smoking 20 cigarettes per year.

Much of the radon shown in Figure 18.12 as a source of radiation is produced in igneous rocks which contain radioisotopes. It then escapes to the atmosphere. The air in houses built of granite in, for example, Aberdeen can have higher than average concentrations of radon.

The origin of the elements

Spectroscopic studies of the light from stars and increasing knowledge of nuclear reactions have made it possible to develop theories about how the elements formed.

There are large quantities of hydrogen in space, but the atoms are widely separated and unlikely to come into contact. In regions between the stars called interstellar gas clouds the hydrogen is more concentrated and there is also dust present. The sheer quantity of matter creates huge gravitational fields and these cause the clouds to contract and the gases to be compressed into dense clumps. At the center of these clumps the compression is such that the gases become very hot, attaining temperatures of several million degrees Celsius. Under these conditions the atoms possess so much energy that fusion can occur and new larger nuclei form since the large repulsive forces between the

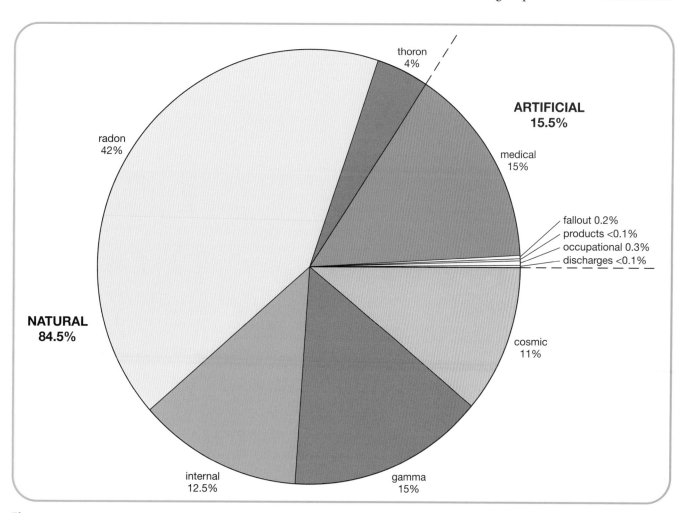

Figure 18.12 Sources of radiation affecting the Scottish population (data from the Scottish Environment Statistics, 1998)

positive nuclei can be overcome. Two of the simpler reactions are:

$$^1_1H + {}^2_1H \rightarrow {}^3_2He$$

$$^2_1H + {}^3_1H \rightarrow {}^4_2He + {}^1_0n$$

Huge quantities of energy are produced and the gas clouds glow. A **nebula** is visible in the sky. Kinetic energy is given to much of the remaining gas and dust, creating winds which carry them away leaving a new star, sometimes with attendant planets, formed from condensed and coalesced material.

In the heaviest stars, with the hottest and most compressed centres, continuing fusion occurs and produces heavier elements.

When hot enough, reactions such as the following can occur.

$$3^4_2He \rightarrow {}^{12}_6C$$

$$^{12}_6C + {}^4_2He \rightarrow {}^{16}_8O$$

Figure 18.13 The Rosette nebula, 2600 light years away

The distribution of the elements is graded with the heaviest nearest the core. When iron is formed energy is no longer released on further fusion (see Figure 18.2), but is absorbed. The core becomes unstable and explodes creating a **supernova**. This disperses the elements as new clouds of dust and gas and the process starts afresh. All naturally-occurring elements have been formed in this way, taking billions of years.

Figure 18.14 The Vela supernova remnant, about 1000 light years from Earth. The explosion happened about 11 000 years ago

Study Questions

In questions 1–4 choose the correct word from the following list to complete the sentence.

> alpha   artificial   fission   fusion   gamma
> neutrons   natural   protons

1   About 85% of the background radiation to which we are exposed arises from _____ sources.

2   Cobalt-60 can be used to treat cancerous tissue as it emits _____ radiation.

3   The formation of elements in the stars occurs by nuclear _____.

4   In nuclear reactors, bombardment by _____ causes uranium nuclei to split.

5   When a radioactive isotope loses an alpha particle

A   both atomic number and mass number change

B   only the atomic number changes

C   neither the atomic number nor the mass number change

D   only the mass number changes.

6   $^{31}P \rightarrow ^{32}P$

Phosphorus-32 can be produced from phosphorus-31 by a process called

A   proton capture

B   beta emission

C   neutron capture

D   gamma emission

7   If a radioactive isotope has a half-life of 2 hours, what fraction of it will remain after 8 hours?

A   ¼

B   ⅙

C   ⅛

D   ¹⁄₁₆

8   $$^{215}_{84}Po \xrightarrow{-\alpha} X \xrightarrow{-\beta} Y$$

From the sequence shown above, it can be deduced that

A   the mass number of **X** is 213

B   **X** is an isotope of astatine

C   the mass number of **Y** is 210

D   **Y** is an isotope of bismuth.

9   Thallium-206 (half-life = 4.2 minutes) emits β particles to form a stable isotope of lead.

The mass of this isotope of lead produced after a 0.4 g sample of thallium-206 has been left for 8.4 minutes would be

A   0.1 g

B   0.2 g

C   0.3 g

D   0.4 g.

10   Write down the symbol of the particle formed when

a) a francium atom loses its outer electron

b) $^{223}_{87}Fr$ loses a beta particle.

11   Complete the following equations:

a) $^{234}_{90}Th \rightarrow \quad + ^{0}_{-1}e$

b) $^{222}_{86}Th \rightarrow \quad + ^{4}_{2}He$

12   Find x, y and z in the following equations:

$$^{6}_{3}Li + ^{1}_{0}n \rightarrow ^{3}_{1}H + x$$

$$^{234}_{92}U \rightarrow y + ^{4}_{2}He$$

$$z \rightarrow ^{212}_{83}Bi + ^{0}_{-1}e$$

13* a)   The radioactive isotope $^{221}_{87}Fr$ decays to form a stable isotope $^{b}_{a}X$ by the following sequence of emissions:

α, α, β, α, β

Identify element X and write values for 'a' and 'b'

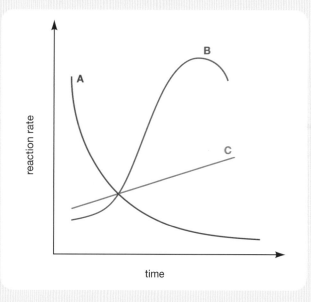

Study Questions (continued)

b) During fission of uranium, one of the reactions which occurs is:

$$^{235}_{92}U + ^{1}_{0}n \rightarrow ^{144}_{56}Ba + ^{90}_{36}Kr + 2^{1}_{0}n$$

i) Which of the curves on the previous page could represent this reaction?

ii) Explain your choice.

14* Smoke detectors use the alpha radiation from americium-241 to ionise the air in a small chamber. When smoke is present, the conductivity of the air is changed and a buzzer is activated.

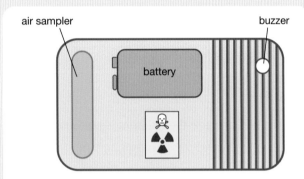

air sampler

buzzer

battery

radioactive source

a) Write a balanced nuclear equation for the alpha decay of americium-241.

b) The half-life of americium-241 is 433 years. Calculate the time taken for the activity of the sample to fall to 12.5% of its original value.

c) Give *two* reasons why americium-241 is a suitable radioisotope for use in an overhead smoke detector.

15* The radioisotope $^{131}_{53}I$ is used in hospitals. It has a half-life of 8 days and decays to give a stable product.

A bag of hospital linen contaminated with $^{131}_{53}I$ was found to give a count rate of 320 s^{-1}

a) Draw a graph to show how the count rate of the linen will change with time.

b) Hospitals are not allowed to dispose of material contaminated with $^{131}_{53}I$ until the count rate has fallen to 30 counts s^{-1}.

Use the graph to determine how long the bag of linen must be stored before disposal.

16* Radioisotopes are used in the treatment of patients suffering from cancer.

a) The isotope $^{60}_{27}Co$ has a half-life of 5.3 years and is used to supply gamma radiation from outside the body of the patient. Give *two* reasons why this isotope would *not* be suitable for use inside the body.

b) $^{32}_{15}P$, a beta-emitting isotope with a half-life of 14 days, is used in the treatment of skin cancer.

i) Show, *by calculation*, how the proton to neutron ratio is changed by the decay of this isotope.

ii) 3 g of the isotope was used to treat cancer over a period of 56 days. Calculate the mass of the isotope which decayed during this time.

17* The radioisotope, sodium-24, can be made in a nuclear reactor by bombarding element X with neutrons.

$$^{a}_{b}X + ^{1}_{0}n \rightarrow ^{24}_{11}Na$$

a) Identify element X and write values for a and b.

b) The graph shows how the mass of a sample of sodium-24 varies with time.

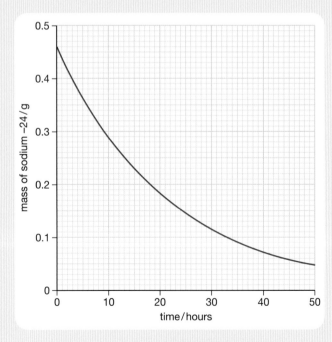

i) What is the half-life of sodium-24?

ii) Calculate the average rate of decay of sodium-24 over the first 10 hours.

Study Questions (continued)

iii) If the temperature of the sodium-24 sample is increased, how would this affect its rate of decay?

c) Two samples of ^{24}Na and ^{24}NaCl have the same mass. Why are their intensities of radiation different?

18* Technetium-99, which has a long half-life, is produced as a radioactive waste product in nuclear reactors. One way of reducing the danger of this isotope is to change it into technetium-100 by bombardment with particles, as shown by the nuclear equation.

$$^{99}_{43}\text{Tc} + \textbf{X} \rightarrow {}^{100}_{43}\text{Tc}$$

a) Identify particle **X**.

b) Technetium-100 decays by beta-emission. Write a balanced equation for this reaction.

c) Technetium-100 has a half-life of 16 s. If a sample of technetium-100 is left for 48 s, what fraction of the sample would remain?

End-of-Course Questions

All of the questions in this section come from past NQ Higher examination papers and are reproduced here with permission from the SQA.

Part 1: Multiple choice questions

1 Which of the following solids has a low melting point and a high electrical conductivity?

   A Iodine

   B Potassium

   C Silicon oxide

   D Potassium fluoride

2 Which covalent gas dissolves in water to form an alkali?

   A HCl

   B CH_4

   C SO_2

   D NH_3

3 Hydrochloric acid reacts with magnesium according to the following equation.

$$Mg(s) + 2H^+(aq) \rightarrow Mg^{2+}(aq) + H_2(g)$$

What volume of $4 \ mol \ l^{-1}$ hydrochloric acid reacts with 0.1 mol of magnesium?

   A $25 \ cm^3$

   B $50 \ cm^3$

   C $100 \ cm^3$

   D $200 \ cm^3$

4 Two identical samples of zinc were added to an excess of two solutions of sulphuric acid, concentrations $2 \ mol \ l^{-1}$ and $1 \ mol \ l^{-1}$ respectively.

Which of the following would have been the same for the two samples?

   A The total mass lost

   B The total time for the reaction

   C The initial reaction rate

   D The average rate of evolution of gas

5 When 3.6 g of butanal (relative formula mass = 72) was burned, 134 kJ of energy was released. From this result, what is the enthalpy of combustion, in $kJ \ mol^{-1}$?

   A -6.7 B $+6.7$

   C -2680 D $+2680$

6

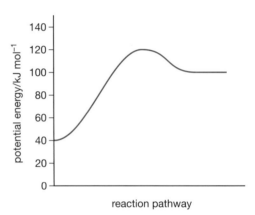

Which of the following sets of data applies to the reaction represented by the above energy diagram?

	Enthalpy change	Activation energy/kJ mol^{-1}
A	Exothermic	60
B	Exothermic	80
C	Endothermic	60
D	Endothermic	80

7 Which equation represents the first ionisation energy of a diatomic element?

   A $\frac{1}{2}X_2(s) \rightarrow X^+(g)$

   B $\frac{1}{2}X_2(g) \rightarrow X^-(g)$

   C $X(g) \rightarrow X^+(g)$

   D $X(s) \rightarrow X^-(g)$

8 Which type of bonding can be described as intermolecular?

   A Covalent bonding

   B Hydrogen bonding

   C Ionic bonding

   D Metallic bonding

9 Which statement may be correctly applied to silicon dioxide?

A It consists of discrete molecules.

B It has a covalent network structure.

C It is similar in structure to carbon dioxide.

D Van der Waals' attractions are important to its structure.

10 The shapes of some common molecules are shown below and each contains at least one polar bond. Which molecule is non-polar?

A H—Cl

B
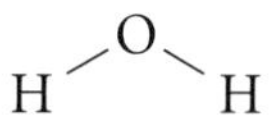
O
H
H

C O=C=O

D
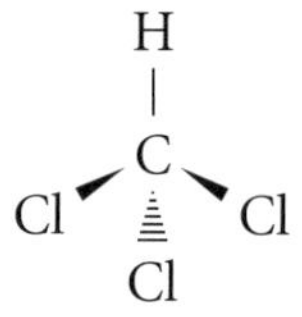
H
C
Cl
Cl
Cl

11

$$2NO(g) + O_2(g) \rightarrow 2NO_2(g)$$

How many litres of nitrogen dioxide gas could theoretically be obtained in the reaction of 1 litre of nitrogen monoxide gas with 2 litres of oxygen gas? (All volumes are measured under the same conditions of temperature and pressure.)

A 1

B 2

C 3

D 4

12 Which of the following gases has the same volume as 128.2 g of sulphur dioxide gas. (All volumes are measured under the same conditions of temperature and pressure.)

A 2.0 g of hydrogen

B 8.0 g of helium

C 32.0 g of oxygen

D 80.8 g of neon

13 The Avogadro Constant is the same as the number of

A atoms in 24 g of carbon

B molecules in 16 g of oxygen

C molecules in 2 g of hydrogen

D ions in 1 litre of sodium chloride solution, concentration 1 mol l^{-1}.

14 Which of the following processes can be used industrially to produce aromatic hydrocarbons?

A Reforming of naphtha

B Catalytic cracking of propene

C Reforming of coal

D Catalytic cracking of heavy oil fractions

15 Which of the following compounds does **not** have isomeric structures?

A C_2HCl_3

B $C_2H_4Cl_2$

C Propene

D Propan-1-ol

16 Which of the following structural formulae represents a tertiary alcohol?

A
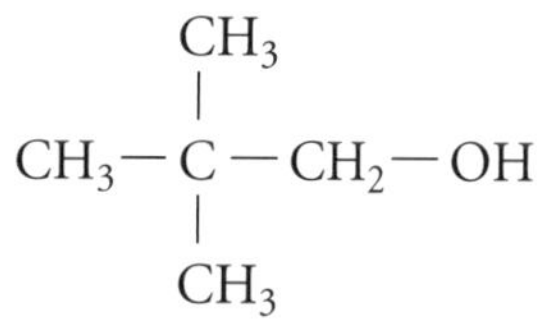
CH3
CH3—C—CH2—OH
CH3

B
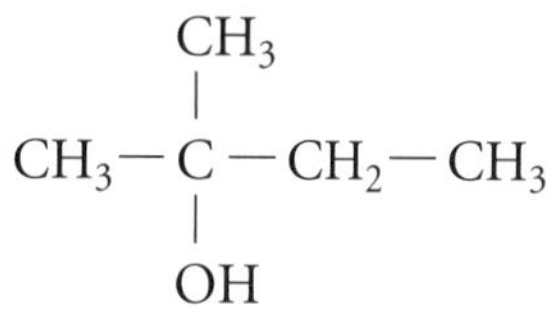
CH3
CH3—C—CH2—CH3
OH

C
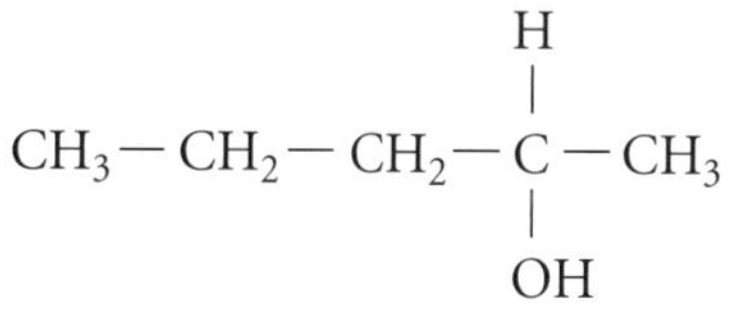
H
CH3—CH2—CH2—C—CH3
OH

D
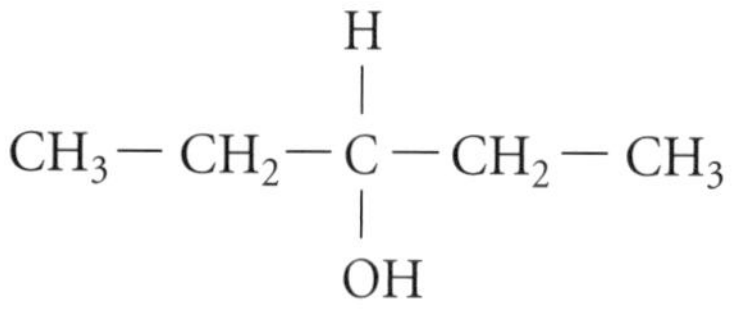
H
CH3—CH2—C—CH2—CH3
OH

17 An ester has the structural formula:

$$CH_3 - CH_2 - \overset{\overset{\displaystyle O}{\|}}{C} - O - \overset{\overset{\displaystyle CH_3}{|}}{\underset{\underset{\displaystyle CH_3}{|}}{C}} - H$$

On hydrolysis, the ester would produce

A ethanoic acid and propan-1-ol

B ethanoic acid and propan-2-ol

C propanoic acid and propan-1-ol

D propanoic acid and propan-2-ol

18 Ozone has an important role in the upper atmosphere because it

A reflects ultraviolet radiation

B reflects certain CFCs

C absorbs ultraviolet radiation

D absorbs certain CFCs.

19 Which reaction can be classified as reduction?

A $CH_3CH_2OH \rightarrow CH_3COOH$

B $CH_3CH(OH)CH_3 \rightarrow CH_3COCH_3$

C $CH_3CH_2COCH_3 \rightarrow CH_3CH_2CH(OH)CH_3$

D $CH_3CH_2CHO \rightarrow CH_3CH_2COOH$

20 Some recently developed polymers have unusual properties.

Which polymer is soluble in water?

A Poly(ethyne)

B Poly(ethenol)

C Biopol

D Kevlar

21 Which statement can be applied to polymeric esters?

A They are used for flavourings, perfumes and solvents.

B They are condensation polymers made by the linking up of amino acids.

C They are manufactured for use as textile fabrics and resins.

D They are cross-linked addition polymers.

22 In the formation of 'hardened' fats from vegetable oils, the hydrogen

A causes cross-linking between chains

B causes hydrolysis to occur

C increases the carbon chain length

D reduces the number of carbon–carbon double bonds.

23 Proteins can be denatured under acid conditions.

During this denaturing, the protein molecule

A changes shape

B is dehydrated

C is neutralised

D is polymerised.

24 Some solid ammonium chloride is added to a dilute solution of ammonia.

Which of the following ions will decrease in concentration as a result?

A Ammonium

B Hydrogen

C Hydroxide

D Chloride

25 Equal volumes of solutions of ethanoic acid and hydrochloric acid, of equal concentrations, are compared.

In which of the following cases does the ethanoic acid give the higher value?

A pH of solution

B Conductivity of solution

C Rate of reaction with magnesium

D Volume of sodium hydroxide solution neutralised

26 If 96 500 C of electricity are passed through separate solutions of copper(II) chloride and nickel(II) chloride, then

A equal masses of copper and nickel will be deposited

B the same number of atoms of each metal will be deposited

C the metals will be deposited on the positive electrode

D different numbers of moles of each metal will be deposited.

27 During a redox process in acid solution, iodate ions, $IO_3^-(aq)$, are converted into iodine, $I_2(aq)$.

$$IO_3^-(aq) \rightarrow I_2(aq)$$

The numbers of $H^+(aq)$ and $H_2O(l)$ required to balance the ion–electron equation for the formation of 1 mol of $I_2(aq)$ are, respectively

A 3 and 6

B 6 and 3

C 6 and 12

D 12 and 6

28 In which of the following reactions is hydrogen gas acting as an oxidising agent?

A $H_2 + C_2H_4 \rightarrow C_2H_6$

B $H_2 + Cl_2 \rightarrow 2HCl$

C $H_2 + 2Na \rightarrow 2NaH$

D $H_2 + CuO \rightarrow H_2O + Cu$

29 When 10 g of lead pellets containing radioactive lead are placed in a solution containing 10 g of lead nitrate, radioactivity soon appears in the solution.

Compared to the pellets the solution will show

A different intensity of radiation and different half-life

B the same intensity of radiation but different half-life

C different intensity of radiation but the same half-life

D the same intensity of radiation and the same half-life.

30 $^2_1H + ^3_1H \rightarrow ^4_2He + ^1_0n$

The above process represents

A nuclear fission

B nuclear fusion

C proton capture

D neutron capture.

Part 2: Extended answer questions

1 The effect of temperature changes on reaction rate can be studied using the reaction between an organic acid solution and acidified potassium permanganate solution.

$$5(COOH)_2(aq) + 6H^+(aq) + 2MnO_4^-(aq) \rightarrow$$
$$2Mn^{2+}(aq) + 10CO_2(g) + 8H_2O(l)$$

The apparatus required is shown in the diagram.

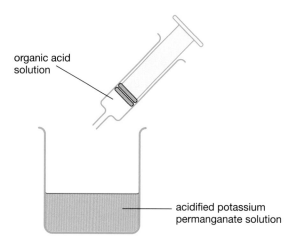

organic acid solution

acidified potassium permanganate solution

a) Name the organic acid. 1

b) Describe how the reaction time can be measured. 1

c) The headings for a set of results are shown below.

Temperature of reaction/°C	Reaction time/s	Reaction rate/

Complete the headings by giving the correct unit for the **third** column. 1

(3)

2 Vinyl acetate is the monomer for the preparation of polyvinylacetate (PVA) which is widely used in the building industry.

Vinyl acetate has the structural formula:

$$CH_3-\overset{\overset{\displaystyle O}{\|}}{C}-O-CH=CH_2$$

a) Draw part of the structure of polyvinylacetate, showing **three** monomer units joined together. 1

b) Vinyl acetate and hexane have the same relative formula mass. Explain why you would expect vinyl acetate to have a higher boiling point than hexane. 2 (3)

3 Calcium hydroxyapatite makes up 95% of tooth enamel.

a) Tooth decay is caused when tooth enamel is attacked by acid in the mouth.

i)   One of the acids which attacks tooth enamel is 2-hydroxypropanoic acid which has the molecular formula $C_3H_6O_3$. Draw a structural formula for this acid.   1

ii) Calcium hydroxyapatite reacts with acid in the mouth as shown by the following balanced equation.

$$Ca_{10}(PO_4)_6(OH)_2 + 8H^+ \rightarrow 10Ca^{2+} + 2H_2O$$
$$\text{calcium hydroxyapatite} \qquad + 6HPO_4{}^{x-}$$

What is the value of x?   1

iii) The pH of a solution in the mouth is 5. What is the concentration of hydrogen ions, in mol l^{-1}, in this solution.   1

b)   Tooth enamel also contains a fibrous protein called collagen.

i)   Describe a difference between a fibrous and a globular protein.   1

ii) Name the **four** elements present in all proteins.   1

(5)

4   Hydrogen sulphide, H_2S, is the unpleasant gas produced when eggs rot.

a) i)   The gas can be prepared by the reaction of iron(II) sulphide with dilute hydrochloric acid. Iron(II) chloride is the other product of the reaction.

Write a balanced chemical equation for this reaction.   1

ii) Iron metal is often present as an impurity in iron(II) sulphide.

Name the other product which would be formed in the reaction with dilute hydrochloric acid if iron metal is present as an impurity.   1

b) The enthalpy of combustion of hydrogen sulphide is $-563\,kJ\,mol^{-1}$.

Use this value and the enthalpy of combustion values in the Data Book to calculate the enthalpy change for the reaction:

$$H_2(g) + S(s) \rightarrow H_2S(g)$$
$$\text{(rhombic)}$$

Show your working clearly.   2

(4)

5   An ester can be prepared by the following sequence of reactions.

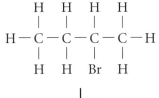

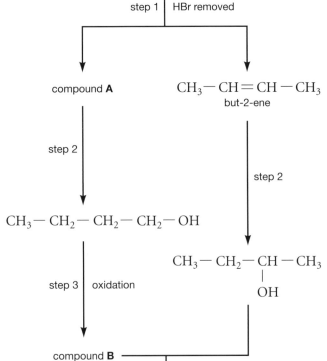

a) i)   Draw a structural formula for compound A.   1

ii) But-2-ene and compound A undergo the same type of reaction in Step 2.

Name this type of reaction.   1

iii) Acidified potassium dichromate solution can be used to carry out Step 3.

What colour change would be observed   1

iv) Name compound B.   1

b) i)   What evidence would show that an ester had been formed in Step 4?   1

ii) Give **one** use for esters.   1

(6)

6 A student added 0.20 g of silver nitrate, $AgNO_3$, to 25 cm³ of water. This solution was then added to 20 cm³ of 0.0010 mol l⁻¹ hydrochloric acid as shown in the diagram.

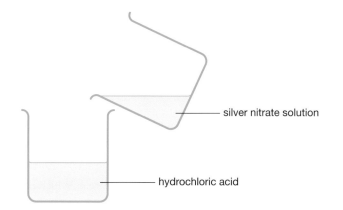

silver nitrate solution

hydrochloric acid

The equation for the reaction which occurs is:

$$AgNO_3(aq) + HCl(aq) \rightarrow AgCl(s) + HNO_3(aq)$$

a) i)   Name the type of reaction which takes place.   1

   ii)   Show by calculation which reactant is in excess. Show your working clearly.   2

b)   The hydrochloric acid in the experiment can be described as a dilute solution of a strong acid.

   i)   What is meant by a strong acid?   1

   ii)   What is the pH of the 0.0010 mol l⁻¹ hydrochloric acid used in the experiment?   1

   (5)

7 When sodium hydrogencarbonate is heated to 112°C it decomposes and the gas, carbon dioxide, is given off:

$$2NaHCO_3(s) \rightarrow Na_2CO_3(s) + CO_2(g) + H_2O(g)$$

The following apparatus can be used to measure the volume of carbon dioxide produced by the reaction.

solid sodium hydrogencarbonate

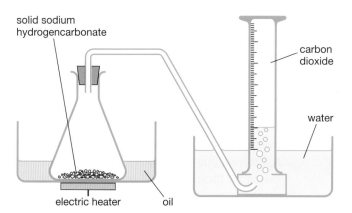

carbon dioxide

water

electric heater   oil

a)   Why is an oil bath used and **not** a water bath?   1

b) i)   Calculate the theoretical volume of carbon dioxide produced by the complete decomposition of 1.68 g of sodium hydrogencarbonate. (Take the molar volume of carbon dioxide to be 23 litres mol⁻¹.) Show your working clearly.   2

   ii)   Assuming that all of the sodium hydrogencarbonate is decomposed, suggest why the volume of carbon dioxide collected in the measuring cylinder would be less than the theoretical value.   1

   (4)

8 If both potassium iodide solution, KI(aq), and liquid chloroform, $CHCl_3(l)$, are added to a test-tube with some iodine, the iodine dissolves in both. Two layers are formed as shown in the diagram.

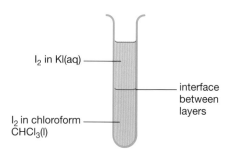

I_2 in KI(aq)

interface between layers

I_2 in chloroform $CHCl_3(l)$

An equilibrium is set up:

$$I_2 \text{ in KI(aq)} \rightleftharpoons I_2 \text{ in } CHCl_3(l)$$

The iodine is always distributed between the two layers in the same ratio:

$$\frac{\text{concentration of } I_2 \text{ in } CHCl_3(l)}{\text{concentration of } I_2 \text{ in KI(aq)}} = \frac{3}{1}$$

a)   What is meant by the term **equilibrium**?   1

b)   When more potassium iodide solution is added to the top layer the equilibrium is disturbed. What happens to restore the equilibrium?   1

c)   0.4 g of I_2 is dissolved in 10 cm³ of KI(aq) and 10 cm³ of $CHCl_3(l)$.

   Calculate the concentration of iodine, in g l⁻¹, contained in $CHCl_3(l)$.   1

   (3)

9 Phosphorus-32 is a radioisotope that decays by beta-emission.

a) Write the nuclear equation for the decay of phosphorus-32. 1

b) i) An 8 g sample of phosphorus-32 was freshly prepared. Calculate the number of phosphorus atoms in the 8 g sample. 1

ii) The half-life of phosphorus-32 is 14.3 days. Calculate the time it would take for the mass of phosphorus-32 in the 8 g sample to fall to 1 g. 1

(3)

10 An experiment using dilute hydrochloric acid and sodium hydroxide solution was carried out to determine the enthalpy of neutralisation.

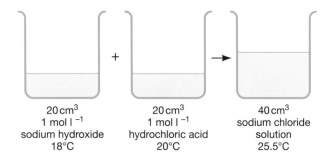

20 cm³	20 cm³	40 cm³
1 mol l⁻¹	1 mol l⁻¹	sodium chloride
sodium hydroxide	hydrochloric acid	solution
18°C	20°C	25.5°C

a) Using the information in the diagram, calculate the enthalpy of neutralisation in kJ mol⁻¹. Show your working clearly. 2

b) Calculate the concentration of hydroxide ions in the 1 mol l⁻¹ hydrochloric acid used in the experiment. 1

(3)

11 Uranium ore is converted into uranium(IV) fluoride, UF_4, to produce fuel for nuclear power stations.

a) In one process, uranium can be extracted from the uranium(IV) fluoride by a redox reaction with magnesium as follows.

$$2Mg + UF_4 \rightarrow 2MgF_2 + U$$

i) Give another name for this type of redox reaction. 1

ii) Write the ion-electron equation for the reduction reaction that takes place. 1

iii) The reaction with magnesium is carried out at a high temperature. The reaction vessel is filled with argon rather than air.

Suggest a reason for using argon rather than air. 1

b) In a second process, the uranium(IV) fluoride is converted into UF_6 as shown.

$$UF_4(s) + F_2(g) \rightarrow UF_6(g)$$

i) Name the type of bonding in $UF_6(g)$. 1

ii) Both UF_4 and UF_6 are radioactive. How does the half-life of the uranium in UF_4 compare with the half-life of the uranium in UF_6? 1

(5)

12 Vitamin C is required by our bodies for producing the protein, collagen. Collagen can form sheets that support skin and internal organs.

a) i) There are two main types of protein. Which of the two main types is collagen? 1

ii) Part of the structure of collagen is shown.

Draw a structural formula for an amino acid that

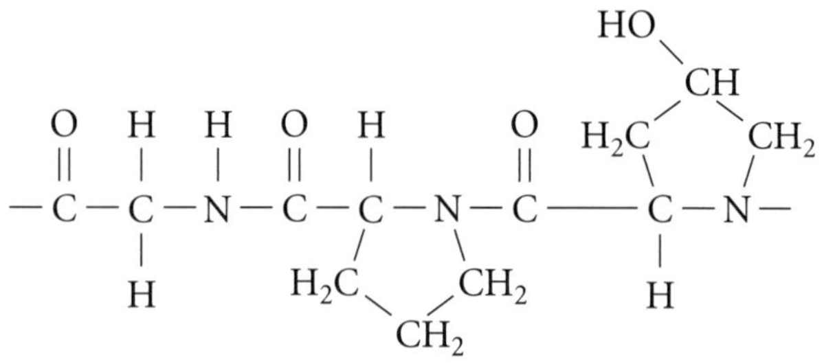

$$\begin{array}{c} O \quad H \quad H \quad O \quad H \qquad\qquad O \\ \| \quad | \quad | \quad \| \quad | \qquad\qquad \| \\ -C-C-N-C-C-N-C \quad\quad C-N- \\ | \qquad\qquad / \quad\quad \backslash \qquad | \\ H \qquad H_2C \quad CH_2 \qquad H \\ \backslash \quad / \\ CH_2 \end{array}$$

could be obtained by hydrolysing this part of the collagen. 1

b) A standard solution of iodine can be used to determine the mass of vitamin C in orange juice.

Iodine reacts with vitamin C as shown by the following equation.

$$C_6H_8O_6(aq) + I_2(aq) \rightarrow C_6H_6O_6(aq) +$$
vitamin C $\qquad\qquad 2H^+(aq) + 2I^-(aq)$

In an investigation using a carton containing 500 cm³ of orange juice, separate 50.0 cm³ samples were measured out. Each sample was then titrated with a 0.0050 mol l⁻¹ solution of iodine.

i) Why would starch solution be added to each 50.0 cm³ sample of orange juice before titrating against iodine solution? 1

ii) Titrating the whole carton of orange juice would require large volumes of iodine solution. Apart from this disadvantage give another reason for titrating several smaller samples of orange juice. 1

iii) An average of 21.4 cm³ of the iodine solution was required for the complete reaction with the vitamin C in 50.0 cm³ of orange juice.

Use this result to calculate the mass of vitamin C, in grams, in the 500 cm³ carton of orange juice. Show your working clearly. **2**

(6)

13 An industrial method for the production of ethanol is outlined in the flow diagram.

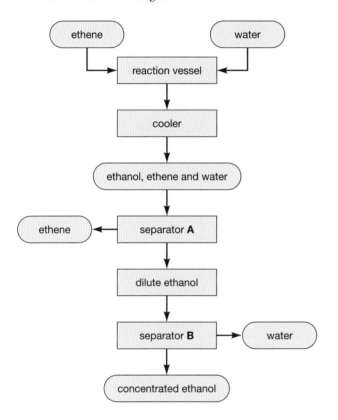

a) The starting materials are ethene and water. Water is a raw material but ethene is not. Why is ethene **not** a raw material? **1**

b) i) Unreacted ethene is removed in separator **A**. Suggest how the separated ethene could be used to increase the efficiency of the overall process. **1**

ii) Name the process that takes place in separator **B**. **1**

c) In the reaction vessel, ethanol is produced in an exothermic reaction.

$$C_2H_4(g) + H_2O(g) \rightleftharpoons C_2H_5OH(g)$$

i) Name the type of chemical reaction that takes place in the reaction vessel. **1**

ii) What would happen to the equilibrium position if the temperature inside the reaction vessel was increased? **1**

iii) If 1.64 kg of ethanol (relative formula mass = 46) is produced from 10.0 kg of ethene (relative formula mass = 28), calculate the percentage yield of ethanol. **2**

(7)

14 Chlorine can be produced commercially from concentrated sodium chloride solution in a membrane cell. Only sodium ions can pass through the membrane. These ions move in the direction shown in the diagram.

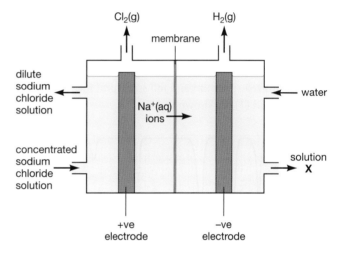

The reactions at each electrode are:

+ve electrode: $2Cl^-(aq) \rightarrow Cl_2(g) + 2e^-$

−ve electrode: $2H_2O(l) + 2e^- \rightarrow H_2(g) + 2OH^-(aq)$

a) Write the overall redox equation for the reaction in the membrane cell. **1**

b) i) Name solution X. **1**

ii) Hydrogen gas is produced at the negative electrode in the membrane cell. Suggest why this gas could be a valuable resource in the future. **1**

c) Calculate the mass of chlorine, in kilograms, produced in a membrane cell using a current of 80 000 A for 10 hours. Show your working clearly. **3**

(6)

15 Infrared spectroscopy can be used to help identify the bonds which are present in an organic molecule.

Different bonds absorb infrared radiation of different wave numbers.

The table below shows the range of wave numbers of infrared radiation absorbed by the bonds indicated with thicker lines.

Bond	Wave number range/cm^{-1}
O—H	3650–3590
C≡C—H	3300
C—C—H	2962–2853
C≡C	2260–2100
C—O	1150–1070

The infrared spectrum and full structural formula for compound ① are shown.

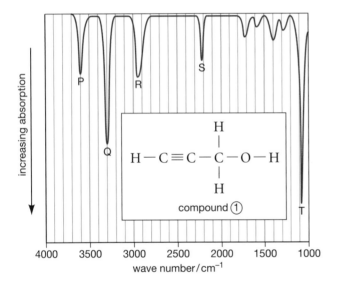

a) In the above full structural formula for compound ①, use an arrow to indicate the bond that could be responsible for the absorption at **T** in its infrared spectrum.   1

b) A series of reactions is carried out starting with compound ①.

$$\text{compound ①} \xrightarrow{\text{1 mol H}_2} \text{compound ②} \xrightarrow{\text{1 mol H}_2} \text{compound ③}$$

i)   Give the letters for the **two** absorptions which would **not** appear in the infrared spectrum for compound ②.   1

ii) Name compound ③.   1

                                                                (3)

Chemical Dictionary

A

Activated complex: An unstable arrangement of atoms formed at the maximum of the potential energy barrier during a reaction.

Activation energy: The energy required by colliding molecules to form an activated complex.

Addition: A reaction in which two or more molecules combine to produce a larger molecule and nothing else.

Addition polymerisation: A process whereby many small unsaturated molecules (monomers) join to form one large molecule (a polymer) and nothing else.

Alcohols: Carbon compounds which contain the hydroxyl functional group, —OH, except when it is linked to a benzene ring.

Aldehydes: Carbon compounds which contain the —CHO functional group. They are reducing agents.

Alicyclic compounds: Carbon compounds having a ring structure which is non-aromatic, e.g. cyclohexane. In most alicyclic compounds the ring is not planar.

Aliphatic compounds: Carbon compounds which have a straight-chain or branched-chain structure, e.g. hexane, 2-methyl propan-1-ol.

Alkanals: A homologous series of aldehydes, general formula $C_nH_{2n}O$. The first member is methanal, HCHO.

Alkanes: A homologous series of saturated hydrocarbons, general formula C_nH_{2n+2}. The first member is methane, CH_4.

Alkanoic acids: A homologous series of carboxylic acids, general formula $C_nH_{2n}O_2$. The first member is methanoic acid, HCOOH.

Alkanols: A homologous series of alcohols, general formula $C_nH_{2n+1}OH$. The first member is methanol, CH_3OH.

Alkanones: A homologous series of ketones, general formula $C_nH_{2n}O$. The first member is propanone, CH_3COCH_3.

Alkenes: A homologous series of unsaturated hydrocarbons, general formula C_nH_{2n}. Each member contains a carbon–carbon double bond. The first member is ethene $CH_2{=}CH_2$.

Alkyl group: A group of carbon and hydrogen atoms forming a branch in a carbon compound, e.g. methyl group, CH_3—; ethyl group, C_2H_5—.

Alkynes: A homologous series of unsaturated hydrocarbons, general formula C_nH_{2n-2}. Each member contains a carbon–carbon triple bond. The first member is ethyne, C_2H_2.

Alpha particle: Charged particle consisting of two protons and two neutrons emitted by some radioisotopes. Identical to a helium nucleus and represented as ^4_2He.

Amide link: Group of atoms formed by condensation polymerisation of amino acids in the formation of protein chains. The amide link is —CO—NH— and occurs between each pair of amino acid residues in the chain. Also called peptide link.

Amino acids: Compounds of general formula, $H_2NCHXCOOH$ – where X is, for example, H, CH_3, $C_6H_5CH_2$ – which polymerise by condensation to form proteins. Essential amino acids cannot be synthesised by an organism and must be present in its diet.

Aromatic compounds: Carbon compounds which contain the characteristic benzene ring of six carbon atoms. The ring is planar and has delocalised electrons.

Atomic number: The number of protons in the nucleus of an atom.

Average rate: The change in mass or concentration of a reactant or product divided by the time interval during which the change occurs.

Avogadro Constant, L: The number of formula units in the gram formula mass of a substance, e.g. the number of molecules in 18 g of water, $L = 6.02 \times 10^{23}\ mol^{-1}$.

B

Beta particle: Charged particle consisting of a single electron emitted by some radioisotopes. Represented as $_{-1}^{0}e$

Binding energy: The energy which must be supplied to an atomic nucleus to cause it to split up into its component protons and neutrons.

Biodegradable: Able to rot away by natural biological processes.

Biogas: Gas produced by anaerobic fermentation of organic waste. Mainly consists of methane with some CO_2.

C

Carbonyl group: The functional group present in ketones, namely $>C=O$ (also present in aldehydes as part of their functional group, —CHO).

Carboxyl group: The functional group present in carboxylic acids, namely —COOH.

Carboxylic acids: Carbon compounds which contain the carboxyl functional group.

Catalyst: A substance which speeds up a reaction without itself being used up. It lowers the activation energy of the reaction.

Catalyst poisoning: This occurs when a substance forms strong bonds with the surface of a catalyst thus reducing its efficiency.

Concentration: The amount of solute dissolved in a given volume of solution. The usual units are moles per litre (mol 1^{-1}).

Condensation polymerisation: A process whereby many small molecules (monomers) join to form a large molecule (a polymer), with water or other small molecules formed at the same time.

Condensation reaction: A reaction in which two (or more) molecules join to produce a single larger molecule, with water or another small molecule formed at the same time.

Covalent atomic radius: A useful measure of atomic size, being half the distance between the nuclei of two covalently-bonded atoms of an element. Covalent bond lengths between any two atoms can be obtained by adding the appropriate covalent atomic radii.

Covalent bonding: Bond formed between two non-metal atoms by the sharing of a pair of electrons.

Cracking: The breaking up of larger hydrocarbon molecules (usually alkanes) to produce a mixture of smaller molecules (usually alkanes and alkenes). Heat alone is used in thermal cracking. The use of a catalyst (in catalytic cracking) allows the process to be carried out at a lower temperature.

Critical mass: The minimum mass of material needed to sustain a chain fission reaction in a nuclear reactor or nuclear weapon.

Cycloalkanes: A homologous series of saturated ring molecules with general formula C_nH_{2n}. The simplest is cyclopropane, C_3H_6.

D

Dehydration: The removal of water from a single compound, e.g. dehydration of ethanol, C_2H_5OH, produces ethene, C_2H_4.

Dehydrogenation: The removal of hydrogen from a single compound, e.g. dehydrogenation of propane, C_3H_8, produces propene, C_3H_6.

Delocalised electrons: Electrons which are not confined to a single orbital between a pair of atoms, e.g. in metallic bonding. The benzene 'ring' also has delocalised electrons.

Denaturing of proteins: The changes resulting in the secondary structure of proteins from increases of temperature or reduction in pH. Loss of enzyme activity is one important consequence.

Dipole–dipole interactions: The attraction between molecules which possess a permanent dipole because of the presence of polar bonds.

Dissociation: The splitting into ions of acids and bases, e.g.

$$HCl \ (aq) \rightarrow H^+ \ (aq) + Cl^- \ (aq)$$

$$NH_3 \ (g) + H_2O \ (l) \rightarrow NH_4^+ \ (aq) + OH^- \ (aq)$$

E

Electrochemical series: A list of chemicals arranged in order of their increasing ability to gain electrons, i.e. in order of increasing oxidising power.

Electrolysis: The process which occurs when a current of electricity is passed through a molten electrolyte (resulting in decomposition) or an electrolyte solution (which results in decomposition of the solute and/or the water).

Electron: A particle which moves around the nucleus of an atom. It has a single negative charge but its mass is negligible compared to that of a proton or neutron.

Electronegativity: The strength of the attraction by an atom of an element for its bonding electrons. If the electronegativities of two atoms sharing electrons is similar, the bond will be almost purely covalent. The greater the difference in electronegativities, the more likely the bond is to be polar covalent, or even ionic.

Empirical formula: Shows the simplest ratio of atoms in a compound, for example CH_3 is the empirical formula for ethane (molecular formula C_2H_6).

Endothermic reaction: A reaction in which heat energy is absorbed from the surroundings. It has a positive enthalpy change ($\Delta H+$).

Enthalpy change: The difference in heat energy between reactants and products in a reaction.

Enthalpy of combustion: The enthalpy change when one mole of a substance is completely burned in oxygen.

Enthalpy of neutralisation of an acid: The enthalpy change when the acid is neutralised to form one mole of water. The enthalpy of neutralisation of a base can be similarly defined.

Enthalpy of solution: The enthalpy change when one mole of a substance is dissolved in sufficient water to form a dilute solution.

Enzyme: A globular protein which is able to catalyse a specific reaction.

Equilibrium: State attained in a reversible reaction when forward and reverse reactions are taking place at the same rate.

Esters: Carbon compounds formed when alcohols react with carboxylic acids by condensation.

Exothermic reaction: A reaction in which heat energy is released to the surroundings. It has a negative enthalpy change ($\Delta H-$).

F

Faraday: The quantity of electricity equal to the total charge of one mole of electrons, i.e. 96 500 coulombs.

Fats: Esters formed from one molecule of glycerol and three molecules of, usually saturated, long-chain carboxylic acids. The compounds have melting points high enough to be solid at room temperature. See also oils.

Feedstock: A substance derived from a raw material which is used to manufacture another substance.

Fermentation: Changes in organic substances brought about by enzymes from living organisms such as yeasts or bacteria.

Fractional distillation of crude oil: A means of separation into groups of hydrocarbons with similar boiling points (fractions).

Fuel: A substance which is used as a source of energy. This is released when the fuel burns.

Functional group: A group of atoms or type of carbon–carbon bond which provides a series of carbon compounds with its characteristic chemical properties, e.g. —CHO, —C=C—.

G

Gamma radiation: High frequency, and high energy, electromagnetic radiation emitted by radioactive substances.

Group: A column of elements in the Periodic Table. The values of a selected physical property show a distinct trend of increase or decrease down the column. The chemical properties of the elements in the group are similar.

H

Haber Process: The industrial production of ammonia from nitrogen and hydrogen using high pressure and temperature, with iron as a catalyst.

Half-life: The time in which the activity of a radioisotope decays by half, or in which half of its atoms disintegrate. Half-lives vary from isotope to isotope. Their length varies, ranging from fractions of seconds to millions of years.

Hess's Law: The enthalpy change of a chemical reaction depends only on the chemical nature and physical state of the reactants and products and is independent of any intermediate steps.

Heterogeneous catalyst: A catalyst which is in a different physical state from the reactants.

Homogeneous catalyst: A catalyst which is in the same physical state as the reactants.

Homologous series: A group of chemically similar compounds which can be represented by a general formula. Physical properties change progressively through the series, for example, the alkanes, general formula C_nH_{2n+2}, show a steady increase in boiling point.

Hormones: Chemicals, often complex proteins, which regulate metabolic processes in the body. An example is insulin which regulates sugar metabolism.

Hydration: The addition of water to an unsaturated compound, e.g. the hydration of ethene, C_2H_4, produces ethanol, C_2H_5OH.

Hydrocarbon: A compound containing the elements carbon and hydrogen only.

Hydrogenation: The addition of hydrogen to an unsaturated compound, e.g. hydrogenation converts alkenes to alkanes and oils into fats.

Hydrogen bonds: Intermolecular forces of attraction. The molecules must contain highly polar bonds in which hydrogen atoms are linked to very electronegative nitrogen, oxygen or fluorine atoms. The hydrogen atoms are then left with a positive charge and are attracted to the electronegative atoms of other molecules. They are a specific, stronger type of dipole–dipole interaction.

Hydrolysis: The breaking down of larger molecules into smaller molecules by reaction with water.

I

Intermolecular bonds: Bonds between molecules e.g. van der Waal's bonds, dipole–dipole interactions and hydrogen bonds. They are much weaker than intramolecular bonds.

Intramolecular bonds: Bonds within molecules, i.e. covalent and polar covalent bonds.

Ion–electron equation: Equation which shows either the loss of electrons (oxidation) or the gain of electrons (reduction).

Ionic bond: Bond formed as a result of attraction between positive and negative ions.

Ionic product: In water, the ionic product, K_w, is $[H^+][OH^-]$. Its value is 10^{-14} mol^2 l^{-2} at approx. 24°C.

Ionisation: The loss or gain of electrons by neutral atoms to form ions, e.g.

$$Na(g) \rightarrow Na^+(g) + e^-$$

$$Cl(g) + e^- \rightarrow Cl^-(g)$$

'Ionisation enthalpy' is usually reserved for enthalpy changes referring to the formation of positive ions. For the formation of negative ions 'electron-gain enthalpy' is used.

Ions: Atoms or groups of atoms which possess a positive or negative charge due to loss or gain of electrons, for example Na^+ and CO_3^{2-}.

Isomers: Compounds which have the same molecular formula but different structural formulae.

Isotopes: Atoms of the same element which have different numbers of neutrons. They have the same atomic number but different mass numbers.

K

Ketones: Carbon compounds which contain the carbonyl group. They are not reducing agents (unlike aldehydes).

L

Lattice: The three-dimensional arrangement of positive and negative ions in the solid, crystalline state of ionic compounds.

Le Chatelier's Principle: If any change of physical or chemical conditions is imposed on any chemical equilibrium then the equilibrium alters in the direction which tends to counteract the change of conditions. For a summary of the application of Le Chatelier's Principle see page 166.

M

Mass number: The total number of protons and neutrons in the nucleus of an atom.

Metallic bonding: The bonding responsible for typically metallic properties such as malleability, ductility and electrical conductivity in metals and alloys. Each atom loses its outer electrons to form positive ions. These ions pack together in a regular crystalline arrangement with the electrons delocalised through the structure binding the ions together.

Miscibility: The ability of liquids to mix perfectly together. In contrast, immiscible liquids form clearly defined layers with the denser liquid forming the lower layer.

Molar volume: The volume of a mole of a gas at a specified temperature and pressure.

Mole: The gram formula mass of a substance. It contains 6.02×10^{23} formula units of the substance.

Molecular formula: Formula which shows the number of atoms of the different elements which are present in one molecule of a substance.

Molecule: A group of atoms held together by covalent bonds.

Monomers: Relatively small molecules which can join together to produce a very large molecule (a polymer) by a process called polymerisation.

N

Neutral solution: Solution in which the concentrations of $H^+(aq)$ and $OH^-(aq)$ ions are equal (pH = 7).

Neutron: A particle found in the nucleus of an atom. It has the same mass as a proton but no charge.

Nucleus: The extremely small centre of an atom where the neutrons and protons are found.

O

Oils: Esters formed from one molecule of glycerol and three molecules of, usually unsaturated, carboxylic acids. Oils have melting points low enough to be liquid at normal room temperature. See fats.

Ostwald Process: The industrial production of nitric acid from ammonia by a process which includes catalytic oxidation.

Oxidation: A process in which electrons are lost.

Oxidising agent: A substance which gains electrons, i.e. is an electron acceptor.

P

Peptide link: See amide link.

Percentage yield: This is the actual yield of substance obtained divided by the theoretical yield calculated from the balanced equation then multiplied by 100.

Period: A horizontal row in the Periodic Table.

Periodic Table: An arrangement of the elements in order of increasing atomic number, with chemically similar elements occurring in the same main vertical columns (groups).

pH: A measure of the acidity of a solution, strictly pH $= -\log_{10}[H^+]$. Since, in any solution in water, $[H^+][OH^-] = 10^{-14} \text{ mol}^2 \text{ l}^{-2}$, pHs also apply to solutions of alkalis. (See Chapter 16.)

Phenyl group: A group of carbon and hydrogen atoms, formula $C_6H_5\text{—}$, found in many aromatic compounds, e.g. phenylethene, $C_6H_5CH=CH_2$.

Polar covalent bonds: Bonds formed between non-metallic atoms by sharing a pair of electrons. If the atoms have considerably different electronegativities, the electrons are not shared equally, the more electronegative atom becoming slightly negative in comparison to the other atom. As a result the bond is 'polar', e.g. $H^{\delta+}\text{—}Cl^{\delta-}$.

Polymer: A very large molecule which is formed by the joining together of many smaller molecules (monomers).

Polymerisation: The process whereby a polymer is formed. (See addition polymerisation and condensation polymerisation.)

Proton: A particle found in the nucleus of an atom. It has a single positive charge and the same mass as a neutron.

Q

Quantitative electrolysis: An experiment in which the quantity of product obtained during electrolysis is related to the quantity of electricity used.

R

Radioactivity: The emission of radiation from some atomic nuclei which disintegrate because their proton–neutron ratio makes them inherently unstable. Three types of radiation, α, β or γ, may be emitted.

Redox reaction: A reaction in which reduction and oxidation take place. Electrons are lost by one substance and gained by another.

Redox titration: An experiment in which the volumes of aqueous solutions of a reducing agent and an oxidising agent, which react together completely, are accurately measured. The concentration of one of the reactants can then be determined provided the concentration of the other reactant is known.

Reducing agent: A substance which loses electrons, i.e. is an electron donor.

Reduction: A process in which electrons are gained.

Reforming: A reaction which alters the structure of hydrocarbon molecules, e.g. straight-chain hydrocarbons may be converted to branched-chain or aromatic hydrocarbons.

Relative atomic mass: The average mass of one atom of an element on a scale where one atom of $^{12}_{6}C$ has a mass of 12 units exactly.

Renewable energy source: One which will not run out in the foreseeable future, e.g. solar, wind and tidal power, fuel obtained from crops.

Reversible reaction: One which proceeds in both directions, for example:

$$N_2 + 3H_2 \rightleftharpoons 2NH_3$$

S

Saturated hydrocarbon: A hydrocarbon in which all carbon–carbon covalent bonds are single bonds.

Screening: The ability of electrons in inner energy levels of an atom to reduce the attraction of the nuclear charge for the electrons of the outermost levels.

Spectator ion: An ion which is present in a reaction mixture but takes no part in the reaction.

Standard solution: A solution of known concentration.

State symbols: Symbols used to indicate the state of atoms, ions or molecules: (s) = solid; (1) = liquid; (g) = gas; (aq) = aqueous (dissolved in water).

Steam cracking: The breakdown of hydrocarbons in the presence of steam, notably used in the production of ethene and propene.

Structural formula: A formula which shows the arrangement of atoms in a molecule or ion. A full structural formula shows all of the bonds, for example, propane:

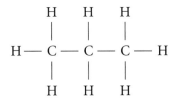

A shortened structural formula shows the sequence of groups of atoms, for example, propane: $CH_3CH_2CH_3$.

Strong acids and alkalis: Acids and alkalis which are fully dissociated into ions in solution. Strong acids include hydrochloric, nitric and sulphuric acids. Strong alkalis include sodium and potassium hydroxides.

Supernova: When all the hydrogen in a star has been converted into heavier elements, the star begins to absorb energy and collapse. Its rotation increases and it breaks up, flinging material into space to be used in the formation of other stars. The event is an explosion much brighter than the sun, the effects being visible in the sky for centuries.

Synthesis gas: A mixture of carbon monoxide and hydrogen made by steam-reforming natural gas, methane. It is used for the synthesis of methanol.

T

Thermoplastic: Refers to polymers which can be softened by heating and then remoulded. They harden again on cooling.

Thermosetting: Refers to polymers which form as a rigid cross-linked structure when their monomers react on heating. They cannot be softened by further heating.

Transition metals: The elements which form a 'bridge' in the Periodic Table between groups II and III, for example, iron and copper.

U

Unsaturated compounds: Compounds in which there are carbon–carbon double or triple bonds, e.g. alkenes, alkynes and vegetable oils.

V

Variable: Something that can be changed in a chemical reaction e.g. temperature, particle size, concentration, etc.

Viscosity: A description of how 'thick' a liquid is, for example, engine oil is 'thicker' (more viscous) than petrol.

W

Weak acids and alkalis: Acids and alkalis which do not dissociate fully into ions in solution. Weak acids include most carboxylic acids, carbonic acid and sulphurous acid. Weak alkalis include ammonia and most amines.

Periodic Table of the Elements

Group I Group II

Group III Group IV Group V Group VI Group VII Group 0

Key

1.0	relative atomic mass
H	symbol
hydrogen	name
1	atomic number

Transition elements

Group I	Group II											Group III	Group IV	Group V	Group VI	Group VII	Group 0
1.0 **H** hydrogen 1																	4.0 **He** helium 2
6.9 **Li** lithium 3	9.0 **Be** beryllium 4											10.8 **B** boron 5	12.0 **C** carbon 6	14.0 **N** nitrogen 7	16.0 **O** oxygen 8	19.0 **F** fluorine 9	20.2 **Ne** neon 10
23.0 **Na** sodium 11	24.3 **Mg** magnesium 12											27.0 **Al** aluminium 13	28.1 **Si** silicon 14	31.0 **P** phosphorus 15	32.1 **S** sulphur 16	35.5 **Cl** chlorine 17	40.0 **Ar** argon 18
39.1 **K** potassium 19	40.0 **Ca** calcium 20	45.0 **Sc** scandium 21	47.9 **Ti** titanium 22	51.0 **V** vanadium 23	52.0 **Cr** chromium 24	54.9 **Mn** manganese 25	55.8 **Fe** iron 26	58.9 **Co** cobalt 27	58.7 **Ni** nickel 28	63.5 **Cu** copper 29	65.4 **Zn** zinc 30	69.7 **Ga** gallium 31	72.6 **Ge** germanium 32	74.9 **As** arsenic 33	79.0 **Se** selenium 34	79.9 **Br** bromine 35	83.8 **Kr** krypton 36
85.5 **Rb** rubidium 37	87.6 **Sr** strontium 38	88.9 **Y** yttrium 39	91.2 **Zr** zirconium 40	92.9 **Nb** niobium 41	95.9 **Mo** molybdenum 42	**Tc** technetium 43	101.1 **Ru** ruthenium 44	102.9 **Rh** rhodium 45	106.4 **Pd** palladium 46	107.9 **Ag** silver 47	112.4 **Cd** cadmium 48	114.8 **In** indium 49	118.7 **Sn** tin 50	121.8 **Sb** antimony 51	127.6 **Te** tellurium 52	126.9 **I** iodine 53	131.3 **Xe** xenon 54
132.9 **Cs** caesium 55	137.3 **Ba** barium 56		178.5 **Hf** hafnium 72	181.0 **Ta** tantalum 73	183.9 **W** tungsten 74	186.2 **Re** rhenium 75	190.2 **Os** osmium 76	192.2 **Ir** iridium 77	195.1 **Pt** platinum 78	197.0 **Au** gold 79	200.6 **Hg** mercury 80	204.4 **Tl** thallium 81	207.2 **Pb** lead 82	209.0 **Bi** bismuth 83	**Po** polonium 84	**At** astatine 85	**Rn** radon 86
Fr francium 87	226.0 **Ra** radium 88		104 **Rf** rutherfordium 104	**Db** dubnium 105	**Sg** seaborgium 106	**Bh** bohrium 107	**Hs** hassium 108	**Mt** meitnerium 109	**Uun** ununnilium 110	**Uuu** unununium 111	**Uub** ununbium 112						

Lanthanides

138.9 **La** lanthanum 57	140.1 **Ce** cerium 58	140.9 **Pr** praseodymium 59	144.2 **Nd** neodymium 60	**Pm** promethium 61	150.4 **Sm** samarium 62	152.0 **Eu** europium 63	157.3 **Gd** gadolinium 64	158.9 **Tb** terbium 65	162.5 **Dy** dysprosium 66	164.9 **Ho** holmium 67	167.3 **Er** erbium 68	168.9 **Tm** thulium 69	173.0 **Yb** ytterbium 70	175.0 **Lu** lutetium 71

Actinides

227.0 **Ac** actinium 89	232.0 **Th** thorium 90	231.0 **Pa** protactinium 91	238.0 **U** uranium 92	237.0 **Np** neptunium 93	**Pu** plutonium 94	**Am** americium 95	**Cm** curium 96	**Bk** berkelium 97	**Cf** californium 98	**Es** einsteinium 99	**Fm** fermium 100	**Md** mendelevium 101	**No** nobelium 102	**Lr** lawrencium 103

Relative atomic masses are shown only for elements which have stable isotopes or isotopes with very long half-life.

Index

Answers to In-Text Questions

Chapter 1: Reaction Rate

1 a) 0.2   b) 1.5   c) 0.1   d) 30

2 a) $0.5 \, mol \, l^{-1}$ b) $2.5 \, mol \, l^{-1}$
  c) $0.05 \, mol \, l^{-1}$ d) $5 \, mol \, l^{-1}$

3 a) i) 0.02   ii) 3.4 g

  b) i) 6   ii) 636 g

  c) i) 0.5   ii) 74.2 g

  d) i) 2.5   ii) 330.3 g

4 a) In the first 30 s the average rate in terms of

   i) mass of CO_2 produced = $0.042 \, g \, s^{-1}$

   ii) decreasing concentration of acid
       = $0.038 \, mol \, l^{-1} \, s^{-1}$.

  b) Between 60 s and 90 s the average rate in terms of

   i) mass of CO_2 produced = $0.016 \, g \, l^{-1}$

   ii) decreasing concentration of acid
       = $0.014 \, mol \, l^{-1} \, s^{-1}$.

5 The average rate in terms of

  a) mass of CO_2 produced = $0.48 \, g \, min^{-1}$

  b) decreasing concentration of acid
     = $0.44 \, mol \, l^{-1} \, min^{-1}$.

6

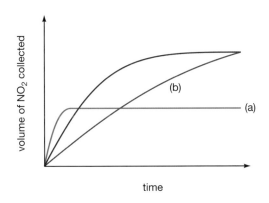

7 a) Decreasing concentration of acid/H^+ ions as the
     reaction proceeds.

b)

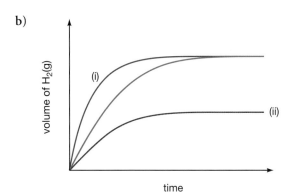

8 a) i.e. the reaction between hydrogen and oxygen.

9 Burning phosphorus: low activation energy;

  burning magnesium: high activation energy.

10 6.62 g

11 798 tonnes

12 a) 0.025 mol of Mg present and this would require 0.05
      mol of HCl.

      0.08 mol of HCl present.

   b) 0.03 mol of HCl in excess.

   c) Titrate the excess acid using a suitable indicator (e.g.
      methyl orange) and NaOH(aq) of known
      concentration.

13 a) 0.04 mol of Zn present and this would require 0.04
      mol of $CuSO_4$.

      0.03 mol of $CuSO_4$ present.

   b) 1.91 g

   c) Add dilute HCl to dissolve the excess Zn (Cu does not
      react). Filter off the copper; wash, dry and weigh it.

14 a) 0.004 mol of $Pb(NO_3)_2$ present and this would
      require 0.008 mol of KI.

      0.006 mol of KI present. Hence, $Pb(NO_3)_2$ is present
      in excess.

   b) 1.383 g

15 a) Heterogeneous, the reactants are gases while the
      catalyst is a solid.

   b) Homogeneous, the reactants and the catalyst are
      liquids.

c) Heterogenous, the reactants are gases while the catalyst is a liquid.

Chapter 2: Enthalpy

1 a) Reaction C

b) i) E_A: Reaction A: 50 kJ; B: 30 kJ; C: 40 kJ

ii) ΔH: Reaction A: -10 kJ; B: -40 kJ; C: 20 kJ

c) i) Reaction A   ii) Reaction B

2 -877.8 kJ mol^{-1}

3 a) -1254 kJ mol^{-1}

b) This is much less than the Data Book value. The main sources of error are likely to be incomplete combustion of propanol and heat losses to the surroundings.

4 34.8 kJ mol^{-1}

5 -40.1 kJ mol^{-1}

6 -54.3 kJ mol^{-1}

7 -56.85 kJ mol^{-1}

Chapter 3: Patterns in the Periodic Table

1 a) Potassium has an extra layer of electrons.

b) Although the atoms have the same number of layers of electrons, the chlorine atom has a larger nuclear charge than sodium. This attracts the outer electrons more strongly and causes a decrease in the atomic radius.

2 The first electron to be removed from each group I element is in the outside layer. The second is in a layer nearer to the nucleus and requires more energy to remove it.

3 a) $596 + 1160 = 1756$ kJ mol^{-1}

b) $584 + 1830 + 2760 = 5174$ kJ mol^{-1}

4 Xe has the largest (non-radioactive) noble gas atoms, and the electron to be removed to create Xe$^+$ will be further from the nucleus than the equivalent electron in any other noble gas atom. Hence the 1st ionisation enthalpy of Xe is the lowest of any of the noble gases.

Chapter 4: Bonding, Structure and Properties of the Elements

1 Electronegativity (the same for atoms of the same element).

2 More electrons in the molecules as group is descended, leading to stronger van der Waals' bonds, leading to higher melting points.

3 Aluminium has three outer electrons compared to sodium's one. These electrons delocalise leading to greater conductivity.

4 Metals

5 Diamond has a network–covalent structure. Sulphur has S_8 molecules which contain covalent bonds, but the molecules are held together to form the lump by weak van der Waals' bonds.

Chapter 5: Bonding, Structure and Properties of Compounds

1 a) 2.0   b) 1.7   c) 2.2 CsCl is most ionic

2 Electronegativity difference is 1.4. In fact it is ionic. You should recognise that it will be, at least, strongly polar.

3 KBr and MgO (ionic ratio 1 : 1)

4

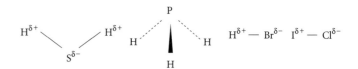

5 $H^{\delta+}$ —$Cl^{\delta-}$ deflected

$O=O$ not deflected

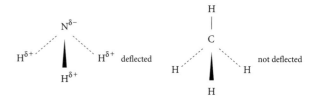

Chapter 6: The Mole

1 a) 159.8 g Br$_2$ b) 16 g O$_2$

2 A, B, D and F are true.

3 a) 39.1 g   b) 34 g   c) 58.1 g

4 25 litres

5 a) $0.84\,g\,l^{-1}$ b) $0.71\,g\,l^{-1}$ c) $1.83\,g\,l^{-1}$

6 a) i) $960\,cm^3$ (or 0.96 litres) ii) 120 litres

   b) i) 3 mol ii) 0.015 mol

7 a) 4.8 litres b) 2.4 litres

8 600 litres

9 $30\,cm^3$ of O_2 required; $60\,cm^3$ of NO_2 produced.

10 $100\,cm^3$ of O_2 required; $60\,cm^3$ of CO_2 and $80\,cm^3$ of water vapour produced.

11 a) i) $400\,cm^3$ of O_2 ii) $200\,cm^3$ of CO_2

   b) i) 32.5 litres of O_2 ii) 20 litres of CO_2

12 a) Aluminium in excess b) 2.4 litres

13 a) i) Oxygen in excess

      ii) $5\,cm^3$ of excess O_2 and $10\,cm^3$ of CO_2 produced.

   b) i) Propane in excess

      ii) $5\,cm^3$ of excess C_3H_8 and $15\,cm^3$ of CO_2 produced.

Chapter 7: Fuels

1 a) $C_6H_{14} \rightarrow C_6H_6 + 4H_2$

   b)

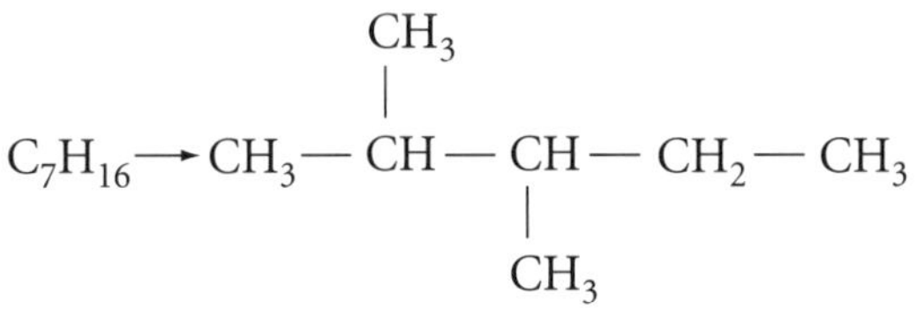

$$C_7H_{16} \rightarrow CH_3-\underset{\underset{CH_3}{|}}{CH}-\underset{\underset{CH_3}{|}}{CH}-CH_2-CH_3$$

   The product of this reaction can have different structures, provided the branches are not attached to C atoms at the end of the chain.

2 a) 29.7 MJ

   b) 41.9 MJ

Chapter 8: Hydrocarbons

1

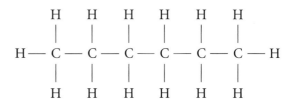

hexane

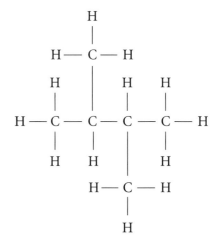

2,3–dimethylbutane

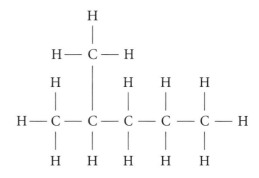

2–methylpentane

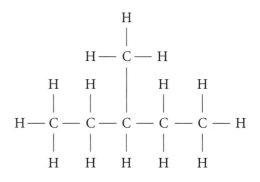

3–methylpentane

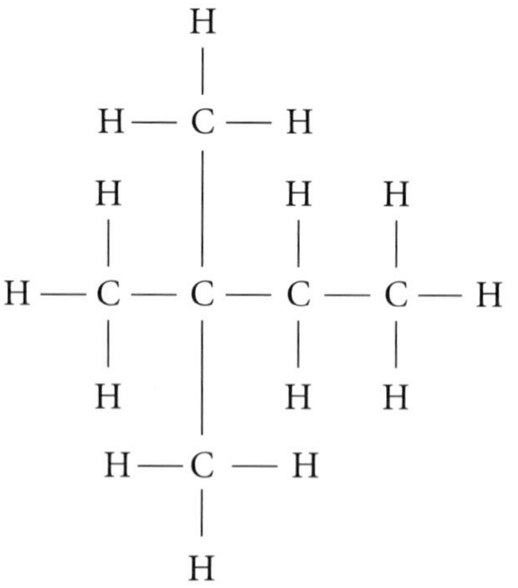

$$
\begin{array}{c}
\quad\quad\quad H \\
\quad\quad\quad | \\
H - C - H \\
\quad\quad\quad | \\
H \quad\quad\quad H \quad\; H \\
| \quad\quad\quad | \quad\; | \\
H - C - C - C - C - H \\
| \quad\quad\quad | \quad\; | \\
H \quad\quad\quad H \quad\; H \\
\quad\quad H - C - H \\
\quad\quad\quad\quad | \\
\quad\quad\quad\quad H
\end{array}
$$

2,2–dimethylbutane

2

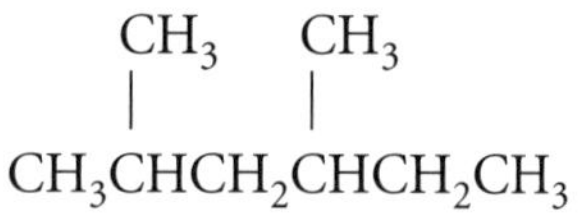

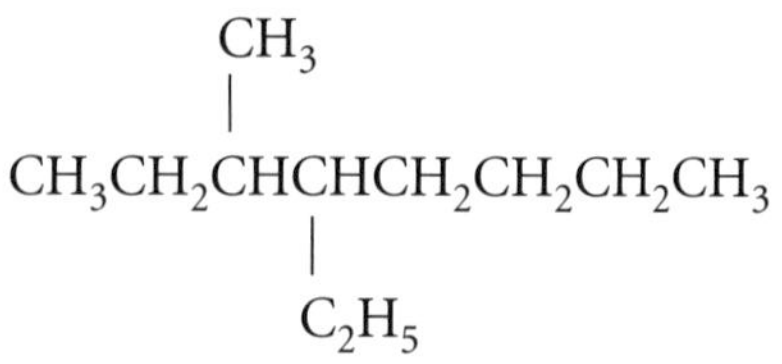

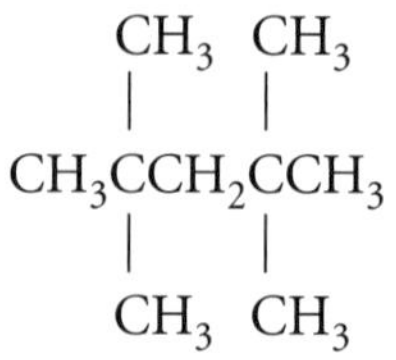

$$
\begin{array}{c}
CH_3 \quad\; CH_3 \\
| \quad\quad\quad | \\
CH_3CHCH_2CHCH_2CH_3
\end{array}
$$

$$
\begin{array}{c}
CH_3 \\
| \\
CH_3CH_2CHCHCH_2CH_2CH_2CH_3 \\
| \\
C_2H_5
\end{array}
$$

$$
\begin{array}{c}
CH_3 \; CH_3 \\
| \quad\; | \\
CH_3CCH_2CCH_3 \\
| \quad\; | \\
CH_3 \; CH_3
\end{array}
$$

3 a) 2,3-dimethylpentane **b)** 3,3-diethylhexane
c) 2,2,6-trimethylheptane

4 a) pent-2-ene **b)** 4,5-dimethylhex-1-ene
c) 2-methylbut-2-ene

5

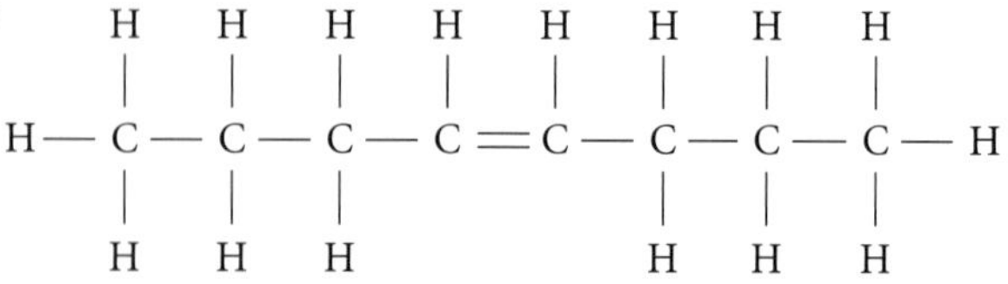

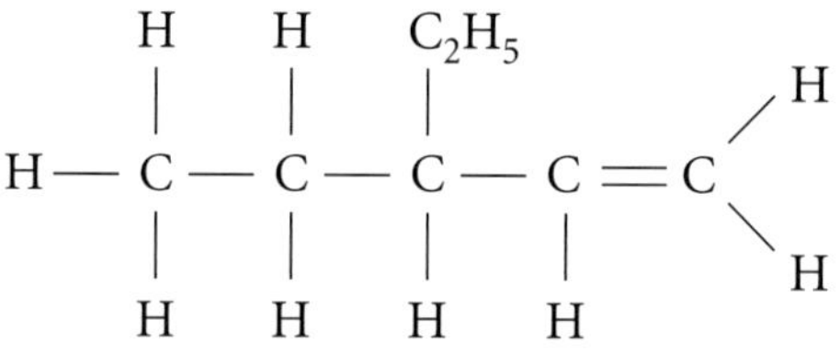

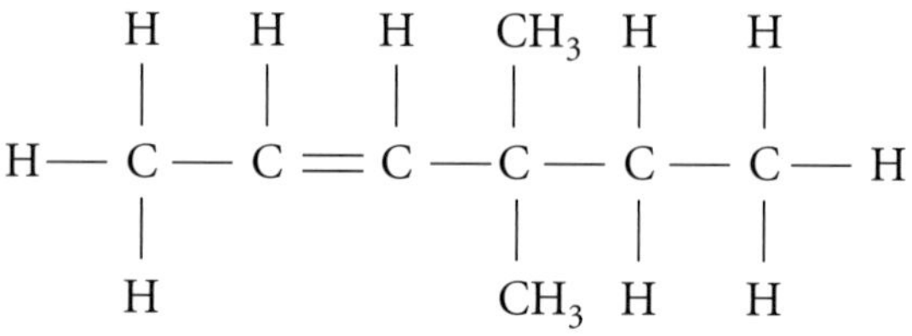

$$
\begin{array}{c}
H \;\; H \;\; H \;\; H \;\; H \;\; H \;\; H \;\; H \\
| \;\; | \;\; | \;\; | \;\; | \;\; | \;\; | \;\; | \\
H - C - C - C - C = C - C - C - C - H \\
| \;\; | \;\; | \quad\quad\quad | \;\; | \;\; | \\
H \;\; H \;\; H \quad\quad\quad H \;\; H \;\; H
\end{array}
$$

$$
\begin{array}{c}
H \;\; H \;\; C_2H_5 \\
| \;\; | \;\; | \quad\quad H \\
H - C - C - C - C = C \\
| \;\; | \;\; | \quad\quad H \\
H \;\; H \;\; H \;\; H
\end{array}
$$

$$
\begin{array}{c}
H \;\; H \;\; H \;\; CH_3 \; H \;\; H \\
| \;\; | \;\; | \;\; | \;\; | \;\; | \\
H - C - C = C - C - C - C - H \\
| \quad\quad\quad\quad\; | \;\; | \;\; | \\
H \quad\quad\quad\quad\; CH_3 \; H \;\; H
\end{array}
$$

Note: When drawing full structural formulae it is usually acceptable to abbreviate any branches as shown in examples **b)** and **c)**.

6

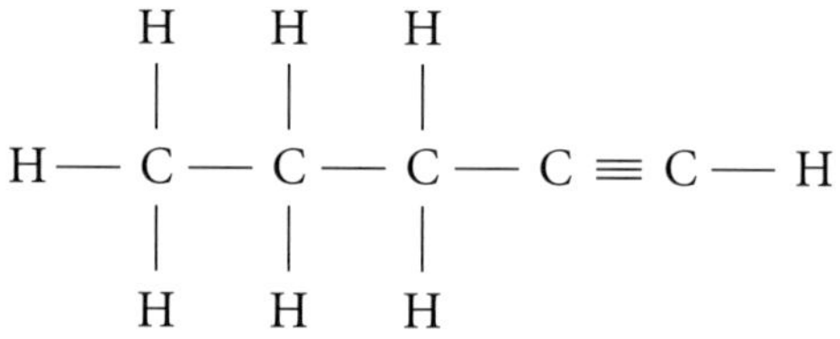

$$
\begin{array}{c}
H \;\; H \;\; H \\
| \;\; | \;\; | \\
H - C - C - C - C \equiv C - H \\
| \;\; | \;\; | \\
H \;\; H \;\; H
\end{array}
$$

pent–1–yne

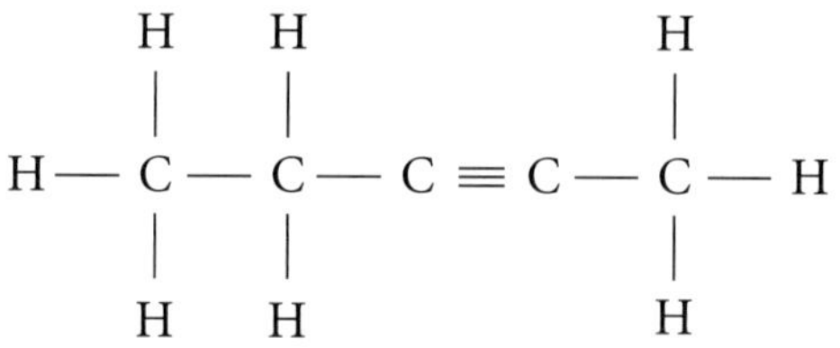

$$
\begin{array}{c}
H \;\; H \quad\quad\quad\quad H \\
| \;\; | \quad\quad\quad\quad | \\
H - C - C - C \equiv C - C - H \\
| \;\; | \quad\quad\quad\quad | \\
H \;\; H \quad\quad\quad\quad H
\end{array}
$$

pent–2–yne

$$
\begin{array}{c}
H \quad\; CH_3 \\
| \quad\quad | \\
H - C - C - C \equiv C - H \\
| \quad\quad | \\
H \quad\; H
\end{array}
$$

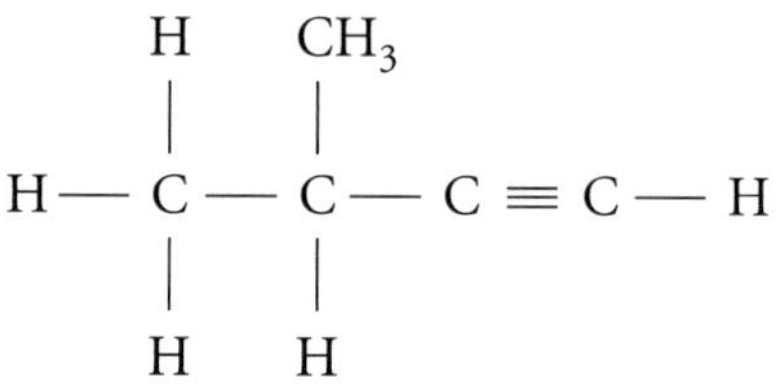

3–methylbut–1–yne

7 a)

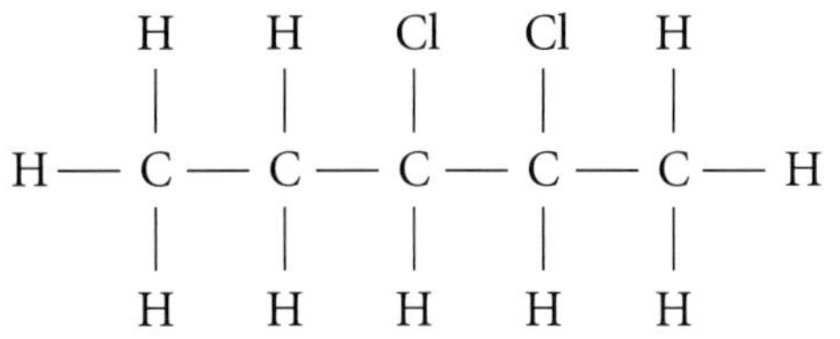

```
     H    H    Cl   Cl   H
     |    |    |    |    |
H —  C —  C —  C —  C —  C — H
     |    |    |    |    |
     H    H    H    H    H
```

2,3–dichloropentane

b)

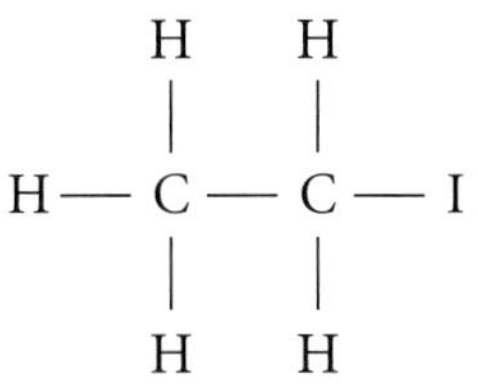

```
     H    H
     |    |
H —  C —  C — I
     |    |
     H    H
```

iodoethane

c)

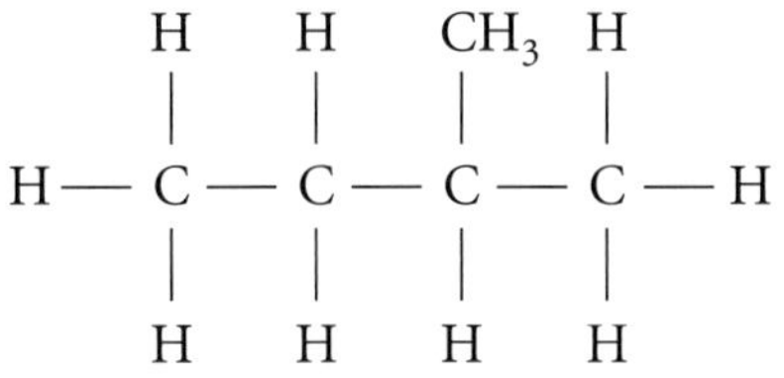

```
     H    H    CH₃  H
     |    |    |    |
H —  C —  C —  C —  C — H
     |    |    |    |
     H    H    H    H
```

2–methylbutane

8

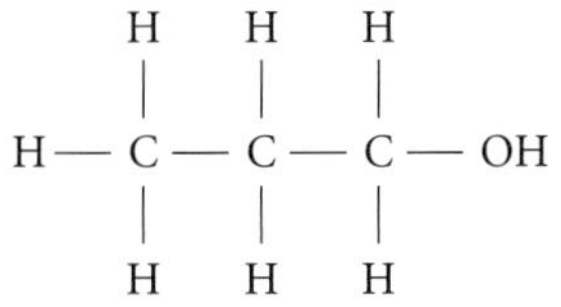

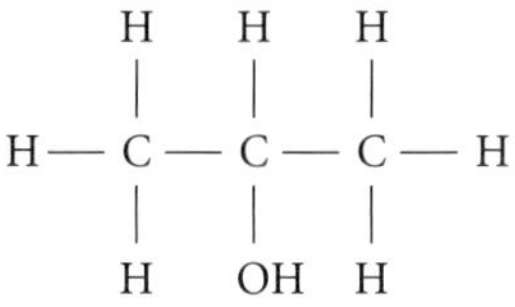

```
     H    H    H                H    H    H
     |    |    |                |    |    |
H —  C —  C —  C — OH     H —  C —  C —  C — H
     |    |    |                |    |    |
     H    H    H                H    OH   H
```

9 a)

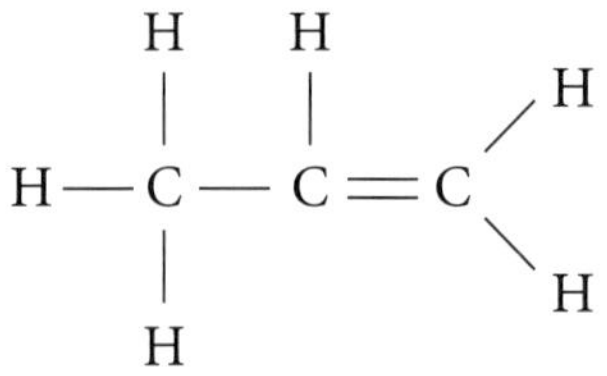

```
     H    H         H
     |    |        /
H —  C —  C  ═  C
     |             \
     H              H
```

propene

b)

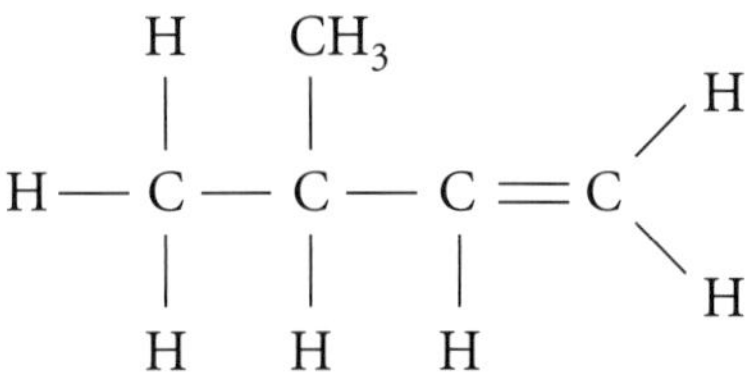

```
     H    CH₃         H
     |    |          /
H —  C —  C —  C  ═  C
     |    |    |      \
     H    H    H       H
```

3–methylbut–1–ene

c)

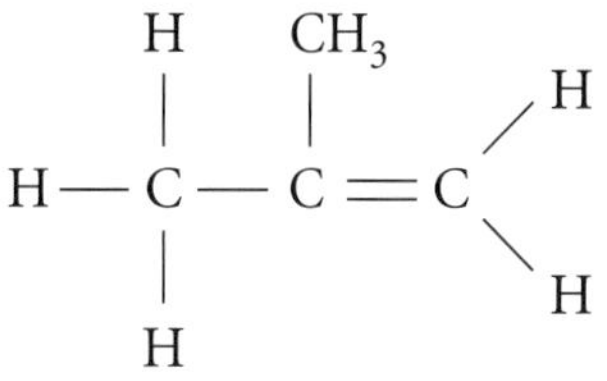

```
     H    CH₃       H
     |    |        /
H —  C —  C  ═  C
     |             \
     H              H
```

2–methylpropene

10 a)

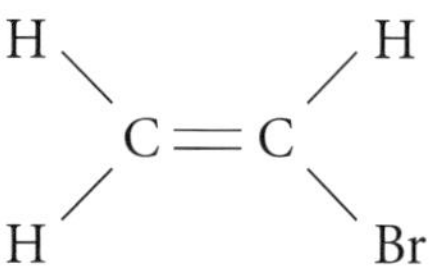

```
H               H
  \            /
    C  ═  C
  /            \
H               Br
```

bromoethene

b)

```
     H    Cl   Cl
     |    |    |
H —  C —  C —  C — H
     |    |    |
     H    Cl   Cl
```

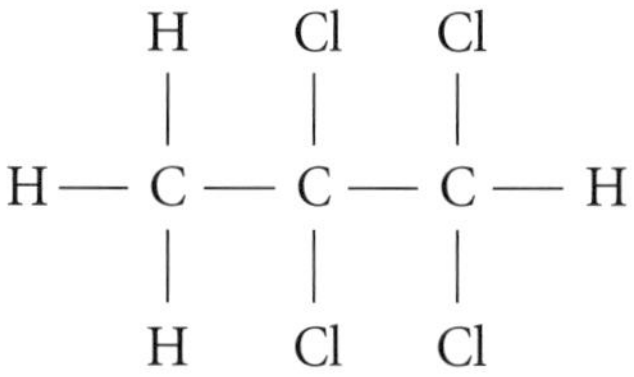

1,1,2,2–tetrachloropropane

c)

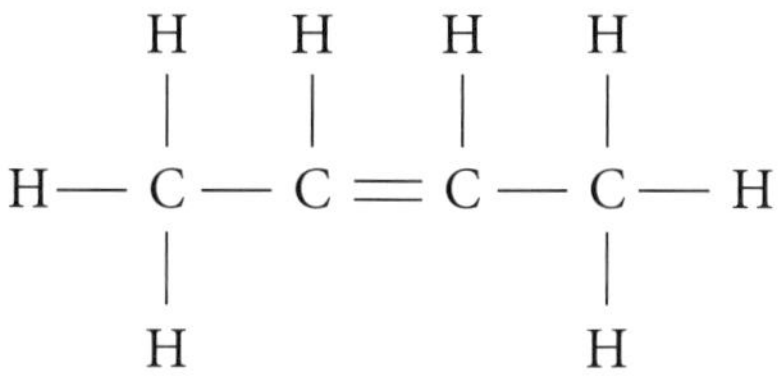

```
     H    H    H    H
     |    |    |    |
H —  C —  C  ═  C —  C — H
     |              |
     H              H
```

but–2–ene

11 The molecular shape of ethane is three-dimensional, ethene is planar and ethyne is linear as shown in the photograph.

12 a) C_6H_6O

  b) $C_7H_6O_2$

  c) C_8H_8

13 a)

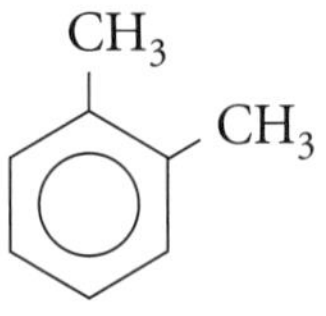

CH₃
CH₃

1,2-dimethylbenzene

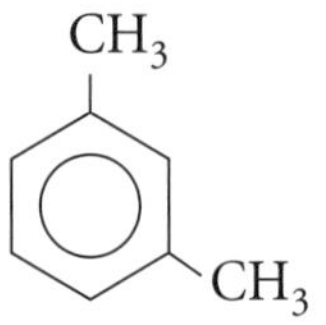

CH₃
CH₃

1,3-dimethylbenzene

b)

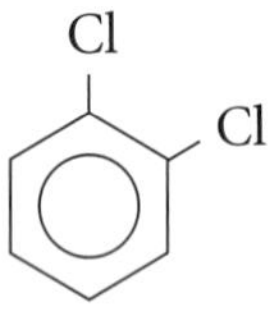

Cl
Cl

1,2-dichlorobenzene

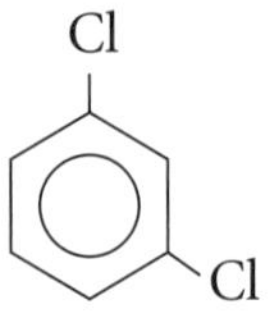

Cl
Cl

1,3-dichlorobenzene

Cl
Cl

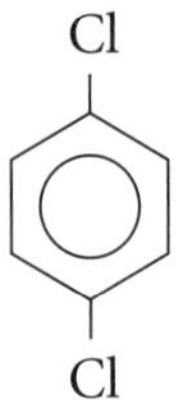

1,4-dichlorobenzene

Chapter 9: Alcohols, Aldehydes and Ketones

1

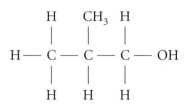

2–methylpropan–1–ol is a primary alkanol

2 a)

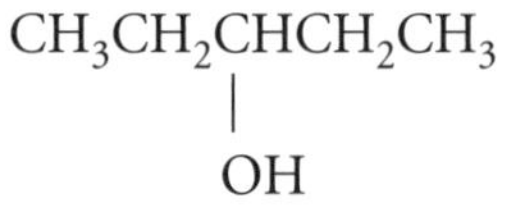

$CH_3CH_2CHCH_2CH_3$
OH

secondary

b)

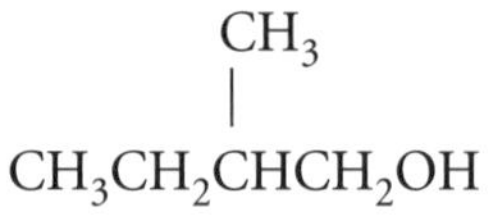

CH_3
$CH_3CH_2CHCH_2OH$

primary

c)

C_2H_5
$CH_3CH_2CCH_2CH_3$
OH

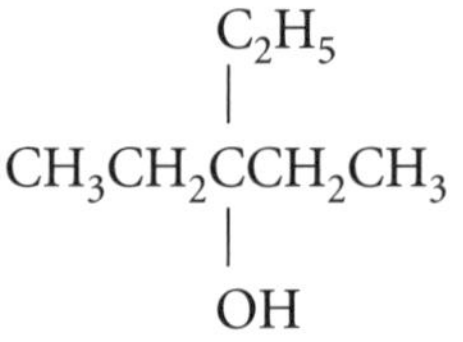

tertiary

3 a) 2-methylbutan-2-ol (tertiary)
b) 2,2-dimethylbutan-l-ol (primary)
c) 4-methylpentan-2-ol (secondary)

4 a)

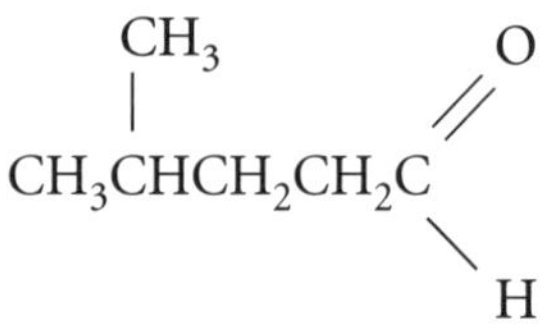

CH_3 O
$CH_3CHCH_2CH_2C$
                 H

made by oxidising 4–methylpentan–1–ol

b)

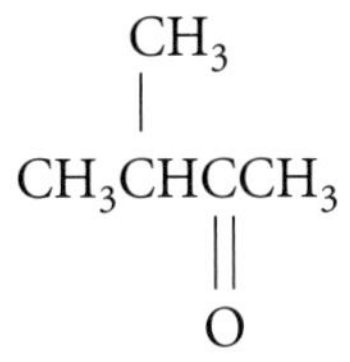

$$CH_3CHCCH_3$$
with CH$_3$ above and O (double bond) below

made by oxidising 3-methylbutan-2-ol.

5 a) 4-ethylhexan-3-one   b) 3,3-dimethylbutanal

6 a) $CH_3OH \rightarrow$ $HCHO \rightarrow$ $HCOOH$

O:H ratio: 1:4 = 0.25:1   1:2 = 0.5:1   2:2 = 1:1

In each step the O:H ratio is increasing, i.e. oxidation is taking place.

b) i) $C_4H_9OH \rightarrow C_3H_7CHO \rightarrow C_3H_7COOH$

butan-1-ol butanal   butanoic acid

O:H ratio: 1:10 = 0.1:1   1:8 = 0.13:1   2:8 = 0.25:1

ii) $C_3H_7CH(OH)CH_3 \rightarrow C_3H_7COCH_3$

pentan-2-ol   pentan-2-one

O:H ratio: 1:12 = 0.08:1   1:10 = 0.1:1

Chapter 10: Carboxylic Acids and Esters

1 a)

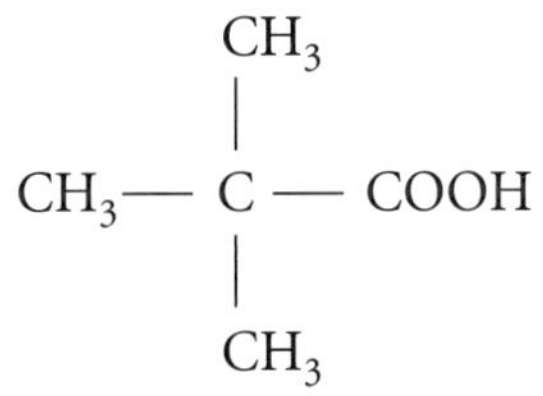

$$CH_3 - C - COOH$$
with CH$_3$ above and CH$_3$ below

b)

$$CH_3CHCH_2CHCH_2COOH$$
with CH$_3$ and C$_2$H$_5$ above

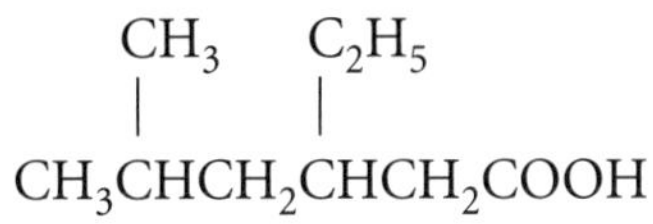

2 2,3-dimethylpentanoic acid, made from:
$CH_3CH_2CH(CH_3)CH(CH_3)CH_2OH$

3

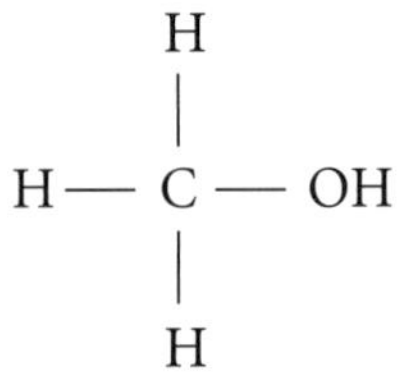

$$H - C - OH$$
with H above and H below

methanol

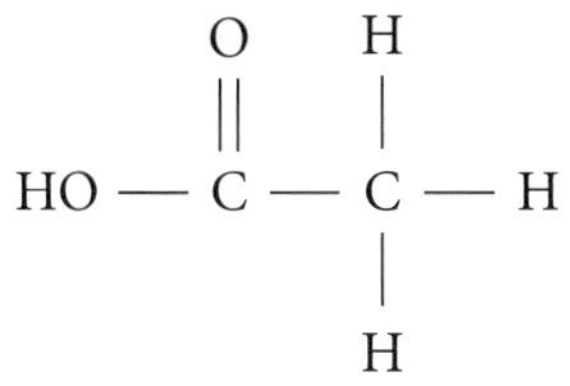

ethanoic acid

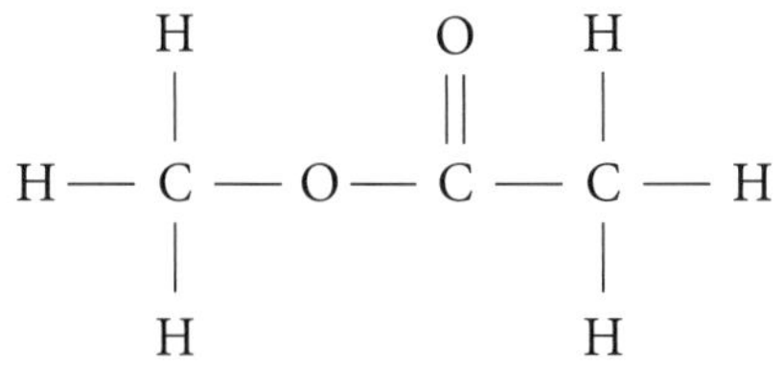

methyl ethanoate (ester produced)

4 $CH_3CH_2CH_2COOH$
butanoic acid
$HOCH_2CH_2CH_3$
propan-1-ol
$CH_3CH_2CH_2COOCH_2CH_2CH_3$
propyl butanoate (ester produced)

5 a) Propyl methanoate, on hydrolysis produces:

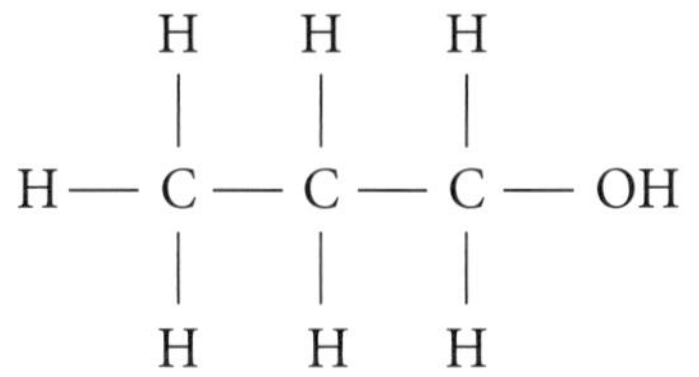

$$H - C - C - C - OH \quad \text{and}$$

$$HO - C - H$$
(O double bond)

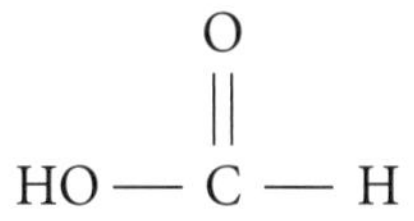

b) Methyl benzoate, on hydrolysis produces:

(benzene ring) $C - OH$ and $HO - C - H$

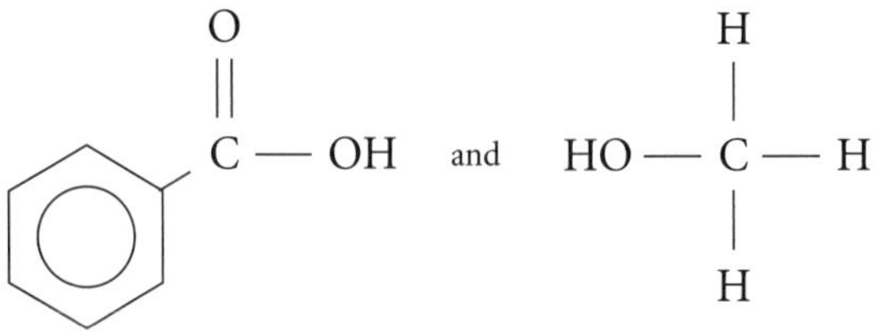

6 a) $CH_3CH_2CH_2COOCH_2CH_3 + KOH \rightarrow$
ethyl butanoate
$CH_3CH_2CH_2COO^-K^+ + HOCH_2CH_3$
potassium butanoate   ethanol

b) $CH_3CH_2CH_2CH_2CH_2OOCCH_3 + H_2O \rightarrow$
pentyl ethanoate
$CH_3CH_2CH_2CH_2CH_2OH + HOOCCH_3$
pentan-1-ol   ethanoic acid

7 82.8%

8 4.33 g

Chapter 11: Polymers

1

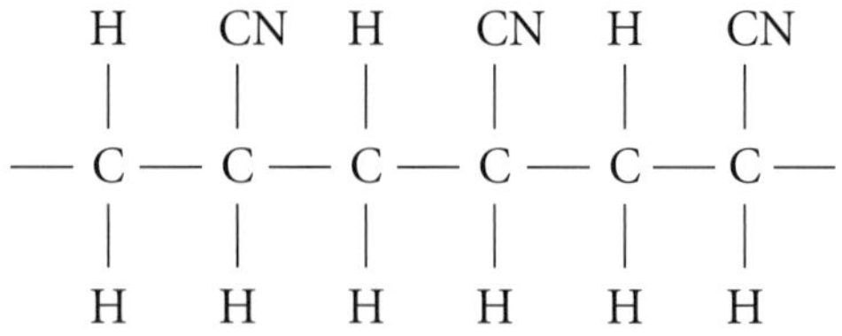

2

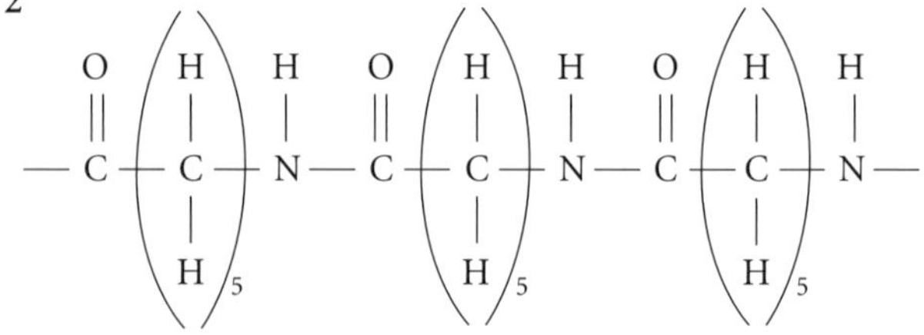

3

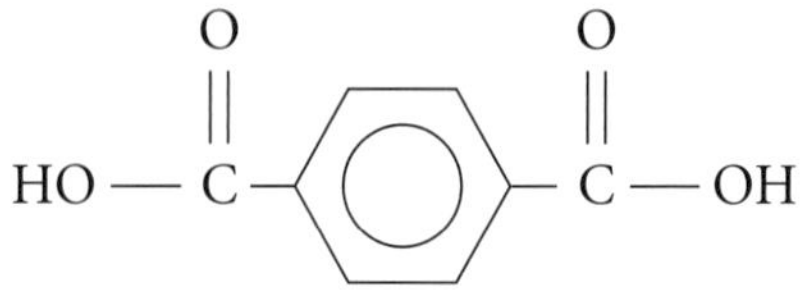

and

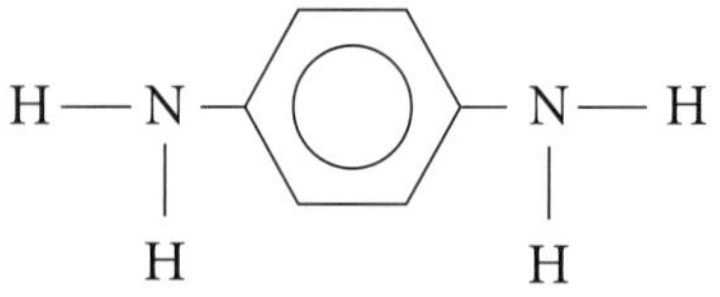

4 $CH_2{=}CHOH$

5 Benzene's electrons are delocalised within individual molecules, not over a long chain.

6

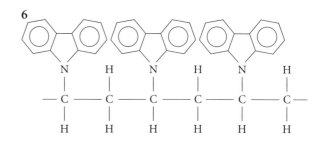

Chapter 12: Natural Products

1

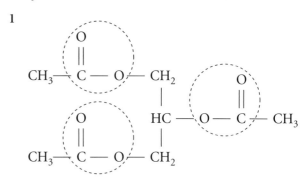

2 a) An amino acid

b)

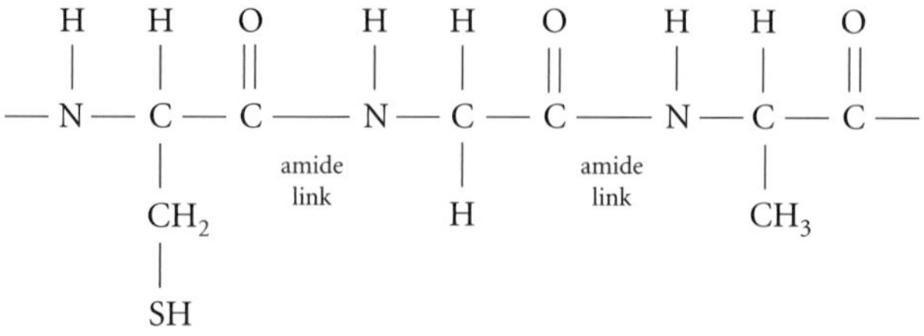

c) The amino acids could be joined in different orders.

d) It contains sulphur.

Chapter 13: The Chemical Industry

1 Advantages include cheap or renewable source of electricity from hydroelectricity, provides employment in areas of lower population.

Disadvantages include distance from port importing raw material and markets which process the product, possible effect on tourism, long term effect on environment such as atmospheric pollution and eventual dereliction.

Chapter 14: Hess's Law

1 $226 \, kJ \, mol^{-1}$

2 $-202 \, kJ \, mol^{-1}$

Chapter 15: Equilibrium

1 Bleaching action

   a) decreases b) decreases c) increases

2 a) Rate increases b) Yield decreases

3 a) Reaction 2 b) Reaction 1

4 a) Equilibrium reached faster since HCl dissolves giving H^+.

   b) Yield increases since dissolving HCl removes water and equilibrium goes to the right.

5 H_2SO_4 dissolves, effectively absorbing H_2O, equilibrium goes to the right, yield of ester increases. As H_2SO_4 dissolves, H^+ is formed, this acts as a catalyst, equilibrium reached faster.

Chapter 16: Acids and Bases

1 $[OH^-] = 10^{-11} \, mol \, l^{-1}$

2 a) $pH = 4$ b) $[H^+] = 10^{-9} \, mol \, l^{-1}$, $pH = 9$

3 $[H^+] = 10^{-2} \, mol \, l^{-1}$, $pH = 2$. For $10^{-2} \, mol \, l^{-1}$ CH_3COOH, the pH is higher.

4 a) $0.0025 \, mol$ b) $0.002 \, mol$ c) $0.002 \, mol$.

5 a) $0.002 \, mol$ b) $0.05 \, mol$.

6 $pH > 7$

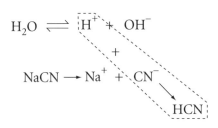

Leaving excess OH^- ions

Chapter 17: Redox Reactions

1 a) $Cu \rightarrow Cu^{2+} + 2e^-$

     $Ag^+ + e^- \rightarrow Ag \; (\times 2)$

     $\overline{2Ag^+ + Cu \rightarrow 2Ag + Cu^{2+}}$

  b) $Cr \rightarrow Cr^{3+} + 3e^- \; (\times 2)$

     $Ni^{2+} + 2e^- \rightarrow Ni \; (\times 3)$

     $\overline{3Ni^{2+} + 2Cr \rightarrow 3Ni + 2Cr^{3+}}$

2 a) $MnO_2 + 2H_2O \rightarrow MnO_4^- + 4H^+ + 3e^-$ Oxidation

  b) $FeO_4^{2-} + 8H^+ + 3e^- \rightarrow Fe^{3+} + 4H_2O$ Reduction

  c) $V^{3+} + 3H_2O \rightarrow VO_3^- + 6H^+ + 2e^-$ Oxidation

3 1) $2Fe^{3+} + 2I^- \rightarrow 2Fe^{2+} + I_2$

Charge: $6+$ $2-$ $4+$ 0

        $\underbrace{\qquad}_{4+}$ $\underbrace{\qquad}_{4+}$

Total charge on each side of the equation is the same.

  2) $Cr_2O_7^{2-} + 14H^+ + 6Fe^{2+} \rightarrow 6Fe^{3+} + 2Cr^{3+} + 7H_2O$

Charge: $2-$ $14+$ $12+$ $18+$ $6+$ 0

        $\underbrace{\qquad\qquad}_{24+}$ $\underbrace{\qquad\qquad}_{24+}$

Total charge on each side of the equation is the same.

4 i) Reducing agent   Oxidising agent   Spectator ion(s)

  a) Sulphite ions, SO_3^{2-} Iodine, I_2 Na^+

  b) Iron(II) ions, Fe^{2+} H_2O_2 SO_4^{2-}

  c) Tin(II) ions, Sn^{2+} $Cr_2O_7^{2-}$ K^+ and Cl^-

  ii) a) $I_2 + SO_3^{2-} + H_2O \rightarrow 2I^- + SO_4^{2-} + 2H^+$

     b) $2Fe^{2+} + H_2O_2 + 2H^+ \rightarrow 2Fe^{3+} + 2H_2O$

     c) $3Sn^{2+} + Cr_2O_7^{2-} + 14H^+ \rightarrow 3Sn^{4+} + 2Cr^{3+} + 7H_2O$

5 a) $2.5 \, mol \, l^{-1}$ b) $0.5 \, mol \, l^{-1}$ c) $0.3 \, mol \, l^{-1}$

6 a) $0.16 \, mol \, l^{-1}$ b) $0.8 \, mol \, l^{-1}$ c) $1.0 \, mol \, l^{-1}$

7 $1.01 \, g$

8 $4 \, A$

9 a) $9.65 \times 10^8 \, C$ b) 18 hours

Chapter 18: Nuclear Chemistry

1 $^{234}_{92}U \rightarrow {}^{230}_{90}Th + {}^{4}_{2}He$

   $^{216}_{84}Po \rightarrow {}^{212}_{82}Pb + {}^{4}_{2}He$

   $^{230}_{90}Th \rightarrow {}^{226}_{88}Ra + {}^{4}_{2}He$

2 $^{214}_{82}Pb \rightarrow {}^{214}_{83}Bi + {}^{0}_{-1}e$

   $^{228}_{89}Ac \rightarrow {}^{228}_{90}Th + {}^{0}_{-1}e$

   $^{210}_{83}Bi \rightarrow {}^{210}_{84}Po + {}^{0}_{-1}e$

3 a) $^{45}_{21}Sc + {}^{1}_{0}n \rightarrow {}^{42}_{19}K + {}^{4}_{2}He$

   b) $^{9}_{4}Be + {}^{4}_{2}He \rightarrow {}^{12}_{6}C + {}^{1}_{0}n$

   c) $^{14}_{7}N + {}^{4}_{2}He \rightarrow {}^{17}_{8}O + {}^{1}_{1}H$

4 $(\frac{1}{2})^8 = \frac{1}{256}$

5 $3t_{\frac{1}{2}} = 7.5$ yrs

6 $3t_{\frac{1}{2}} = 37.2$ yrs

Answers to Study Questions

Chapter 1: Reaction Rates

1 temperature

2 heterogeneous

3 divided

4 sulphur

5 B   6 B   7 D

8 a) The total volume will not be constant so that the concentrations of all reactants will vary.

  b) Any 2 from:

  Use more accurate measuring cylinders, especially for smaller volumes.

  Start timing while adding the H_2O_2 solution.

  Place a white tile or paper below the beaker.

9 a) $0.65\,cm^3\,s^{-1}$

  b)

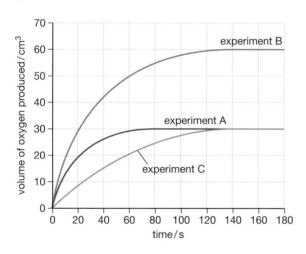

10 a) i) $PbCO_3$ in excess   ii) $3.312\,g$

  b) i) HCl in excess   ii) $40.05\,g$

Chapter 2: Enthalpy

1 lower

2 endothermic

3 water

4 unstable

5 B   6 B

7 a) $190\,kJ$ b) $-20\,kJ$, exothermic

  c)

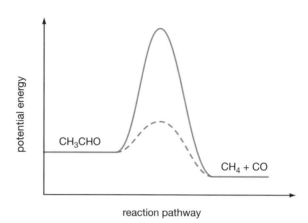

8 Reaction is highly exothermic; both products are gaseous and expand rapidly to be expelled at high velocity or both reactants are liquid so mix easily.

9 $-50.2\,kJ\,mol^{-1}$

10 $17.8°C$

11 a) 0.218 b) $15.8\,g$

12 a) i) $H^+(aq)$ are not used up.

    ii) Homogeneous, same physical state as reactant

  b)

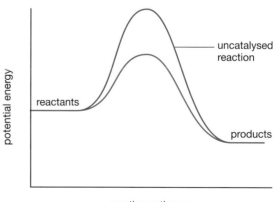

Chapter 3: Patterns in the Periodic Table

1 D   2 B   3 B   4 C

5 a) Stronger van der Waals' forces due to increasing numbers of electrons.

   b) They have covalent network structures.

   c) Outermost electron is further from the nucleus due to an increasing number of inner energy levels which also have a screening effect on the attraction by the nucleus.

   d) Group VII or halogens.

6 a) A hydrogen atom has only one electron.

   b) i) $He(g) \rightarrow He^+(g) + e^-$

       ii) Helium has a greater nuclear charge.

7 a) Electronegativity

   b) It decreases

   c) K has one more energy level than Na giving more screening between the nucleus and the outermost electron. Hence less energy is needed to remove it even though there are more protons in the nucleus.

8 a) Ionic radius for element 13 is less than that of Mg^{2+} but more than that of B^{3+}. Ionic radius for element 15 is greater than that of N^{3-}.

   b) i)   It is only a proton **or** it has no electrons.

       ii) Li^+ has more protons than H^- and therefore greater nuclear attraction for the same number of electrons.

   c) N^{3-} has one more electron energy level.

Chapter 4: Bonding, Structure and Properties of Elements

1 molecular

2 protons

3 van der Waals' forces

4 second

5 A   6 C   7 B   8 C

9 a) Table should show:

   Lithium: m. pt. 181 °C, metallic bonds broken.

   Boron: m. pt. 2300 °C, covalent bonds broken.

   Nitrogen: m. pt. −210 °C, van der Waals' forces broken.

   b) $O(g) \rightarrow O^+(g) + e^-$

   c) They increase due to increasing nuclear charge.

10 a) Fullerene consists of discrete molecules, diamond is a covalent network.

    b) Helium is an inert gas. If air is present, graphite will burn.

    c) If it decolourises bromine water, it is unsaturated. If not, it is saturated.

11 a) In descending order: Types e, b, d, c, a

    b) Type d

Chapter 5: Bonding, Structure and Properties of Compounds

1 network

2 polar

3 large

4 hydrogen bonding

5 A   6 B   7 C

8 A   9 C   10 D

11 a) i)   It decreases due to increasing nuclear charge.

        ii) It increases due to more energy levels.

    b) Caesium and fluorine

12 a) Both have (polar) covalent bonding.

    b) i) −10°C   ii) 2230°C

    c) This is due to weak bonds (van der Waals' forces) between molecules.

13 a) No, as ions are not free to move.

    b) Yes, as outer electrons are delocalised.

    c) No, as it is covalent.

14 a) Methanol and hydrazine

    b) Hydrogen bonds are stronger than van der Waals' forces so molecules are more strongly attracted to each other.

    c) Heavier molecules have more electrons and therefore have stronger van der Waals' forces between molecules.

15 a) Group 0 **or** Noble gases

    b) C and S have the same electronegativity.

c) Going down the group, atoms have more energy levels. The inner levels have a screening effect so the ability to attract bonding electrons decreases despite the greater nuclear charge.

16 a) More electrons present so the van der Waals' forces increase in strength.

   b) This is due to hydrogen bonds between molecules.

Chapter 6: The Mole

Section A: The Avogadro Constant

1 B 2 C 3 B

4 D 5 C 6 C

7 a) 180 g b) 7.5 g

8 a) 103.3 g b) 155 g

9 0.5

10 1.204×10^{24} (i.e. 2 L)

11 a) One mole of acid present, so only 0.5 mol of Zn reacts.

    b) Statements B and C are true.

Section B: Gas Volumes

1 B 2 B 3 D

4 2.32 g

5 a) 0.5 mol HCl requires 0.25 mol FeS, but 30 g FeS = 0.342 mol.

    b) 6 litres

6 a) 45.9

    b) These liquids would not vapourise.

7 a) $2CO + O_2 \rightarrow 2CO_2$

    b) Carbon dioxide

    c) $80 \, cm^3$ d) $110 \, cm^3$

8 a) $2C_2H_2 + 5O_2 \rightarrow 4CO_2 + 2H_2O$

    b) $95 \, cm^3 \, O_2$ (excess); $100 \, cm^3 \, CO_2$

9 a) It has a larger surface.

    b) 10

    c) i) The volume increases.

       ii) No effect

    d) 1.2 litres

Chapter 7: Fuels

1 increases

2 absence

3 more

4 ethanol

5 A 6 C 7 C

8 a) Methane

    b) i) Sodium hydroxide **or** other alkali

       ii) Diagram should show gas collected in a gas syringe **or** over water in a measuring cylinder.

    c) Ethanol

9 a) i) Methane

       ii) Methanol has hydrogen bonds between molecules.

    b) It can be made from sugars produced by photosynthesis in crops such as grapes or sugar cane.

10 a) The only product of combustion is water vapour.

    b) More kerosine will be burned producing more carbon dioxide.

11 B and C contribute to global warming since both ultimately involve the combustion of fossil fuels.

Chapter 8: Hydrocarbons

1 branched

2 triple

3 hydration

4 benzene

5 B 6 A 7 D

8 a) Reforming

    b) 3,3-dimethylhexane

    c) It has a branched chain structure.

9 a) Addition

    b)

X:
```
    H   H
    |   |
H — C — C — H
    |   |
    Br  Br
```

Y:
```
    H   Br
    |   |
H — C — C — H
    |   |
    H   Br
```

X is 1,2-dibromoethane, Y is 1,1-dibromoethane.

c) i) Hydrogen bromide, HBr

ii) HBr is acidic and so reacts with the alkali, KOH.

10 a) 0.4

b) Cyclohexane

c) i) about 1.0

ii) Benzene

11 a) 2,2,4-trimethylpentane

b)

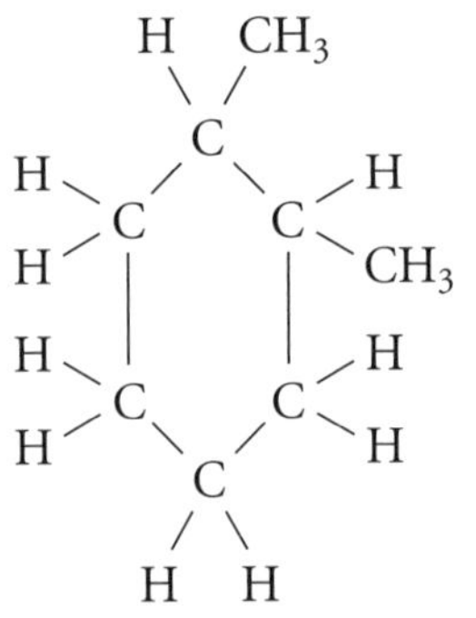

c) oct-1-ene (or other straight chain isomer)

d)

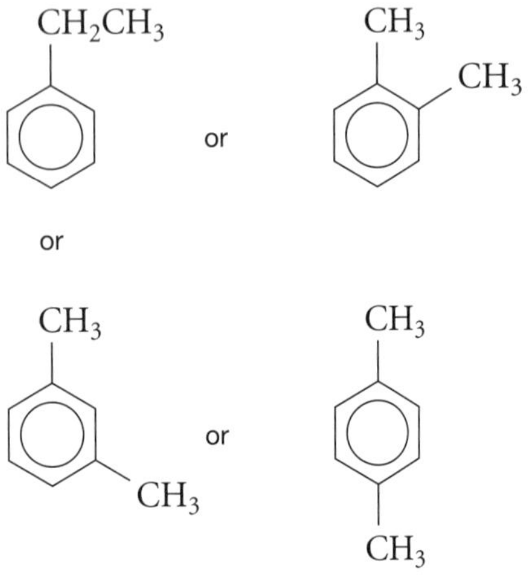

12 a) 4-methylpent-1-ene

b) i) Benzene   ii) Reforming

c) Nitrogen and carbon dioxide

Chapter 9: Alcohols, Aldehydes and Ketones

1 secondary

2 increased

3 carbonyl

4 blue-green

5 C   6 D   7 B   8 D

9 a)

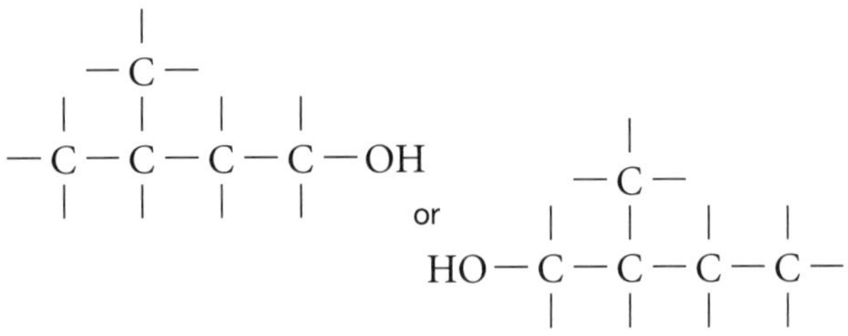

b)

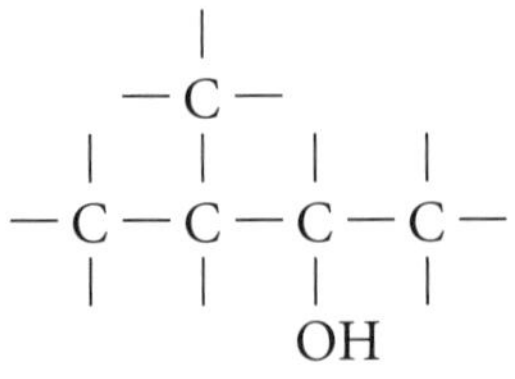

c)

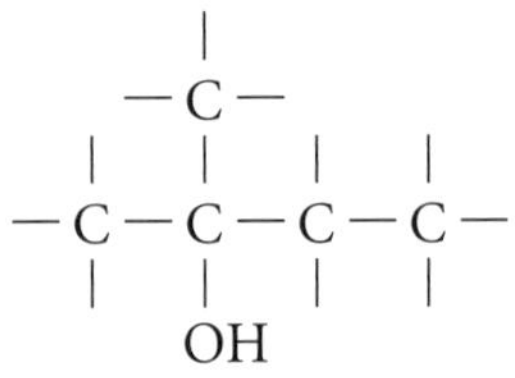

10 a) 3-methylbutan-1-ol **or** 2-methylbutan-1-ol

b) 3-methylbutan-2-ol

c) 2-methylbutan-2-ol

11 a) I is butan-1-ol, II is butan-2-ol, III is 2-methylpropan-2-ol, IV is 2-methylpropan-1-ol.

b) I and IV are primary, II is secondary, III tertiary.

c) i) D is the tertiary alcohol, ie D is **III**.

ii) A & B are primary; C is secondary, ie C is **II**.

iii) D, when dehydrated, gives a branched alkene.

I cannot do so but IV will, so B is **I** and A is **IV**.

12 a) A is but-2-ene, B is 2-chlorobutane, C is butan-2-ol.

b) Hydrogen chloride (*not* hydrochloric acid)

c) i) Orange to blue-green

ii) Oxidation

iii)

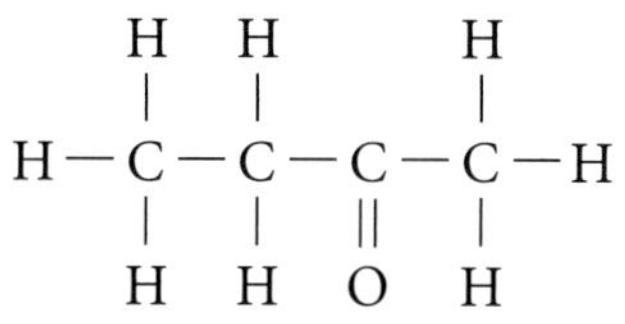

H H H
| | |
H − C − C − C − C − H
| | || |
H H O H

D is butanone.

iv) No, since D is a ketone (or D is not an aldehyde)

Chapter 10: Carboxylic Acids and Esters

1 A **2** B **3** D **4** C

5 a) i) Methyl ethanoate **ii)** Condensation

iii)

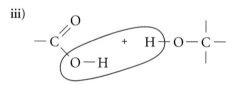

b) i) An oily layer forms on the surface of the solution.

ii) Any 2 from: extinguish any nearby flames, careful addition of conc. H_2SO_4, wrap wet paper towel round top of test tube.

6 a) i) Dehydration **ii)** Condensation

b) Pent-1-ene

c) Pentan-2-one, acidified potassium dichromate solution or copper(II) oxide

d)

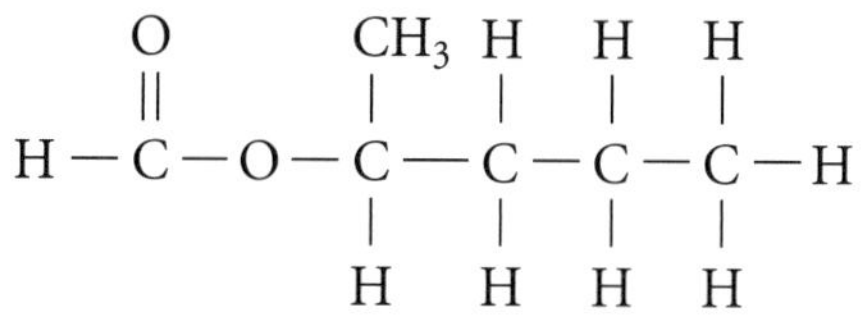

O   CH₃ H H H
|| | | | |
H − C − O − C − C − C − C − H
        | | | |
        H   H H H

or

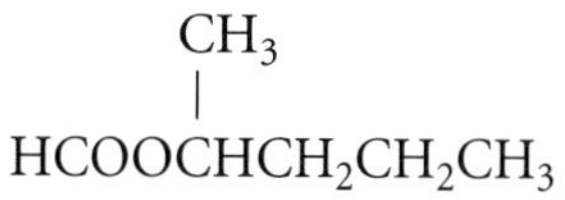

CH₃
|
HCOOCHCH₂CH₂CH₃

7 a) Primary, aromatic

b) i)

⬡ CH₂CH₂OOCCH₂CH₃

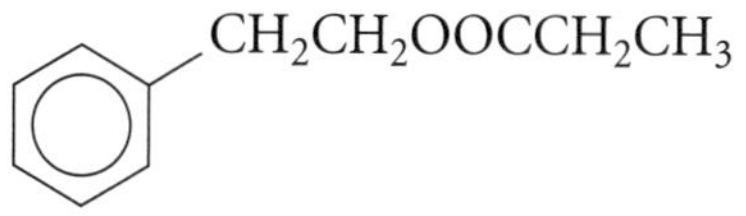

ii) 1.48 tonnes of propanoic acid requires only 2.44 tonnes of phenylethanol for complete reaction.

iii) 2.492 tonnes

8 a) 50% **b)** 4.27 g **c)** 35%

9 a) Step 3: Carefully add a few drops of conc. H_2SO_4 to the test tube.

Step 4: Place the test tube in a beaker containing hot water.

b) This is done to condense any volatile reactant or product.

c) 2-methylpropan-1-ol

Chapter 11: Polymers

1 condensation

2 kevlar

3 diols

4 water-soluble

5 A **6** C **7** D **8** A

9 a) Circle any one of the 3 –CH_2OH groups.

b) Remove H from any of the —O—H groups and attach the following group instead.

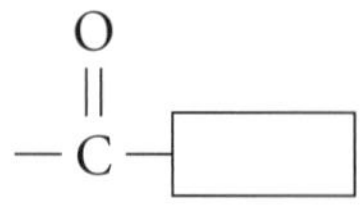

c) Its molecular shape is incompatible with the shape of our enzyme molecules.

d) i) As the monomers have only 2 functional groups per molecule, they form a linear polymer when they react **or** they are unable to form cross-links.

   ii) It is neither linear nor cross-linked **or** it is a discrete molecule.

10 a) Condensation

   b) Polyamide

   c) Hydrogen bonds

11 a) Addition

   b)

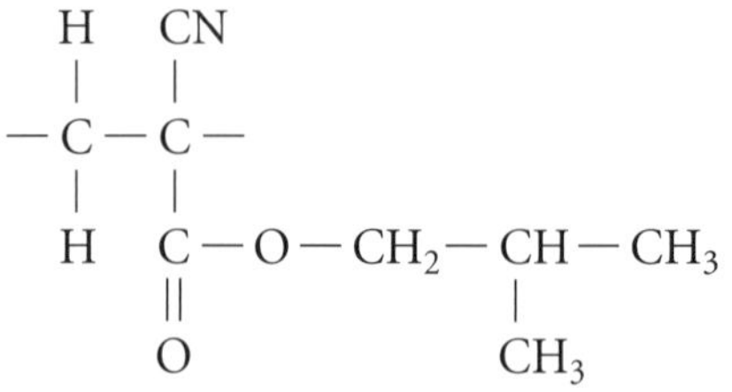

   H   CN
   | |
 — C — C —
   | |
   H   C — O — CH₂ — CH — CH₃
       ‖ |
       O   CH₃

12 a)

   H   H   H   H
   | | | |
 — C — C = C — C —
   | |
   H   H

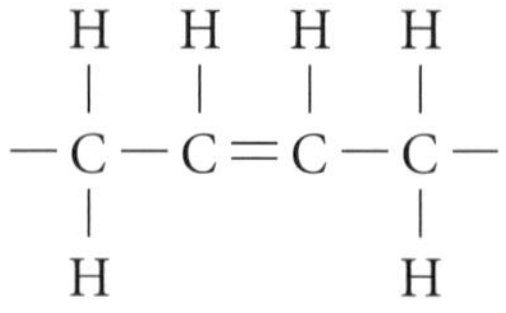

   b)   CH₂ = CHCH = CH₂

13 a) C₈H₆O₄

   b) i) Condensation

      ii)

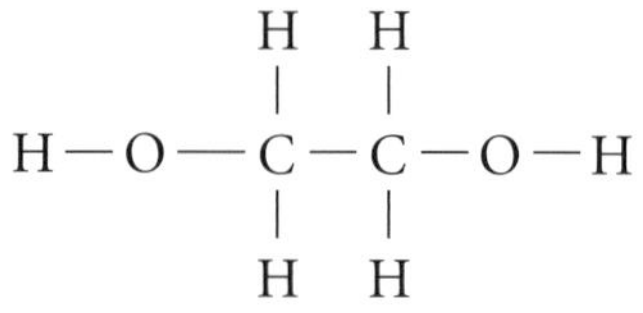

      H   H
      | |
H — O — C — C — O — H
      | |
      H   H

      iii) It has a cross-linked or 3-dimensional structure.

Chapter 12: Natural Products

1 greater

2 hydrolysis

3 globular

4 denatured

5 B 6 D 7 A 8 C

9 a) Each glycerol molecule has 3 —OH groups, which condense with 3 acid molecules to form a triglyceride molecule.

   b) They become saturated **or** hydrogen adds across the carbon–carbon double bonds.

   c) Carbon–carbon double bonds

10 a) 2 b) C₁₈H₃₆O₂

   c) Glycerol **or** propane-1,2,3-triol

   CH₂OH
   |
   CHOH
   |
   CH₂OH

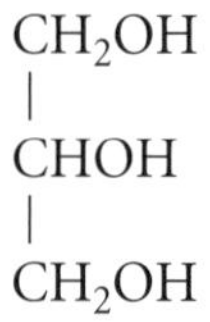

11 a) Glycerol **or** propane-1,2,3-triol

   b) Carbon-carbon double bond

   c) Hydrogenation **or** addition (of hydrogen) **or** hardening

12 a)

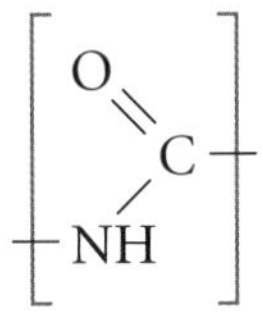

b)

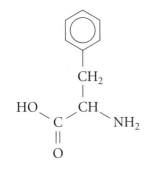

c)

d) It may become hydrolysed.

13 a)

Any

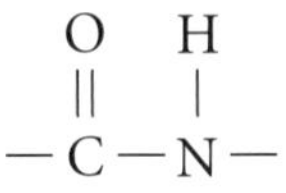

$$\begin{matrix} O & H \\ \| & | \\ -C - N- \end{matrix}$$

b)

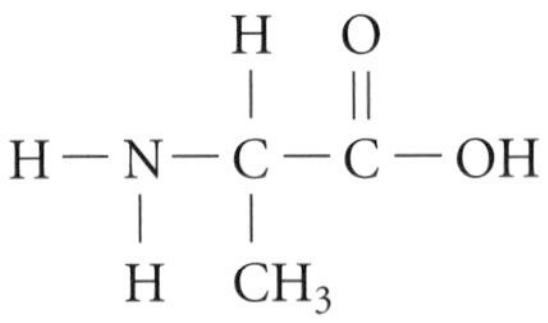

$$\begin{matrix} H & O \\ | & \| \\ H-N-C-C-OH \\ | & | \\ H & CH_3 \end{matrix}$$
or
$$\begin{matrix} H & H & O \\ | & | & \| \\ H-N-C-C-OH \\ | \\ CH_2 \\ | \\ H-C-CH_3 \\ | \\ CH_3 \end{matrix}$$

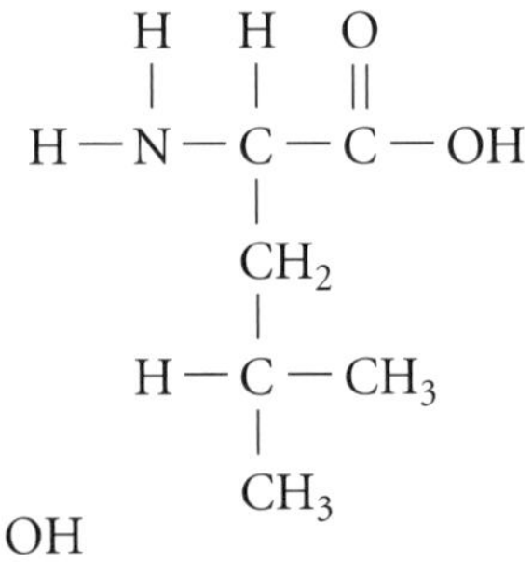

or
$$\begin{matrix} H & H & O \\ | & | & \| \\ H-N-C-C-OH \\ | \\ CH_2 \\ | \\ \text{(phenol ring)} \\ | \\ OH \end{matrix}$$

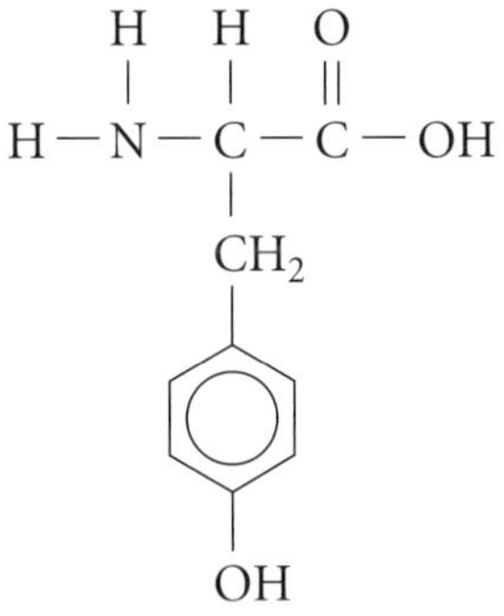

c) i) Sucrose molecules have a different shape from those of maltose.

ii) Maltase is denatured (or its shape changes) at low pH.

14 a) Fibrous

b) i) Hydrogen peroxide

ii) Count the number of bubbles of gas given off during the next 3 minutes.

Chapter 13: The Chemical Industry

1 raw material

2 continuous

3 capital

4 B   **5** D

6 a)

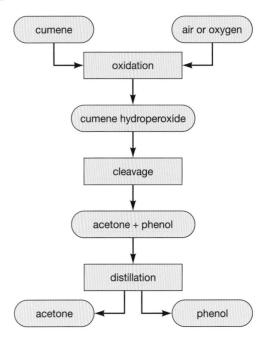

b) 16.2 tonnes

c) Encircle any 2 of: energy costs, availability of feedstocks, number of stages involved, market for by-product(s), waste disposal or emissions to the air

7 a) i) $CH_4(g) + H_2O(g) \rightleftharpoons CO(g) + 3H_2(g)$

730°C, 30 atmospheres pressure, nickel oxide catalyst

ii) $CO(g) + H_2O(g) \rightleftharpoons CO_2(g) + H_2(g)$

330°C, iron oxide catalyst

b) It reacts with some hydrogen to produce more steam.

c) This is to avoid poisoning of the catalysts used in the subsequent stages.

d) $K_2CO_3 + H_2O + CO_2 \rightarrow 2KHCO_3$

8 a) i) Monochloroethene

   ii) To make polyvinyl chloride, PVC

b) $HCl, C_2H_4, CH_2ClCH_2Cl$

c) $CH_2ClCH_2Cl \rightarrow CH_2=CHCl + HCl$

d) Distillation

e) Neutralisation

9 a) Ammonia, carbon dioxide, seawater

b) First equation: ΔH +, second: ΔH −

c) $Ca(OH)_2 + 2NH_4Cl \rightarrow CaCl_2 + 2NH_3 + 2H_2O$

d) It forms insoluble $MgCO_3$ (which can be removed by filtration.)

e) Any 2 from: recycling of NH_3 and/or CO_2, energy saving by burning coke in **A**, seawater is a raw material **or** is cheap, continuous process, $CaCl_2$ byproduct

Chapter 14: Hess's Law

1 B   2 A

3 a) i) Acid is in excess, 0.05 mol of HCl to 0.04 mol NaOH and mole ratio in the equation is 1:1.

   ii) −42.8 kJ

b) 6.7°C

c) Determine the enthalpy of solution of NaOH(s) in water **or** measure temperature change when 1 g of NaOH(s) is added to 50 cm³ of water.

d) Carry out the reactions in a polystyrene cup.

4 a) $46\,kJ\,mol^{-1}$

b) Benzene is less stable than methane but more stable than ethyne.

5 $-312\,kJ\,mol^{-1}$

6 $-1076\,kJ\,mol^{-1}$

7 $-333\,kJ\,mol^{-1}$

8 $-484\,kJ\,mol^{-1}$

Chapter 15: Equilibrium

1 constant

2 left

3 decreases

4 endothermic

5 B   6 B   7 D   8 A

9 a) Haber process

b) This occurs when a substance forms strong bonds with the catalyst's surface, so keeping new reactants from the surface.

c) It would increase as a temperature fall favours the exothermic reaction.

d) Stronger industrial plant would be needed, costing more to build and maintain.

10 a) The enthalpy change is positive **or** the reaction is endothermic.

b) The position of equilibrium moves to counteract the applied change. An increase in pressure pushes the equilibrium in the direction of fewer moles of gas, i.e. towards reactant side, therefore less product.

c) It increases.

d) It is unchanged.

11 a) i) It moves to the right.

   ii) It moves to the left.

   iii) No effect

b) Nitric acid

12 a) This occurs when the rate of the forward reaction equals the rate of the reverse reaction.

b) Increasing the temperature lowers the yield of ammonia, i.e. it favours the reverse reaction. Therefore this is endothermic and the forward reaction exothermic.

c) i) 40%

   ii) Unreacted gases are recycled.

13 a) $0.05\,mol\,l^{-1}\,s^{-1}$

b) 3

c) i) It is in the same physical state as the reactants.

   ii) No effect

14 a) Advantage: higher yield of ammonia; disadvantage: more expensive to build/maintain plant.

b) i)

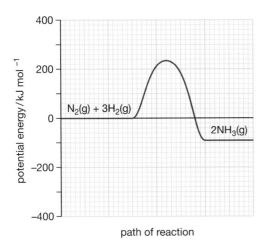

ii) $328 \, \text{kJ mol}^{-1}$

c) 13%

Chapter 16: Acids and Bases

1 slower

2 partially

3 strong

4 lower

5 B   6 C   7 D   8 B   9 A

10 a) i) $1 \times 10^{-6} \, \text{mol l}^{-1}$

ii) $1 \times 10^{-8} \, \text{mol l}^{-1}$

b) It is partially dissociated into ions in aqueous solution.

c) i)   Sodium ethanoate

ii)   Phenolphthalein. The end-point occurs at a pH>7 since sodium ethanoate is the salt of a strong base and a weak acid.

11 a) Acid A has a lower pH (or a higher concentration of $H^+(aq)$ ions) than the equimolar acid B.

b) i)   It will move to the right.

ii)   No effect

iii)   It will move to the right.

c) The equilibrium moves to the right to compensate for the loss of H^+ ions which have combined with ethanoate ions, CH_3COO^-, to form CH_3COOH molecules.

12 a) i)   Any 2 of the following tests. To each acid add Mg, or a carbonate; the weak acid releases a gas more slowly than the strong acid. Test the pH; the weak acid has a higher value.

ii)   Equimolar solutions and temperature

b) The number of double bonds (to oxygen)

c)

$$\begin{array}{c} O \\ \diagdown \\ Cl-O-H \\ \diagup \\ O \end{array}$$

d) Carbonate ions combine with H^+ ions to form carbonic acid molecules. The equilibrium: $H^+ + OH^- \rightleftharpoons H_2O$ moves to the left, thus increasing the concentration of OH^- ions.

Chapter 17: Redox Reactions

1 metals

2 oxidising

3 current

4 more

5 B   6 C   7 A   8 D

9 a) $2IO_3^- + 12H^+ + 10e^- \rightarrow I_2 + 6H_2O$

b) It is an oxidising agent since it accepts electrons.

10 First reaction: $2I^- \rightarrow I_2 + 2e^-$

Second reaction: $2S_2O_3^{2-} \rightarrow S_4O_6^{2-} + 2e^-$

11 a) $2Ce^{4+} + H_2O_2 \rightarrow 2Ce^{3+} + O_2 + 2H^+$

b) Cerium(IV) ions; sulphate ions

c) $0.036 \, \text{mol l}^{-1}$

12 a) $H_2O_2 + 2H^+ + 2I^- \rightarrow I_2 + 2H_2O$

b) $0.00945 \, \text{g}$

13 a) At the negative electrode

b) Cu^{2+}

Working:

$Cu^{n+} + ne^- \rightarrow Cu$

$Q = 1 \times 16 \times 60 = 960\,C$

so 63.5 g Cu produced by $n \times 96\,500\,C$

so 0.32 g Cu produced by $n \times 96\,500 \times \dfrac{0.32}{63.5} = 960\,C$

so $n = \dfrac{960 \times 63.5}{0.32 \times 96\,500} = 2$

14 7.94 g

15 a) 967 kg   **b)** 322.3 kg

16 a) $4Au + 8NaCN + O_2 + 2H_2O \rightarrow$
$\qquad\qquad\qquad 4NaAu(CN)_2 + 4NaOH$

b) Displacement

c) 3+

17 a) 643.3 s

b) i) 20 cm³

ii) No change, as the same number of H^+ ions are produced at the positive electrode as react at the negative electrode.

18 a) 156 684 litres

b) redox reaction or displacement

c) i) $2NaOH + Cl_2 \rightarrow NaClO + NaCl + H_2O$

ii) NaOH reacts with H^+ ions, so the equilibrium moves to the right and the concentration of ClO^- ions increases.

iii) $ClO^- + 2H^+ + 2e^- \rightarrow Cl^- + H_2O$

Chapter 18: Nuclear Chemistry

1 natural

2 gamma

3 fusion

4 neutrons

5 A   **6** C   **7** D   **8** D   **9** C

10 a) Fr^+ **b)** $^{223}_{88}Ra$

11 a) $^{234}_{91}Pa$ **b)** $^{218}_{84}Po$

12 $x = {}^4_2He$, $y = {}^{230}_{90}Th$, $z = {}^{212}_{82}Pb$

13 a) X is bismuth, a = 83, b = 209

b) i) B

ii) More neutrons are produced than are used and initiate more reactions at an ever increasing rate (hence the steep rise) until the target atoms are so few that the reaction ceases rapidly (hence the downturn.)

14 a) $^{241}_{95}Am \rightarrow {}^{237}_{93}Np + {}^4_2He$

b) 1299 years

c) It has a long half-life so activity does not decline significantly in the lifetime of the house. It emits α particles which are 'stopped' by air.

15 a)

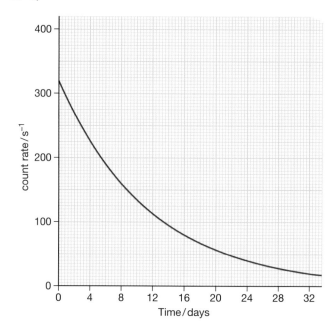

b) 28 days

16 a) Long term effect due to long half-life, gamma radiation damaging to vital organs. Patient would also be a hazard to the public since gamma radiation is not easily screened.

b) i) Initial p:n = 15:17 (or 1:1.13), final p:n = 16:16 (or 1:1)

ii) 2.81 g

17 a) X is sodium, a = 23, b = 11

b) i) 15 hours   **ii)** 0.015 g hr⁻¹   **iii)** No effect

c) They have different numbers of moles (or different numbers of particles) of ^{24}Na.

18 a) A neutron or ^1_0n

b) $^{100}_{43}Tc \rightarrow {}^{100}_{44}Ru + {}^0_{-1}e$

c) 1/8 or 0.125 or 12.5%

Answers to End-of-Course Questions

Part 1: Multiple choice questions

1 B	11 A	21 C
2 D	12 B	22 D
3 B	13 C	23 A
4 A	14 A	24 C
5 C	15 A	25 A
6 D	16 B	26 B
7 C	17 D	27 D
8 B	18 C	28 C
9 B	19 C	29 C
10 C	20 B	30 B

Part 2: Extended answer questions

1 a) Oxalic acid or ethanedioic acid

   b) Start timing while the acid is being added and stop as soon as the reaction mixture is decolourised.

   c) s^{-1}

2 a)

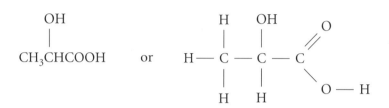

   b) Hexane is non-polar with only van der Waals' forces between molecules. Vinyl acetate is polar and so has stronger intermolecular bonding, i.e. permanent dipole to permanent dipole attractions as well as van der Waals' forces.

3 a) i)

$$CH_3CHCOOH \quad \text{or} \quad H-\overset{\overset{\displaystyle H}{|}}{\underset{\underset{\displaystyle H}{|}}{C}}-\overset{\overset{\displaystyle OH}{|}}{\underset{\underset{\displaystyle H}{|}}{C}}-C\overset{\displaystyle O}{\underset{\displaystyle O-H}{}}$$

      ii) 2   iii) 10^{-5}

   b) i) Fibrous proteins are long and thin and are the main structural materials of animal tissue **or** globular proteins have folded spiral chains and are involved in many biochemical reactions.

      ii) Carbon, hydrogen, nitrogen and oxygen

4 a) i) $FeS + 2HCl \rightarrow FeCl_2 + H_2S$

      ii) Hydrogen

   b) $-20\,kJ\,mol^{-1}$

5 a) i)

$$CH_3CH_2CH = CH_2 \quad \text{or} \quad H-\overset{\overset{\displaystyle H}{|}}{\underset{\underset{\displaystyle H}{|}}{C}}-\overset{\overset{\displaystyle H}{|}}{\underset{\underset{\displaystyle H}{|}}{C}}-\overset{\overset{\displaystyle H}{|}}{\underset{\underset{\displaystyle H}{|}}{C}}=\overset{\overset{\displaystyle H}{|}}{\underset{\underset{\displaystyle H}{|}}{C}}$$

      ii) Hydration **or** addition (of water)

      iii) Orange to blue-green

      iv) Butanoic acid

   b) i) Fruity smell **or** oily layer on water

      ii) Flavourings **or** perfumes **or** solvents

6 a) i) Precipitation

      ii) $AgNO_3$ is in excess, 1.18×10^{-3} mol compared with only 2×10^{-5} mol of HCl.

   b) i) It is fully dissociated into ions in aqueous solution

ii) 3

7 a) A water bath is not hot enough. It can only reach 100°C.

b) i) 0.23 litres **or** 230 cm^3

ii) Carbon dioxide is slightly soluble in water.

8 a) At equilibrium, the rate of the forward reaction equals the rate of the reverse reaction or the concentrations of reactants and products are constant.

b) The equilibrium moves to the left or some iodine goes from the bottom layer to the top layer to re-establish equilibrium.

c) 30

9 a) $^{32}_{15}P \rightarrow\ ^{32}_{16}S +\ ^{0}_{-1}e$

b) i) 1.5×10^{23} **ii)** 42.9 days

10 a) -54.3

b) 1×10^{-14} mol l^{-1}

11 a) i) Displacement

ii) $U^{4+} + 4e^- \rightarrow U$

iii) It prevents formation of metal oxides.

b) i) Covalent   **ii)** They are the same.

12 a) i) Fibrous

ii)

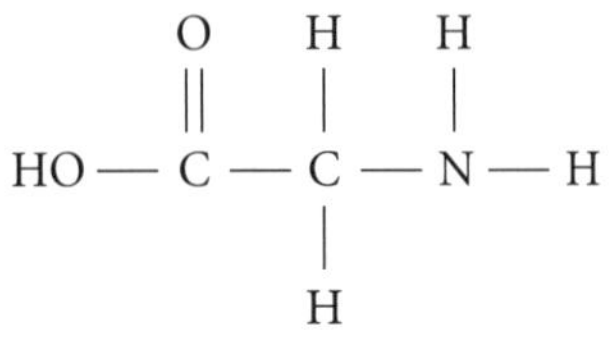

or

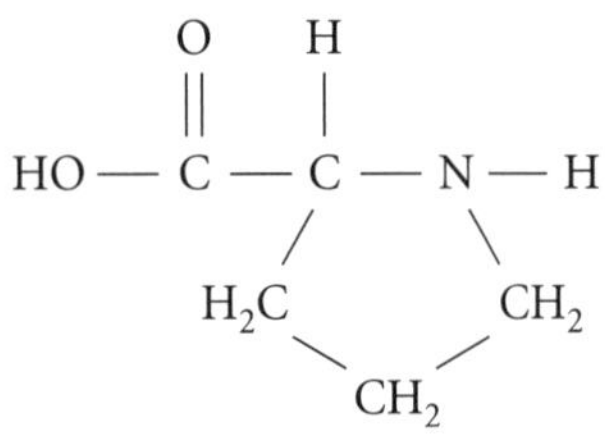

or

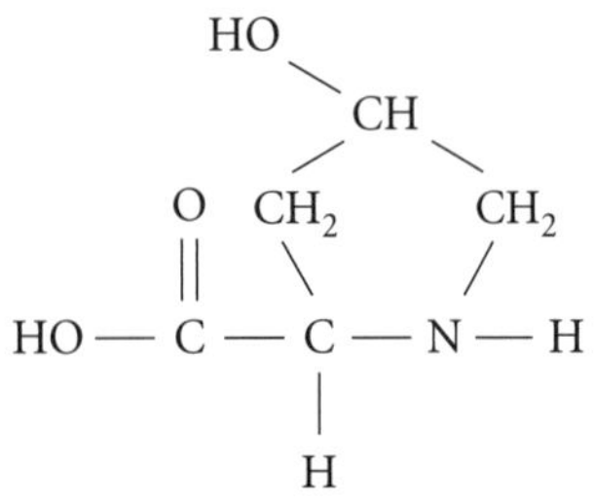

b) i) It is the indicator, a permanent blue colour appearing at the end-point.

ii) The titration can be repeated to obtain an average titre.

iii) 0.188 g

13 a) Ethene is produced from other chemicals.

b) i) It can be recycled to the reaction vessel.

ii) Distillation

c) i) Hydration **or** addition (of water)

ii) It moves to the left.

iii) 10%

14 a) $2Cl^-(aq) + 2H_2O(l) \rightarrow Cl_2(g) + H_2(g) + 2OH^-(aq)$

b) i) Sodium hydroxide

ii) It may be used as a fuel **or** it is a renewable material for energy transfer.

c) 1059

15 a)

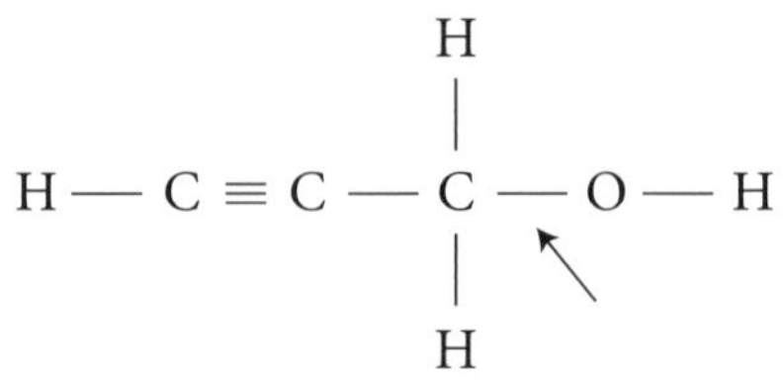

b) i) S and Q   **ii)** Propan-1-ol